SPREADSHEET TOOLS
FOR ENGINEERS
USING EXCEL

THIRD EDITION

BYRON S. GOTTFRIED

PROFESSOR EMERITUS
UNIVERSITY OF PITTSBURGH

Higher Education

Boston   Burr Ridge, IL   Dubuque, IA   Madison, WI   New York   San Francisco   St. Louis
Bangkok   Bogotá   Caracas   Kuala Lumpur   Lisbon   London   Madrid   Mexico City
Milan   Montreal   New Delhi   Santiago   Seoul   Singapore   Sydney   Taipei   Toronto

Higher Education

SPREADSHEET TOOLS FOR ENGINEERS USING EXCEL, THIRD EDITION

1 2 3 4 5 6 7 8 9 0 DOC/DOC 0 9 8 7 6 5

ISBN-13 978–0–07–297184–2
ISBN-10 0–07–297184–3

Publisher: *Suzanne Jeans*
Senior Sponsoring Editor: *Bill Stenquist*
Editorial Assistant: *Megan Hoar*
Outside Developmental Services: *Lachina Publishing Services*
Executive Marketing Manager: *Michael Weitz*
Senior Project Manager: *Kay J. Brimeyer*
Lead Production Supervisor: *Sandy Ludovissy*
Associate Media Producer: *Christina Nelson*
Cover Designer: *Rick D. Noel*
(USE) Cover Image: *McGraw-Hill Digital Archive, image # TA10025*
Compositor: *Lachina Publishing Services*
Typeface: *10.5/12 Times Roman*
Printer: *R. R. Donnelley Crawfordsville, IN*

Library of Congress Cataloging-in-Publication Data

Gottfried, Byron S., 1934–
    Spreadsheet tools for engineers using Excel / Byron S. Gottfried. — 3rd ed.
       p. cm.   — (McGraw-Hill's BEST—basic engineering series and tools)
    Includes index.
    ISBN 978–0–07–297184–2 — ISBN 0–07–297184–3 (hard-copy : alk. paper)
    1. Engineering—Data procession. 2. Microsoft Excel (Computer file). 3. Electronic spreadsheets. I. Title. II. Series.

  TA345.G67   2007
  620'.00285'536—dc22   2005021256
                                                                         CIP

www.mhhe.com

McGraw-Hill continues to bring you the *BEST* (**B**asic **E**ngineering **S**eries and **T**ools) approach to introductory engineering education:

Bertoline, *Introduction to Graphics Communications for Engineers,* 3/e ISBN 0073048364

Chapman, *Fortran 90/95 for Scientists and Engineers*, 2/e ISBN 0072922389

Donaldson, *The Engineering Student Survival Guide*, 3/e ISBN 0073019259

Eide/Jenison/Northup, *Introduction to Engineering Design and Problem Solving*, 2/e
    ISBN 0072402210

Eisenberg, *A Beginner's Guide to Technical Communication* ISBN 0070920451

Finkelstein, *Pocket Book of English Grammar for Engineers and Scientists* ISBN 007352946X

Finkelstein, *Pocket Book of Technical Writing for Engineers and Scientists,* 2/e
    ISBN 0072468491

Gottfried, *Spreadsheet Tools for Engineers Using Excel,* 3/e ISBN 0072971843

Palm, *Introduction to MatLab 7 for Engineers,* ISBN 0072922427

Pritchard, *Mathcad: A Tool for Engineering Problem Solving* ISBN 0070121893

Schinzinger/Martin, *Introduction to Engineering Ethics* ISBN 0072339594

Smith, *Teamwork and Project Management*, 3/e ISBN 0073103675

Tan/D'Orazio, *C Programming for Engineering and Computer Science* ISBN 0079136788

Additional Titles of Interest:

Andersen, *Just Enough Unix*, 5/e ISBN 0072952970

Deacon Carr/Herman/Keldsen/Miller/Wakefield, *Team Learning Assistant Workbook*
    ISBN 0073043893

Eide/Jenison/Mashaw/Northup, *Engineering Fundamentals and Problem Solving*, 4/e
    ISBN 0072430273

Eds, *I-DEAS® Student Guide,* 2/e ISBN 0072525444

Holtzapple/Reece, *Foundations of Engineering*, 2/e ISBN 0072480823

Holtzapple/Reece, *Concepts in Engineering,* ISBN 0073011770

Martin/Schinzinger, *Ethics in Engineering*, 4/e ISBN 0072831154

PREFACE

Excel, developed by the Microsoft Corporation, is the world's most widely used spreadsheet program. It is frequently used for budgeting, financial planning, and record-keeping activities. However, it also includes special features for solving many of the problems that typically arise in engineering analysis, such as determining the roots of algebraic equations, fitting curves through data sets, analyzing data statistically, carrying out studies in engineering economic analysis, and solving complicated optimization problems. Excel can also be used to solve other types of technical problems, such as the evaluation of integrals and the solution of interpolation problems, even though it lacks special features that automate these tasks. It is especially well suited for displaying data in various graphical formats. Armed with these tools, Excel thus becomes the modern-day equivalent of the engineer's classical sliderule.

The success of the earlier editions of *Spreadsheet Tools* has prompted yet another revision of this popular book. This most recent edition has been revised in many ways, in response to the suggestions of several reviewers. In particular:

- The sequence of chapters has been changed so that the chapters are grouped more logically and are introduced in (approximately) the order of increasing complexity.

- New problems and examples have been added in several chapters.

- The material on macros has been expanded into a full chapter.

- Many passages have been rewritten to expand the material and to enhance clarity. The underlying basic ideas always precede the Excel implementation.

- The material on linear interpolation has been combined with the chapter on curve fitting, and the material on polynomial interpolation has been removed.

- Greater emphasis is now placed on the use of trendlines to carry out interpolation.

- A brief appendix has been added, summarizing the Excel functions that are commonly used in engineering applications.

- Descriptive captions now accompany all figures and tables.

- The book is compatible with Excel 2003 (also called Excel 11) and all earlier versions of Excel.

The book is intended primarily as a supplementary textbook for use in introductory engineering courses, although it may also be of interest to more advanced students and many practicing engineers. Chapter 1 sets the tone for engineering analysis in general by presenting a brief approach to problem solving and introducing the role of spreadsheets. Chapters 2 and 3 describe the rudiments of Excel. Although these chapters are by no means exhaustive, they contain enough background material so that the reader can understand and solve the problems in any of the subsequent chapters. Experienced Excel users may choose to review these chapters lightly.

Chapter 4 discusses several different types of graphs that are commonly used in engineering. The material in this chapter provides the graphical background required for some of the later chapters, particularly Chapters 8 and 9. The chapter is primarily concerned with *x-y* graphs (called *XY Charts* or *Scatter Charts* in Excel), including semi-log and log-log graphs. However, line graphs (called *Line Charts* in Excel), bar graphs (called *Column Charts* in Excel), and pie charts are also discussed.

Chapters 5 and 6 are concerned with the transfer of data into and out of Excel, and the organization of data within an Excel worksheet. Chapter 7 discusses unit conversions within Excel, and Chapters 8 through 16 address various analytical techniques that are commonly used by engineers.

The last 12 chapters are essentially independent of one another and can be read in any order. Instructors using this book for a course can pick and choose among these chapters freely, in accordance with their own preferences. Each chapter includes several examples and lots of problems. Instructors can easily supplement these problems with other problem sets, reflecting their own disciplinary interests.

Answers to many of the problems can be downloaded from McGraw-Hill's dedicated website at www.mhhe.com/gottfried3e. Some lengthy data sets are also available for download, allowing you to avoid a great deal of tedious typing.

Corrections and suggestions for the next edition are always welcome. Please send e-mail to bsg@engr.pitt.edu.

In closing, I wish to thank the many readers of the earlier editions for their many helpful comments and suggestions, and I wish to express my gratitude to the editorial staff at McGraw-Hill for their close support and cooperation.

The author would like to acknowledge with appreciation the numerous and valuable comments, suggestions, constructive criticisms, and praise from the following reviewers:

Edward R. Evans, Jr., Penn State, Erie; Hector Gutierrez, Florida Institute of Technology; Chad T. Jafvert, Purdue University; James N. Jensen, University at Buffalo; Christi Patton Luks, University of Tulsa; Ramzi J. Mahmood, California State University, Sacramento; Larry Simonson, South Dakota School of Mines and Technology; J. Steven Swinnea, University of Texas at Austin

Byron S. Gottfried

TABLE OF CONTENTS

Part 1	Excel Fundamentals	1
Chapter 1	**Engineering Analysis and Spreadsheets**	**3**
	1.1 A Spreadsheet Overview	3
	1.2 General Problem-Solving Techniques	8
	1.3 Applicable Engineering Fundamentals	9
	1.4 Mathematical Solution Procedures	16
Chapter 2	**Creating an Excel Worksheet**	**19**
	2.1 Entering and Leaving Excel	20
	2.2 Getting Help	25
	2.3 Moving around the Worksheet	30
	2.4 Entering Data	32
	2.5 Correcting Errors	36
	2.6 Using Formulas	37
	2.7 Using Functions	50
	2.8 Saving and Retrieving a Worksheet	57
	2.9 Printing a Worksheet	60
Chapter 3	**Editing an Excel Worksheet**	**65**
	3.1 Editing the Worksheet	65
	3.2 Undoing Changes	71
	3.3 Copying and Moving Formulas	72
	3.4 Inserting and Deleting Rows and Columns	73
	3.5 Inserting and Deleting Individual Cells	76
	3.6 Smart Tags	78
	3.7 Adjusting Column Widths	79
	3.8 Formatting Data Items	81
	3.9 Editing Shortcuts	81
	3.10 Hyperlinks	82
	3.11 Displaying Cell Formulas	86
	3.12 Closing Remarks	89
Chapter 4	**Graphing Data**	**90**
	4.1 Characteristics of a Good Graph	90
	4.2 Creating a Graph in Excel	93
	4.3 *X-Y* Graphs (Excel *XY Charts*, or *Scatter Charts*)	97
	4.4 Adding Data to an Existing Data Set	102
	4.5 Semi-Log Graphs	110
	4.6 Log-Log Graphs	116

4.7   Line Graphs (Excel *Line Charts*)   122
4.8   Bar Graphs (Excel *Column Charts*)   126
4.9   Pie Charts   131
4.10   Closing Remarks   137

Chapter 5   Organizing Data   138

5.1   Creating a List in Excel   139
5.2   Sorting Data in Excel   142
5.3   Filtering Data in Excel   147
5.4   Pivot Tables   154

Chapter 6   Transferring Data   167

6.1   Importing Data from a Text File   167
6.2   Exporting Data to a Text File   173
6.3   Transferring HTML Data   179
6.4   Transferring Data to Microsoft Word   182
6.5   Transferring Data to Microsoft PowerPoint   194

Part 2   Engineering Applications   201

Chapter 7   Converting Units   202

7.1   Simple Conversions   202
7.2   Simple Conversions in Excel   203
7.3   Converting Temperatures   209
7.4   Complex Conversions   211
7.5   Complex Conversions in Excel   212

Chapter 8   Analyzing Data Statistically   217

8.1   Data Characteristics   217
8.2   Histograms   225
8.3   Cumulative Distributions   233

Chapter 9   Fitting Equations to Data   242

9.1   Linear Interpolation   243
9.2   The Method of Least Squares   249
9.3   Fitting a Straight Line to a Set of Data   251
9.4   Least Squares Curve Fitting in Excel   256
9.5   Fitting Other Functions to a Set of Data   264
9.6   Selecting the Best Function for a Given Data Set   279

Chapter 10   Solving Single Equations   297

10.1   Characteristics of Nonlinear Algebraic Equations   298

10.2 Solving Equations Graphically   300
10.3 Solving Equations Numerically   302
10.4 Solving Equations in Excel Using Goal Seek   309
10.5 Solving Equations in Excel Using Solver   316

Chapter 11   Solving Simultaneous Equations   327

11.1 Matrix Notation   328
11.2 Matrix Operations in Excel   338
11.3 Solving Simultaneous Equations in Excel Using Matrix Inversion   340
11.4 Solving Simultaneous Equations in Excel Using Solver   344

Chapter 12   Evaluating Integrals   360

12.1 The Trapezoidal Rule   361
12.2 Simpson's Rule   374
12.3 Integrating Measured Data   383

Chapter 13   Making Logical Decisions (IF-THEN-ELSE)   389

13.1 Logical (Boolean) Expressions   389
13.2 The IF Function   390
13.3 Nested IF Functions   391

Chapter 14   Recording and Running Macros   399

14.1 Recording a Macro   399
14.2 Executing a Macro   401
14.3 Cell Addressing within a Macro   404
14.4 Saving a Macro   409
14.5 Viewing a Macro   413
14.6 Editing a Macro   418

Chapter 15   Comparing Economic Alternatives   424

15.1 Compound Interest   424
15.2 The Time Value of Money   437
15.3 Uniform, Multipayment Cash Flows   439
15.4 Irregular Cash Flows   447
15.5 Internal Rate of Return   455

Chapter 16   Finding Optimum Solutions   463

16.1 Optimization Problem Characteristics   464
16.2 Solving Optimization Problems in Excel   476

Appendix   494

Index   497

EXAMPLES

Example 1.1	Spreadsheet Analysis of a Projectile's Trajectory	4
Example 1.2	Preparing to Solve a Problem	9
Example 1.3	Assessing the Accuracy of a Solution	12
Example 1.4	A Spreadsheet Solution to the Electrical Circuit Problem	13
Example 2.1	Writing an Excel Formula	39
Example 2.2	A Simple Spreadsheet Application	42
Example 2.3	Naming Cells	44
Example 2.4	Student Exam Scores	52
Example 2.5	Evaluating Trigonometric Functions	53
Example 2.6	Saving a Worksheet	58
Example 2.7	Printing a Worksheet	62
Example 3.1	Selecting a Block of Cells	66
Example 3.2	Preparing a Table by Copying Cells	68
Example 3.3	Editing a Worksheet	83
Example 4.1	Creating an X-Y Graph in Excel	98
Example 4.2	Adding Data to an X-Y Graph	102
Example 4.3	Adding Dependent Variables to an X-Y Graph	105
Example 4.4	Creating a Semi-Log Graph in Excel	114
Example 4.5	Creating a Log-Log Graph in Excel	118
Example 4.6	Creating a Line Graph in Excel	122
Example 4.7	Creating a Bar Graph in Excel	127
Example 4.8	Creating a Pie Chart in Excel	132
Example 5.1	Creating a List in Excel	139
Example 5.2	Sorting a List in Excel	142
Example 5.3	Filtering a List in Excel	147
Example 5.4	Creating a Pivot Table in Excel	157
Example 6.1	Importing Data from a Text File	169
Example 6.2	Exporting Data to a Text File	174
Example 6.3	Exporting Data to an HTML File	180
Example 6.4	Copying Data to Microsoft Word	182
Example 6.5	Embedding Data within Microsoft Word	187
Example 6.6	Linking Data between Excel and Microsoft Word	190
Example 6.7	Transferring Data to Microsoft PowerPoint	195
Example 7.1	A Simple Conversion (Feet to Meters)	203
Example 7.2	A Simple Conversion (Feet to Meters) in Excel	207
Example 7.3	Converting Feet to Millimeters	208
Example 7.4	Converting Celsius to Fahrenheit	210
Example 7.5	Converting a Temperature Difference	210
Example 7.6	A Complex Conversion (psi to Pascals)	212
Example 7.7	A Complex Excel Conversion (psi to Pascals)	213
Example 8.1	Analyzing a Data Set	221
Example 8.2	Constructing a Histogram	226
Example 8.3	Generating a Histogram in Excel	227
Example 8.4	Constructing a Cumulative Distribution	233
Example 8.5	Generating a Cumulative Distribution in Excel	234
Example 8.6	Plotting the Cumulative Distribution	236
Example 8.7	Drawing Inferences from the Cumulative Distribution	237
Example 9.1	Linear Interpolation	244
Example 9.2	Linear Interpolation in Excel	245
Example 9.3	Fitting a Straight Line to a Set of Data	252
Example 9.4	Assessing a Curve Fit	255
Example 9.5	Fitting a Straight Line to a Set of Data in Excel	256
Example 9.6	Use of the Regression Feature in Excel	261
Example 9.7	Fitting an Exponential Function to a Set of Data	265
Example 9.8	Fitting an Exponential Function to a Set of Data in Excel	266
Example 9.9	Fitting a Power Function to a Set of Data	270
Example 9.10	Fitting a Polynomial to a Set of Data	274

Example 9.11	Obtaining a Straight-Line Plot Using Different Coordinate Systems	280
Example 9.12	Fitting Multiple Functions to a Set of Data	283
Example 9.13	Variable Substitution	284
Example 9.14	Scaling a Data Set	287
Example 10.1	Identifying the Real Roots of a Polynomial	299
Example 10.2	Solving a Polynomial Equation Graphically	300
Example 10.3	Solving a Polynomial Equation Using the Method of Bisection	305
Example 10.4	Solving a Polynomial Equation Using the Newton-Raphson Method	307
Example 10.5	Solving a Polynomial Equation in Excel Using Goal Seek	310
Example 10.6	Convergence Considerations	312
Example 10.7	Solving a Polynomial Equation in Excel Using Solver	318
Example 11.1	Writing a System of Simultaneous Equations in Matrix Form	329
Example 11.2	Matrix Multiplication	330
Example 11.3	Reconstructing a System of Simultaneous Equations	331
Example 11.4	Matrix Addition, Matrix Subtraction, and Scalar Multiplication	332
Example 11.5	Properties of the Inverse Matrix	333
Example 11.6	Solving Simultaneous Linear Equations Using Matrix Inversion	335
Example 11.7	Matrix Addition, Matrix Subtraction, and Scalar Multiplication in Excel	339
Example 11.8	Solving Simultaneous Equations in Excel Using Matrix Inversion	341
Example 11.9	Solving Simultaneous Linear Equations in Excel Using Solver	346
Example 11.10	Solving Simultaneous Nonlinear Equations in Excel Using Solver	349
Example 12.1	The Trapezoidal Rule for Unequally Spaced Data	364
Example 12.2	The Trapezoidal Rule for Unequally Spaced Data in Excel	366
Example 12.3	The Trapezoidal Rule for Equally Spaced Data	367
Example 12.4	The Trapezoidal Rule for Equally Spaced Data in Excel	369
Example 12.5	Accuracy and the Number of Intervals	371
Example 12.6	Comparing Simpson's Rule with the Trapezoidal Rule	375
Example 12.7	Simpson's Rule in Excel	376
Example 12.8	Simpson's Rule in Excel – Another Approach	379
Example 12.9	Integrating Measured Data	383
Example 13.1	Student Grades	391
Example 13.2	Student Grades Revisited	392
Example 13.3	Simpson's Rule Revisited	393
Example 14.1	Formatting a Number within a Cell	402
Example 14.2	Evaluating a Polynomial	405
Example 14.3	Analyzing Student Exam Scores	409
Example 14.4	Viewing the Class_Standing Macro	415
Example 14.5	Editing a Macro	418
Example 15.1	Accumulating Compound Interest	426
Example 15.2	A Single-Payment Loan	427
Example 15.3	Compound Interest: Frequency of Compounding	428
Example 15.4	A Single-Payment Loan with Monthly Compounding	430
Example 15.5	A Single-Payment Loan with Monthly Compounding in Excel	431
Example 15.6	Present Value of a Future Sum of Money	437
Example 15.7	Comparing Two Economic Alternatives	437
Example 15.8	Determining the Cost of a Loan	440
Example 15.9	Present Value of a Proposed Investment in Excel	441
Example 15.10	Future Value of a Series of Uniform Payments	443
Example 15.11	Present Value of an Irregular Cash Flow in Excel	447
Example 15.12	Use of the NPV Function in Excel	449
Example 15.13	Comparing Two Investment Opportunities	451
Example 15.14	Calculating the Internal Rate of Return	456
Example 15.15	Comparing Two Investment Opportunities Using IRR	458
Example 15.16	Internal Rate of Return as the Root of a Polynomial	460
Example 16.1	Scheduling Production to Maximize Profit	465
Example 16.2	A Minimum-Weight Structure	468
Example 16.3	Solving the Production Scheduling Problem in Excel	477
Example 16.4	Solving a Nonlinear Minimization Problem in Excel	482

PART 1

EXCEL FUNDAMENTALS

CHAPTER 1

ENGINEERING ANALYSIS AND SPREADSHEETS

Engineering analysis is a systematic process for analyzing and understanding problems that arise in the various fields of engineering. To carry out this process successfully you must be familiar with general problem-solving techniques, you must have an overall understanding of the engineering fundamentals that apply to your particular problem, and you must have a working knowledge of the required mathematical solution procedures. It is also very helpful to have available a computer-based *spreadsheet program* that will solve your problem quickly and easily, once you have defined the problem and set it up properly.

This book discusses the use of spreadsheets to solve a variety of introductory engineering problems. Our emphasis will be on the use of Microsoft *Excel* – a popular spreadsheet program – to carry out these procedures. Examples are provided illustrating the use of these procedures for simple but representative engineering applications. The associated mathematical solution procedures are also presented clearly, so that you have an understanding of what the spreadsheet does and how it goes about its business. In this chapter we begin by explaining what a spreadsheet is, in general terms, and we explain what we mean by *general problem-solving techniques, applicable engineering fundamentals*, and *mathematical solution procedures*.

1.1 A SPREADSHEET OVERVIEW

Before the arrival of personal computers, beginning engineering students were required to learn a number of complicated mathematical procedures in order to solve the equations involved in many engineering calculations. Then they were often required to program these procedures for a computer using a general-purpose programming language. Or they might have written programs that would access a pre-written library of mathematical routines. In either case, the students were required to go through lengthy and tedious procedures. This provided a thorough indoctrination to the use of numerical methods in engineering, but many students would grow impatient and lose interest along the way.

As personal computers became commonplace during the 1980s, *spreadsheets* emerged as one of the principal types of personal computer applications. Though originally intended for carrying out financial calculations, today's spreadsheet programs include provisions for implementing many of the commonly used mathematical procedures used by engineers.

In this book we focus our attention on *Excel* because of its popularity, its widespread availability, and its broad range of features. Excel (and several other competing spreadsheet programs) permits engineers to carry out lengthy calculations very easily, without getting bogged down in complicated mathematical procedures. In addition, Excel allows you to organize your results in a clear and logical manner.

Excel offers other advantages as well. For example, Excel has built-in features that allow you to:

- Import, export, store, process, and sort data.

- Display data graphically.

- Analyze data statistically.

- Fit algebraic equations through data sets.

- Solve single and simultaneous algebraic equations.

- Solve optimization problems.

Moreover, many other mathematical procedures can easily be implemented within Excel, simply by making use of its basic features. Thus, Excel allows you to easily solve many of the problems that commonly arise in engineering analysis. How this is accomplished is one of the major themes of this book.

Before going any further, let's look at a typical Excel spreadsheet with an eye toward understanding what spreadsheets are all about.

Example 1.1 Spreadsheet Analysis of a Projectile's Trajectory

In this example we present an Excel spreadsheet that solves a common physics problem: namely, determining the trajectory of a projectile whose initial velocity is v_0 and initial angle θ. To do so, we will make use of the following well-known equations:

$$x = v_x t \tag{1.1}$$

$$y = v_y t - (gt^2)/2 \tag{1.2}$$

where x = horizontal displacement from the original position, ft

   y = vertical displacement from the original position, ft

   t = time, sec

v_x = initial horizontal velocity, determined as $v_x = v_0 \cos(\theta)$, ft/sec

v_y = initial vertical velocity, determined as $v_y = v_0 \sin(\theta)$, ft/sec

v_0 = initial velocity acting at angle θ, ft/sec. (Note that θ is expressed in radians, not degrees.)

g = acceleration due to gravity, 32.2 ft/sec^2.

You should understand that v_0 and θ represent known information (*input data*). We will use Equations (1.1) and (1.2) to *process* the data, resulting in calculated values for x and y as a function of t (*output data*).

Figure 1.1 shows an Excel spreadsheet (called a *worksheet*) for this problem.

	H3		▼	f_x	100					
	A	B	C	D	E	F	G	H	I	J
1	**Trajectory of a Projectile**									
2										
3	Time (sec)	x (ft)	y (ft)			Initial velocity (v_o) =		100	ft/sec	
4	0.0	0.00	0.00							
5	0.2	16.38	10.77			Angle (θ) =		35	degrees	
6	0.4	32.77	20.14							
7	0.6	49.15	28.11		Gravitational acceleration (g) =			32.2	ft/sec^2	
8	0.8	65.53	34.69							
9	1.0	81.92	39.86				v_x =	81.9	ft/sec	
10	1.2	98.30	43.63							
11	1.4	114.68	46.00				v_y =	57.4	ft/sec	
12	1.6	131.06	46.97							
13	1.8	147.45	46.54							
14	2.0	163.83	44.72							
15	2.2	180.21	41.49							
16	2.4	196.60	36.86							
17	2.6	212.98	30.83							
18	2.8	229.36	23.40							
19	3.0	245.75	14.57							
20	3.2	262.13	4.34							
21	3.4	278.51	-7.28							
22										
23										

Figure 1.1 – A sample Excel worksheet

Note that the worksheet is subdivided into rows and columns, with each row assigned a number and each column identified by a letter. The intersection of a row and column defines a *cell*, with a unique *address*. For example, H3 identifies a cell that contains the numerical value 100; thus H3 is a cell address. Note that the given *parameters* (v_0, θ, and g) are entered into the right portion of the worksheet (in cells H3, H5, and H7, respectively), along with the calculated values of v_x and v_y. The corresponding trajectory is shown as a table in the left portion of the worksheet (columns A through C). Notice that everything is well labeled and units are provided, following good spreadsheet practice.

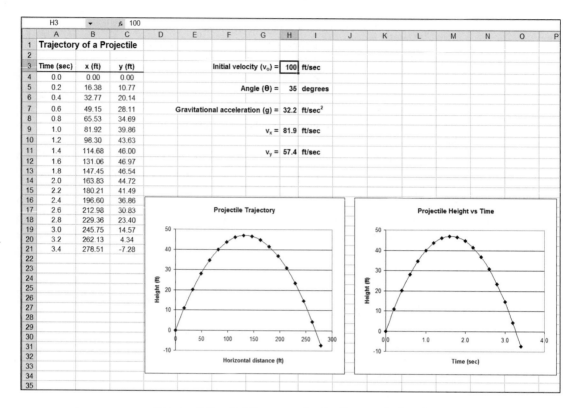

Figure 1.2 – Adding graphs to the worksheet

	F3	▼	ƒₓ 100				
	A	B	C	D	E	F	G
1	Trajectory of a Proj						
2							
3	Time (sec)	x (ft)	y (ft)		Initial velocity (v₀) =	100	ft/sec
4	0	=$F$9*A4	=$F$11*A4-($F$5*A4^2)/2				
5	=A4+0.2	=$F$9*A5	=$F$11*A5-($F$5*A5^2)/2		Angle (θ) = 35		degrees
6	=A5+0.2	=$F$9*A6	=$F$11*A6-($F$5*A6^2)/2				
7	=A6+0.2	=$F$9*A7	=$F$11*A7-($F$5*A7^2)/2		ravitational acceleration (g) = 32.2		ft/sec²
8	=A7+0.2	=$F$9*A8	=$F$11*A8-($F$5*A8^2)/2				
9	=A8+0.2	=$F$9*A9	=$F$11*A9-($F$5*A9^2)/2		vₓ =	=$F$3*COS(RADIANS($F$5))	ft/sec
10	=A9+0.2	=$F$9*A10	=$F$11*A10-($F$5*A10^2)/2				
11	=A10+0.2	=$F$9*A11	=$F$11*A11-($F$5*A11^2)/2		v_y =	=$F$3*SIN(RADIANS($F$5))	ft/sec
12	=A11+0.2	=$F$9*A12	=$F$11*A12-($F$5*A12^2)/2				
13	=A12+0.2	=$F$9*A13	=$F$11*A13-($F$5*A13^2)/2				
14	=A13+0.2	=$F$9*A14	=$F$11*A14-($F$5*A14^2)/2				
15	=A14+0.2	=$F$9*A15	=$F$11*A15-($F$5*A15^2)/2				
16	=A15+0.2	=$F$9*A16	=$F$11*A16-($F$5*A16^2)/2				
17	=A16+0.2	=$F$9*A17	=$F$11*A17-($F$5*A17^2)/2				
18	=A17+0.2	=$F$9*A18	=$F$11*A18-($F$5*A18^2)/2				
19	=A18+0.2	=$F$9*A19	=$F$11*A19-($F$5*A19^2)/2				
20	=A19+0.2	=$F$9*A20	=$F$11*A20-($F$5*A20^2)/2				
21	=A20+0.2	=$F$9*A21	=$F$11*A21-($F$5*A21^2)/2				
22							
23							

Figure 1.3 – Formulas used to generate the values shown in Fig. 1.1

Tabulated data can also be displayed *graphically* within a worksheet, as illustrated in Fig. 1.2. Here we see two different ways to plot the data, even though they have the same shape in this example: y vs x (the actual trajectory) and y vs time.

Note that the entire worksheet requires only three numerical values as input – the values of v_0, θ and g. The remaining values were all generated using *formulas*. Fig. 1.3 shows the formulas that were used to generate the worksheet shown in Fig. 1.1. Note that the formulas shown in columns B and C correspond to equations (1.1) and (1.2), respectively.

It is important to understand that *the entire worksheet will be automatically recalculated if any of the input parameters is assigned a different value.* This important concept is illustrated in Fig. 1.4, where v_0 has been changed to 120 ft/sec and θ has been changed to 40 degrees. Thus, the user can play "what if" by changing various input parameters and immediately seeing the full effects of the changes. This is one of the primary benefits of using a spreadsheet program.

We will discuss the details of entering data and formulas in later chapters of this book. For now, you should focus only on the "big picture" – that is, on what an Excel spreadsheet is, and why it can be useful to engineers and engineering students.

	H3	▼	ƒₓ	120							
	A	B	C	D	E	F	G	H	I	J	
1	**Trajectory of a Projectile**										
2											
3	Time (sec)	x (ft)	y (ft)			Initial velocity (vₒ) =		120	ft/sec		
4	0.0	0.00	0.00								
5	0.2	18.39	14.63			Angle (θ) =		40	degrees		
6	0.4	36.77	27.65								
7	0.6	55.16	39.08		Gravitational acceleration (g) =			32.2	ft/sec²		
8	0.8	73.54	48.91								
9	1.0	91.93	57.13				vₓ =	91.9	ft/sec		
10	1.2	110.31	63.76								
11	1.4	128.70	68.79				v_y =	77.1	ft/sec		
12	1.6	147.08	72.22								
13	1.8	165.47	74.04								
14	2.0	183.85	74.27								
15	2.2	202.24	72.90								
16	2.4	220.62	69.92								
17	2.6	239.01	65.35								
18	2.8	257.39	59.18								
19	3.0	275.78	51.40								
20	3.2	294.16	42.03								
21	3.4	312.55	31.06								
22	3.6	330.93	18.48								
23	3.8	349.32	4.31								
24	4.0	367.70	-11.46								
25											

Figure 1.4 – Effect of changing two parameters

1.2 GENERAL PROBLEM-SOLVING TECHNIQUES

It is very important that you develop good problem-solving habits early in your career. Here are some general suggestions that will help you in the problem-solving process. These suggestions apply to all engineering problems, irrespective of any particular application or any special mathematical procedure.

1. Your perspective of the problem will change over time. Therefore, you should *set aside some time to think about the problem before you attempt to solve it*. This will help you to understand the problem more clearly.

2. Many engineering problems can be represented graphically or pictorially. Therefore, when solving a problem of this type, *draw a sketch of the problem* before you begin to solve it. This will assist you in visualizing the problem.

3. Be sure that you understand the *overall purpose* of the problem and its *key points*. Don't allow yourself to become sidetracked by peripheral or irrelevant information.

4. Ask yourself *what information is known* (input data) and *what information must be determined* (output data). List the input and output information in general terms. (Try to do this in terms of specific variables; do not simply write down numbers.)

5. Ask yourself *what fundamental engineering principles* apply to the problem (see Sec. 1.3). Be certain that you understand how these principles apply to the particular problem you are trying to solve.

6. *Think about how you will solve the problem before you begin the actual solution*. Many problems can be solved several different ways. What mathematical method will you use? Will a computer be required? How will you present the results? Some advance planning will save you a lot of time and grief.

7. Take your time when actually solving the problem. *Develop your solution in an orderly and logical manner*. Be sure that the work is clearly labeled, particularly if you are solving the problem by hand.

8. Once you have obtained a solution, *think about it. Does it make sense?* What assurance do you have that it is correct? Incorrect answers can often be detected in this manner.

9. Be sure that your solution is *clear and complete*. Is the solution presented in an orderly manner? Are the results labeled? Are units included with the numerical answers? Is the logic used to obtain the solution clear? (Most professors are interested in how you obtained your answers as well as the actual numerical results.) Is a table or graph required to present the results in a clear and concise manner?

Remember that *problem solving is a skill that takes time and practice to acquire.* You will become better at it as you acquire more experience with a greater variety of increasingly complex problems.

1.3 APPLICABLE ENGINEERING FUNDAMENTALS

Many engineering problems are based upon one of the following three underlying fundamental principles:

1. *Equilibrium.* Most *steady-state* problems (i.e., problems in which things remain constant with respect to time) are based upon some type of equilibrium. The following are some common forms of equilibrium that you are likely to see in elementary engineering problems:

    (*a*) Force equilibrium.

    (*b*) Flux equilibrium (see item 3 below).

    (*c*) Chemical equilibrium.

2. *Conservation laws.* The two common conservation laws are *conservation of mass* and *conservation of energy.* A great many problems in all fields of engineering are based upon one or both of these principles. (Certain problems are based upon a conservation of momentum principle, though you are unlikely to encounter these in a beginning-level course.)

3. *Rate phenomena.* There are many physically different rate phenomena, though all are represented in the same manner: i.e., a *potential* drives a *flux.* One common example is *Ohm's law* of electrical current flow ($i = \Delta V/R$), where i represents an electrical current (a flux) and ΔV represents a voltage difference (a potential). Here the voltage difference (potential) drives the current flow (flux).

    Another common example of a rate phenomenon is *Fourier's law* of heat conduction ($q = k\Delta T/\Delta L$), where q represents a heat flux (expressed as heat flow per unit area per second) and ΔT represents a temperature difference (a potential). Thus, the temperature difference (potential) drives the heat flux.

Example 1.2 Preparing to Solve a Problem

Suppose you wish to analyze the electrical circuit shown in Fig. 1.5.

(*a*) What is the overall purpose of the problem?

(*b*) What information is known?

(*c*) What information must be determined?

(*d*) What fundamental engineering principles apply to the problem?

(*e*) What will be the overall solution strategy?

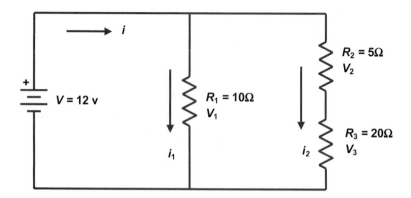

Figure 1.5 – An electrical circuit containing series and parallel resistors

Answers:

(a)   The purpose of the problem is to determine all of the unknown parameters that characterize the behavior of the circuit. These include the current flowing through each path and the voltage drop across each resistance.

(b)   The known information consists of the circuit configuration, the source voltage ($V = 12$ volts), and the values of the individual resistances ($R_1 = 10\ \Omega$, $R_2 = 5\ \Omega$, and $R_3 = 20\ \Omega$).

(c)   The following items must be determined: The currents i, i_1, and i_2 and the voltage drops V_1, V_2, and V_3. (Note that V_1, V_2, and V_3 are the voltage drops across R_1, R_2, and R_3, respectively.)

(d)   The fundamental principles that apply are Ohm's law (rate phenomena) and Kirchhoff's laws (equilibrium).

1.   *Ohm's law*: The voltage drop across a resistor is equal to the product of the current flowing through the resistor and the resistance; i.e., $V = iR$

2.   *Kirchhoff's law – Resistors in series*: The total resistance is equal to the sum of the individual resistances; i.e., $R_T = R_2 + R_3$

   The overall voltage drop is equal to the sum of the voltage drops across the individual resistors; i.e., $V_T = V_2 + V_3$

3.   *Kirchhoff's law – Resistors in parallel*: The voltage boost provided by the source is equal to the voltage drop across each parallel path; i.e., $V = V_1 = V_T$.

   The current provided by the source is equal to the sum of the current flow through each parallel path; i.e., $i = i_1 + i_2$.

   The reciprocal of the overall (equivalent) resistance is equal to sum of the reciprocals of the total resistances for each of the parallel paths; i.e.,

   $$1/R_{eq} = 1/R_1 + 1/R_T = 1/R_1 + 1/(R_2 + R_3)$$

(e)   The overall strategy will be to apply Kirchhoff's laws to the construction of a new circuit that is equivalent to the original circuit but easier to analyze. Ohm's law will be used to determine the unknown currents and voltages along the way. The order of the calculations must be determined carefully. At each step, an unknown quantity must be determined in terms of known information. Thus, the calculations can be carried out in the following manner:

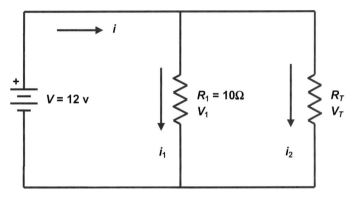

Figure 1.6 – Replacing two series resistors with one overall resistor

(i)   Determine the overall resistance R_T of the rightmost path as $R_T = R_2 + R_3$. This allows us to replace the original network with the simpler network shown in Fig. 1.6.

(ii)   Recognize that $V_1 = (V_2 + V_3)$ = the source voltage V.

(iii)   Determine the current i_1 flowing through the leftmost path (i.e., across R_1) as $i_1 = V_1/R_1$.

(iv)   Determine the current i_2 flowing through the rightmost path as $i_2 = V/R_T$.

(v)   Determine the total current flow i as $i = i_1 + i_2$.

(vi)   Determine V_2 as $V_2 = i_2 \times R_2$.

(vii)   Determine V_3 as $V_3 = i_2 \times R_3$.

(viii)   *Check*: Does $V_2 + V_3 = V$, as it should? (If not, something is wrong. Go back and find the error.)

(ix)   Determine R_{eq} as $1/R_{eq} = 1/R_1 + 1/R_T$. This allows us to replace the network shown in Fig. 1.6 with the simpler network shown in Fig. 1.7.

(x)   *Check*: Does $i = V/R_{eq}$, as it should? (If not, go back and find the error.)

Though this problem is simple enough to solve by hand, a spreadsheet solution is shown in Example 1.4.

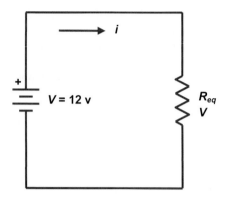

Figure 1.7 – Replacing two parallel resistors with an equivalent resistor

Example 1.3 Assessing the Accuracy of a Solution

A group of students have obtained the following solution for the problem presented in Example 1.2:

$i = 1.68$ amp $i_1 = 0.48$ amp $i_2 = 1.20$ amp
$V_1 = 12.0$ volts $V_2 = 9.6$ volts $V_3 = 2.4$ volts

(*a*) Is the solution clear and complete?

(*b*) Does the solution appear to be correct?

Answers:

(*a*) The solution is clear and complete, though the detailed calculations are not shown. The unknown quantities are labeled with the proper units.

(*b*) When the students thought about the solution, they realized it was not correct. First, of the two branch currents, i_1 and i_2, the larger current (1.20 amps) is flowing through the rightmost path, which has the higher resistance (25 Ω). This does not make sense. Furthermore, within the rightmost path, the larger voltage drop (9.6 volts) occurs across the smaller resistance. This also does not make sense (because Ohm's law states that the *voltage drop is proportional to the resistance*).

Once the students examined their results more carefully, they realized that their calculations were correct but that their results had been transposed. The correct solution is

$i = 1.68$ amp $i_1 = 1.20$ amp $i_2 = 0.48$ amp
$V_1 = 12.0$ volts $V_2 = 2.4$ volts $V_3 = 9.6$ volts

Example 1.4 A Spreadsheet Solution to the Electrical Circuit Problem

Let us examine an Excel spreadsheet solution to the electrical circuit presented in Example 1.2. Our focus in this example will again be an overview of the use of a spreadsheet program in solving a simple engineering problem. We will defer the "how to" details until later chapters.

Figure 1.8 shows an Excel worksheet of the entire problem. In general terms, it is similar to the worksheet shown in Fig. 1.1. Note once again that the worksheet is subdivided into *rows* and *columns*. Each row is numbered, and each column is identified with a letter. The intersection of a row and column defines a *cell*, with a unique *address*. Thus, cell B12 contains the numerical value 25. The heavy outline around cell B12 indicates that it is the *currently active cell*.

	B12	▾	f_x =B7+B8					
	A	B	C	D	E	F	G
1	**Electrical Circuit Analysis**						
2							
3	**Known Information:**						
4							
5	Source voltage V =	12	volts				
6	Resistance R1 =	10	ohms				
7	Resistance R2 =	5	ohms				
8	Resistance R3 =	20	ohms				
9							
10	**Solution:**				**Redundant Checks:**		
11							
12	Resistance Rt =	25	ohms		V = (V1 + V2) =	12	volts
13	Voltage drop V1 =	12	volts		Req =	7.142857	ohms
14	Current i1 =	1.2	amperes		i = V/Req =	1.68	amperes
15	Current i2 =	0.48	amperes				
16	Current i =	1.68	amperes				
17	Voltage drop V2 =	2.4	volts				
18	Voltage drop V3 =	9.6	volts				
19							

Figure 1.8 – A well-labeled worksheet for the electrical circuit problem

Figure 1.9 shows a more abbreviated version of the same worksheet (the units accompanying the numerical values are not included, and the labels in column A are less descriptive). Most students (and many practicing engineers) will develop a worksheet in this manner, though your professor will most likely encourage you to follow the more descriptive style shown in Fig. 1.8. A well-labeled spreadsheet always helps to avoid confusion and minimize communication errors.

Within Fig. 1.9, columns A and D contain descriptive text (*labels*) that identify the numerical data in columns B and E. Cells B5 through B8 contain numerical constants representing the known (given) information. However, cells B12 through B18 and E12 through E14 contain numerical values generated by *formulas*. These values represent the steps used to obtain the problem solution. Each cell (beginning with cell B12 and proceeding down, then moving to column E) corresponds to one of the steps listed under part (*e*) in Example 1.2.

Notice that cell B12 is outlined by a rectangular block. This is the currently active cell. Its value (25) is generated by the formula (=B7+B8) shown at the top of the worksheet. Thus, the value shown in this cell is obtained by adding the numerical values in cells B7 and B8 (the values are 5 and 20, respectively).

	B12	▼	*fx*	=B7+B8		
	A	B	C	D	E	F
1	**Electrical Circuit Analysis**					
2						
3	Known Information:					
4						
5	V =	12				
6	R1 =	10				
7	R2 =	5				
8	R3 =	20				
9						
10	Solution:			Redundant Checks:		
11						
12	Rt =	25		V = (V1 + V2) =	12	
13	V1 =	12		Req =	7.142857	
14	i1 =	1.2		i = V/Req =	1.68	
15	i2 =	0.48				
16	i =	1.68				
17	V2 =	2.4				
18	V3 =	9.6				
19						

Figure 1.9 – A more abbreviated form of the same worksheet

Figure 1.10 shows a corresponding worksheet containing all of the cell formulas used in Fig. 1.7. Cell B12 is again currently active. Notice that the quantities referenced by the active cell formula (i.e., the constants within cells B7 and B8) are also outlined, though these cells have a different type of border than that surrounding the currently active cell.

	B12	▼	*fx*	=B7+B8	
	A	B	C	D	E
1	**Electrical Circuit Analysis**				
2					
3	Known Information:				
4					
5	V =	12			
6	R1 =	10			
7	R2 =	5			
8	R3 =	20			
9					
10	Solution:			Redundant Checks:	
11					
12	Rt =	=B7+B8		V = (V1 + V2) =	=B17+B18
13	V1 =	=B5		Req =	=1/(1/B6+1/B12)
14	i1 =	=B5/B6		i = V/Req =	=E12/E13
15	i2 =	=B5/B12			
16	i =	=B14+B15			
17	V2 =	=B15*B7			
18	V3 =	=B15*B8			
19					

Figure 1.10 – Worksheet showing cell formulas used to obtain calculated values

Finally, in Fig. 1.11 we see what happens when one of the given values is changed. In particular, we have changed the value of R_2 from 5 to 10 ohms. As a result, the values of R_t, i_2, i, V_2, and V_3 have all changed (because their respective cell formulas are all dependent on the value of R_2). (The new values are shown in boldface, for emphasis.) Thus, we see one of the major benefits associated with the use of spreadsheets; namely, if you change one or more of the given input values, the entire worksheet will be automatically and immediately recalculated.

	B7	▼	f_x	10		
	A	B	C	D	E	F
1	**Electrical Circuit Analysis**					
2						
3	Known Information:					
4						
5	V =	12				
6	R1 =	10				
7	R2 =	10	(R2 has been changed from 5 to 10 ohms)			
8	R3 =	20				
9						
10	Solution:			**Redundant Checks:**		
11						
12	Rt =	**30**		V = (V1 + V2) =	12	
13	V1 =	12		Req =	**7.5**	
14	i1 =	1.2		i = V/Req =	**1.6**	
15	i2 =	**0.4**				
16	i =	**1.6**				
17	V2 =	**4**				
18	V3 =	**8**				
19						

Figure 1.11 – Effect of increasing R_2 from 5 to 10 ohms

The details regarding the use of labels, numerical constants, and cell formulas, as well as the method used to display all of the cell formulas, are presented in Chapters 2 and 3. You should focus only on the overall details for now.

When solving a problem with a spreadsheet program, note that you must still think through the problem and apply the applicable engineering fundamentals. The spreadsheet simply assists you by performing the calculations for you (thus eliminating a frequent source of many errors), and allowing you to display the results in an orderly manner.

1.4 MATHEMATICAL SOLUTION PROCEDURES

Once a problem has been defined and properly formulated, it must, of course, be solved for the desired unknown quantities. This generally requires a particular mathematical procedure that is appropriate for the problem under consideration. Thus, you must know the commonly used mathematical procedures, and you must be able to identify which type of procedure to use with each specific problem.

The following types of mathematical procedures are used frequently in introductory-level (and more advanced) engineering analysis problems.

1. *Data analysis* techniques. These are simple statistical techniques that are used to analyze data (e.g., to calculate means, medians, modes, standard deviations, and histograms). Their use enables an engineer to draw meaningful conclusions about information that is hidden within a set of data.

2. *Curve-fitting* techniques. Here we are concerned with passing a curve through an *aggregate* of data rather than through individual data points. Think of this as a systematic way to "eyeball" a curve through a set of data. The resulting curve is generally of more value than the individual data points, particularly when the data points are subject to some *scatter* (as is usually the case when working with real data).

3. *Interpolation* techniques. These techniques allow an engineer to obtain accurate values for the dependent variable when the corresponding independent variable falls within a set of tabulated data points. Problems of this type arise frequently when working with measured data.

4. Techniques for solving *single algebraic equations*. Such equations appear with great frequency in virtually all areas of engineering. It is important to be able to solve them quickly and efficiently.

5. Techniques for solving *simultaneous linear algebraic equations*. Most of the applications that give rise to single algebraic equations also result in simultaneous algebraic equations once the problem conditions become somewhat more complicated. Moreover, many very complicated problems in engineering analysis can be represented as a set of simultaneous linear algebraic equations.

6. Techniques for *evaluating integrals*. In calculus, the classical interpretation of an integral is the area under a curve. (If you haven't learned this yet, you will very soon.) Many engineering applications require the evaluation of integrals in order to determine averages or to determine the cumulative effect of some process that varies with time or distance.

7. Techniques for carrying out *engineering economic analysis*. Most realistic engineering problems have more than one solution. (There are many ways to design a bridge.) The choice among them is often based upon economic considerations. Engineering economic analysis provides criteria and techniques for comparing one solution with another.

8. *Optimization* techniques. If a problem has multiple solutions, finding the best among them may involve lengthy and tedious search procedures. Optimization techniques provide efficient, systematic methods for carrying out these searches.

There are different approaches to each set of solution procedures. *Classical methods*, which are taught in beginning college math courses, are based upon the use of algebra and calculus. These methods tend to be very elegant but generally apply only to relatively simple problems.

Computers can solve more complicated problems, often using *numerical methods* based upon successive approximations. Unfortunately, solutions obtained in this manner are expressed simply as numerical values, rather than algebraic equations showing the relationship between dependent and independent variables. Thus, numerical methods do not provide the same *functional relationships* that result from the use of classical methods – a distinct disadvantage. On the other hand, numerical methods can provide solutions quickly and easily, even for large and complicated problems. Therefore, the use of computer-based numerical solutions has become very common, both for engineering students and for practicing engineers.

Problems

Each of the following problems can be solved individually or by a small group of students.

1.1 Suggest two practical situations that might require calculating the trajectory of an object, such as that shown in Example 1.1.

1.2 *Fick's law* applies to the diffusion of molecules across a membrane. Fick's law is often written as $q = D(\Delta C/\Delta L)$, where q represents the flow of molecules per unit area per second, D is the diffusion coefficient (or diffusivity), ΔC is the difference in the concentration across the membrane, and ΔL is the thickness of the membrane.

(*a*) In this process, what is the flux? What are its units?

(*b*) What is the potential? What are its units?

(*c*) What is the proportionality factor relating flux and potential? What are its units?

1.3 Cite an example of a problem that is based upon force equilibrium. Draw a sketch. Indicate what is likely to be known and what is likely to be unknown.

1.4   Cite an example of a problem that is based upon chemical equilibrium. Draw a sketch of the problem. Indicate what fundamental physical and/or chemical principles apply to your problem. Suggest how the problem might be solved (but do not actually try to solve it).

1.5   Cite an example of an industrial process that is based upon conservation of mass. Indicate what other physical and/or chemical processes might apply.

1.6   Cite an example of an industrial process that is based upon conservation of energy. (Choose a different process from that cited in the previous problem.) Indicate what other physical and/or chemical processes might also apply.

1.7   Suppose you are asked to analyze the truss shown in Fig. 1.12.

   (*a*)   What is the overall purpose of the problem?

   (*b*)   What information is known?

   (*c*)   What information must be determined?

   (*d*)   What fundamental engineering principles apply to the problem?

   (*e*)   What will be the overall solution strategy?

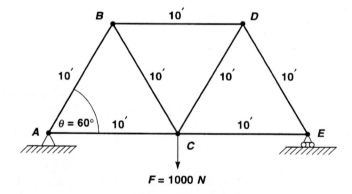

Figure 1.12 – A simple truss

1.8   Can you suggest several fundamental engineering principles that apply to the launching of a NASA space shuttle?

CHAPTER 2

CREATING AN
EXCEL WORKSHEET

To recap what we learned in Chap. 1, a *spreadsheet* is basically a table containing numerical and/or alphanumeric values. The individual elements within the spreadsheet are known as *cells*. A cell can contain two different kinds of data: a *numerical constant* (a *number*) or a *text constant* (also called a *label* or a *string*). Each cell is referenced by its column heading (typically a single letter) and its row number. This is called a *cell address* or *cell reference*. Thus, B3 refers to the cell in column B, row 3. A tabular collection of cells is referred to as a *worksheet*.

If a cell contains a numerical value, the number may have been entered directly, or it may be the result of a *formula evaluation*. Formulas express interdependencies among cells within a worksheet. For example, suppose the numerical value in cell C7 is generated by the formula =(C3+C4+C5). This formula states that the value in cell C7 is obtained by adding the values in cells C3, C4, and C5. Thus, changing the content of one cell (say C4) will change the contents of other cells (cell C7, and any other cells whose values are generated by formulas that refer to cell C4). This important feature allows the user to carry out *what-if* studies: i.e., *what* happens *if* the content of a particular cell is altered? What will be the effect of this change on other values within the worksheet?

Contemporary spreadsheet programs also provide many other features including formatting, editing, file management and database capabilities, a wide variety of special functions, and the ability to represent data graphically. In addition, the newer versions of most popular spreadsheet programs include special mathematical procedures that are useful for solving problems that arise in engineering analysis. The use of many of these features is discussed in this book.

Excel is a popular spreadsheet program developed and supplied by the Microsoft Corporation, either as a stand-alone product or as a part of Microsoft's *Office* suite. It contains many different features, including multiple command paths (different ways to do the same thing), capabilities for creating a variety of graphs and drawings, data management capabilities, various shortcut menus, and extensive online help. In the next two chapters we will consider only those bare-bones features that are essential for the material described later in this book. For more information on Excel, you are encouraged to consult Excel's online help, or any of the many tradebooks describing its use.

Versions of Excel are available both for Intel-compatible PCs using various Windows operating environments and for the Apple family of computers. This discussion assumes you are familiar with at least one of these environments.

2.1 ENTERING AND LEAVING EXCEL

To enter Excel, select the appropriate icon from the Desktop or from the Microsoft Office Shortcut Bar. (An *icon* is a small graphical symbol.) Typically, this will involve clicking the mouse on an easily recognizable Excel icon. Once you do this, you will see the opening Excel window, consisting of several lines of information surrounding an empty Excel *worksheet*, as shown in Fig. 2.1. The principal items within the window are illustrated in Fig. 2.1 and described below. (Your screen may look slightly different, depending on your particular version of Excel.)

Title Bar

The top line is called the *Title Bar*. It includes the spreadsheet name, an icon that closes Excel at the left, and icons that either change the size of the active window or close Excel at the right. We will discuss these icons later, as the need arises. For now, however, note that you can exit from Excel by clicking on the left icon and then selecting Close from the resulting drop-down menu, or by clicking on the rightmost icon (the × symbol within a small box).

Menu Bar

The second line is called the *Menu Bar*. Selecting one of the choices (File, Edit, View, Insert, . . . , Help) causes one of Excel's *drop-down menus* to appear. The File menu, for example, includes selections for opening new or existing worksheet files, closing files, saving files, printing files, and exiting Excel. The Edit menu includes selections for deleting, moving, or copying portions of a worksheet, and for inserting or deleting objects within the worksheet. And so on.

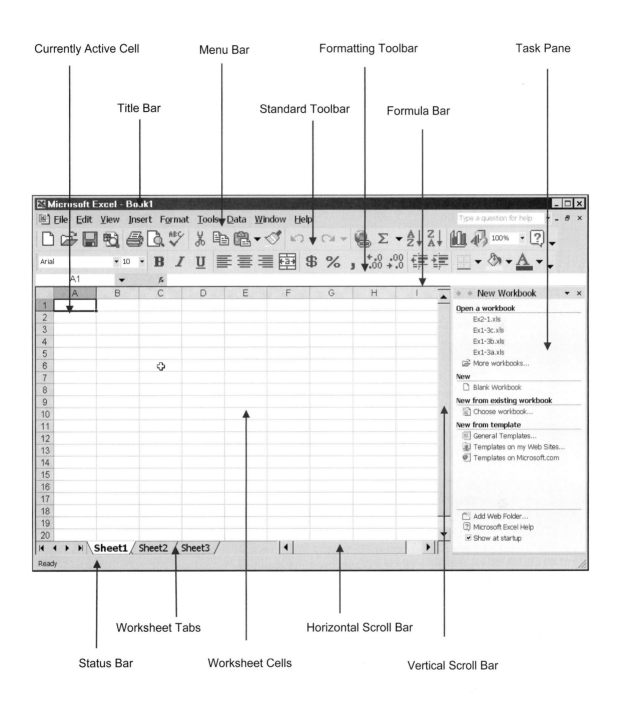

Figure 2.1 – An Excel window

We will discuss the more commonly used menu selections within several of the drop-down menus within the next two chapters, as the need arises. For now, however, note that each drop-down menu in the more recent versions of Excel may appear in two different forms. A short menu, showing the most commonly used features, may appear initially. This may be replaced automatically by a longer menu after a few seconds of inactivity. The long menu may also be obtained by clicking on the double arrow at the bottom of the short menu. Figure 2.2(*a*) shows the short form of the Edit menu, whereas Fig. 2.2(*b*) shows the long form.

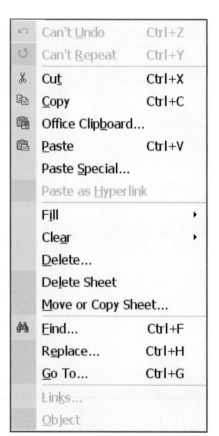

Figure 2.2(*a*) – A short menu   **Figure 2.2(*b*) – A long menu**

You can leave Excel at any time by selecting Exit from the File menu. (This is an alternative to clicking on the rightmost icon in the title bar, as described previously.) Or you can close the current worksheet, while remaining in Excel, by clicking on the *leftmost* icon in the *Menu Bar* and then selecting Close from the resulting menu. Note also that you can obtain detailed assistance by selecting Help from the Menu Bar or by pressing function key F1 (more about this later).

Toolbars

The third and fourth lines are called *toolbars*. The icons in these lines duplicate several of the more commonly used menu selections that are available from the Menu Bar. For example, the upper toolbar (the *Standard Toolbar*) contains icons that will open a new workbook, open an existing workbook, save a workbook, print a workbook, etc. (a *workbook* is a collection of one or more worksheets, as described below). This toolbar also contains icons that allow you to carry out certain common operations, such as summing the values in a row or column of numbers. The lower toolbar (the *Formatting Toolbar*) contains icons that allow you to control the appearance of the items shown on the worksheet.

The toolbars also contain drop-down menus whose choices control the portion of the worksheet that is displayed, the current font, and the current font size. These choices also duplicate selections that are available from the Menu Bar. Thus, the toolbars contain repetitions of selections available elsewhere. They are present only for convenience; in fact, they can be removed from the screen if you wish (by selecting Toolbars from the View menu and then adding or removing the available choices from the resulting dialog box).

The Excel Worksheet

Most of the Excel window is occupied by a worksheet consisting of a grid of *cells*. Each cell has its own column letter and row number. These cells will contain the actual problem description; hence, their contents will comprise the detailed worksheet. We will see how information is entered and manipulated within these cells later in this chapter (see Sec. 2.3).

The Formula Bar

The *Formula Bar* appears just above the worksheet. It contains information about the *active worksheet cell* (see Sec. 2.2). The Formula Bar includes the large formula display area on the right, and a smaller *Name Box* on the left. In Fig. 2.1 the formula display area contains a reference to the currently active cell A1. We will say more about the Formula Bar later in this chapter (see Sec. 2.3).

Worksheet Tabs

Beneath the worksheet is a set of *worksheet tabs,* similar to the tabs that separate the sections within a standard three-ring notebook. Each of these tabs accesses a different worksheet. Any of these worksheets can be selected simply by clicking on the appropriate tab. Collectively, these worksheets are referred to as a *workbook*. Thus, Fig. 2.1 indicates that Sheet1 is the active worksheet in the workbook Book1 (note that the current workbook name appears in the title bar). The underlying assumption is that all of the worksheets within a workbook are related to one another. Each workbook is saved and stored as a single file.

Scroll Bars

The worksheet window also contains horizontal and vertical *scroll bars*. The horizontal scroll bar appears in the second last row, to the right of the sheet tabs, and the vertical scroll bar appears at the right side of the screen. The scroll bars allow rapid movement through the worksheet. They are particularly useful for those worksheets that extend beyond the confines of the screen.

To move through the worksheet, click the mouse on the *scroll button* that appears within each scroll bar and drag the button in the desired direction of movement. For example, to move vertically downward through a long worksheet, drag the vertical scroll button (shown just beneath the upward-pointing arrow in Fig. 2.1) downward. You may also move through the worksheet by clicking on one of the arrows at either end of each scroll bar. Or you may click on the scroll bar itself, on either side of the scroll button.

The Status Bar

The bottom line is called the *Status Bar*. It provides information about the current state of the spreadsheet model (e.g., Ready) and the current command or menu selection. It also shows the current input status, with flags such as CAPS (for uppercase), EXT (extended cell selection mode), NUM (numeric keypad activated), and so on.

The Task Pane

Newer versions of Excel may display a *Task Pane* to the right of the worksheet when Excel is first opened, as shown in Fig. 2.1. The Task Pane offers a convenient way to open a new or recently used worksheet. It can be removed by clicking on the × symbol in its upper right corner, and it can be made to reappear by selecting Task Pane from the View menu. Excel can be configured so that the Task Pane does not appear when Excel is first opened, as discussed below.

Adding/Removing Items from the Worksheet Window

Some of the items discussed above can be removed from the worksheet window, resulting in a larger worksheet display (more rows and columns) with less clutter. For example, the Standard Toolbar and the Formatting Toolbar can be removed by selecting Toolbars from the View menu. In addition, many items, such as the Formula Bar, the worksheet tabs, the scroll bars, the Task Pane, and the Status Bar, can be removed by selecting Options/View from the Tools menu and then deselecting these items from the resulting dialog box, which is shown in Fig. 2.3. The toolbars can also be removed by selecting Full Screen from the View menu. These worksheet items can be restored by reversing these procedures. Other items (for example, other toolbars) can also be added or deleted in this manner.

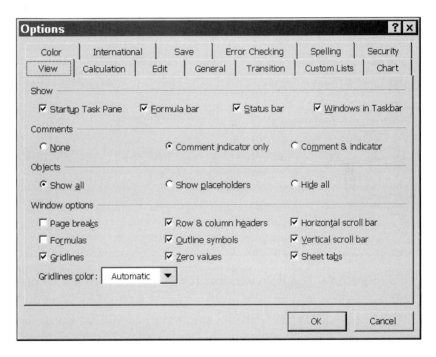

Figure 2.3 – The Tools/Options/View dialog box

2.2 GETTING HELP

Excel includes a very thorough online help feature. To access online help, select Help from the Menu Bar or press function key F1. In either case, a Help window will appear, similar to that shown in Fig. 2.4. The appearance of the Help window may vary from one version of Excel to another, though they are all similar.

Notice the three tabs above the left pane of Fig. 2.4. Selecting the Contents tab will result in a list of overall topics, as shown in the figure. You may select any topic simply by clicking on it (or by clicking on the accompanying plus sign, causing a list of sub-topics to appear). Clicking on a topic or a sub-topic will result in a detailed explanation of that topic or sub-topic in the right pane, as shown in Fig. 2.4 for Excel XP.

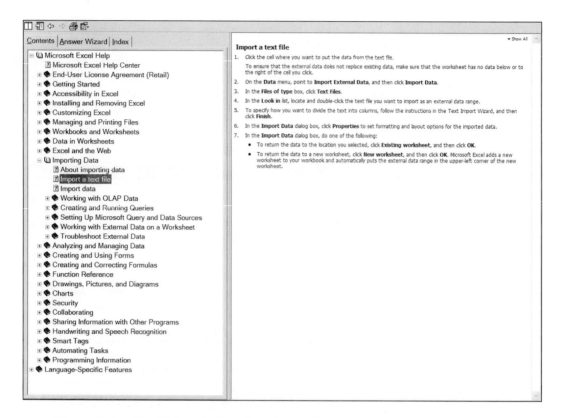

Figure 2.4 – The Help window for Excel XP, with the Contents tab active

Selecting the Answer Wizard tab results in a dialog box that allows you to enter a question in plain text, such as "How can I set the page margins?" Excel will then attempt to provide an appropriate answer within the right pane. Figure 2.5 illustrates how this works in Excel XP.

Clicking on the Index tab results in a lengthy, detailed list of topics, similar to the index in a book. Many people find this the most useful online help feature. Simply find a topic and click on it, resulting in a detailed explanation of that topic within the right pane. Use of the Index feature with Excel XP is illustrated in Fig. 2.6 for Excel XP.

Other versions of Excel may display the Help window differently. For example, Fig. 2.7(a) shows the Help window that is obtained by pressing F1 or selecting Help from the Menu Bar when using Excel 2003 (a version of Excel that is more recent than Excel XP). The window first appears with the Assistance (Search for:) feature active, as in Fig. 2.7(a). Once you enter a search item, you may specify either Online (internet) or Offline (local) help.

If you click on Table of Contents (beneath the Search for: box), the Contents feature will become active, as shown in Fig. 2.7(b).

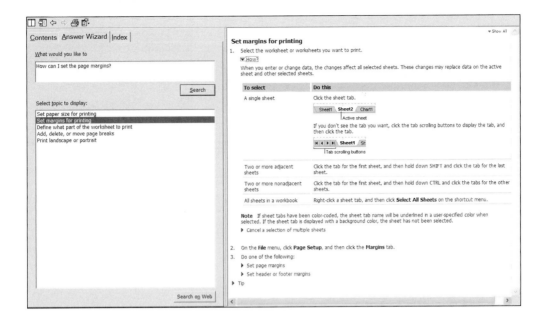

Figure 2.5 – The Help window for Excel XP, with the Answer Wizard tab active

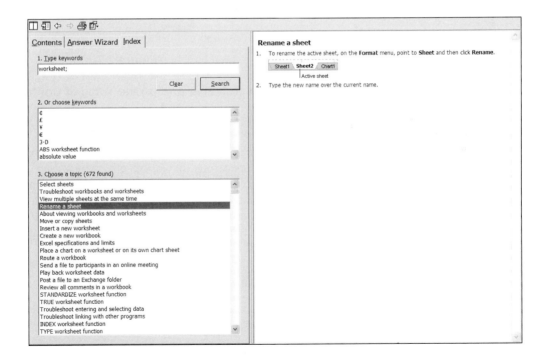

Figure 2.6 – The Help window for Excel XP, with the Index tab active

Figure 2.7(*a*)

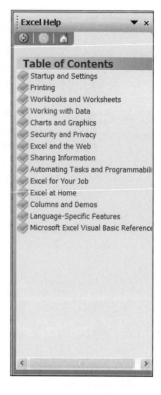

Figure 2.7(*b*)

The Excel 2003 Help window, showing (*a*) the Search area and (*b*) the Table of Contents

Some versions of Excel also include a What's This? feature, which explains the meaning of individual items appearing in worksheets, menus, and dialog boxes. To activate this feature, select What's This? from the Help menu, or hold down the Shift key and press function key F1. Then place the mouse pointer over a specific item, such as a toolbar button or a user-response area within a dialog box, and click the left mouse button. This will generate an information box, typically explaining what the item is and how it can be used.

Help may also be obtained on the Internet by selecting Office on the Web or Microsoft Office Online from the Help menu.

Leaving Excel

We have already seen several different ways to leave Excel. To recap, you can double-click on the left icon in the Title Bar; you can click once on the left icon and then select Close from the resulting menu; or you can click on the rightmost icon (containing the ×) in the Title Bar. You may also leave by selecting Exit from the File menu. If your worksheet is new or if it has been altered in any way, an attempt to exit will generate a dialog box, asking you whether or not to save the current version of the worksheet. If you respond by selecting Yes, another dialog box will appear, prompting you for the name and location of the worksheet file. These two dialog boxes safeguard against your inadvertently leaving Excel without first saving your work. (Some other save options, which allow you to save your work periodically before leaving Excel, are described in Sec. 2.8.)

The use of the features found in the Menu Bar and the various toolbars becomes self-evident with just a little practice. We will discuss certain of these features later in this book, as the need arises.

Problem

2.1 When working through this and the next few problems, feel free to experiment on your own after you have worked through the suggested problems.

(*a*) Enter Excel from the Macintosh or Windows desktop.

(*b*) Click on the middle icon within the group of three at the right side of the Title Bar a few times. Be sure you understand what this icon does.

(*c*) Minimize Excel by clicking on the leftmost icon within the group of three in the Title Bar. Then restore the Excel window by clicking on the Excel button within the *Task Bar* at the bottom of the desktop.

(*d*) Close Excel (i.e., return to the Desktop) by double-clicking on the left icon in the Title Bar (or by clicking once on the left icon and then selecting Close from the resulting menu).

(*e*) Reenter Excel. Then select File from the Menu Bar. Examine the selections in the File menu. Then remove the File menu by pressing the Escape key or by clicking on some other portion of the worksheet.

(*f*) Select Edit from the Menu Bar. Examine the selections in the Edit menu. Then remove the Edit menu using the method described above.

(*g*) Select Format from the Menu Bar. Examine the selections in the Format menu. Choose the Cells selection, then examine the Number, Alignment, Font, and Border tabs (one at a time). Be sure that you understand what each of these tabs is for. (You may exit from any of the resulting dialog boxes by selecting Cancel.) When you have examined each of these Format selections, remove the Format menu using the method described in (*e*).

(*h*) Select Help from the Menu Bar. (On most computers you can also obtain the Help menu by pressing function key F1.) Then choose Contents from the Help menu. After examining Contents, choose Index. Be sure you understand how to use each of these features. Open a few specific help items. (Exit from each help item in the usual manner, by double-clicking on the icon in the upper left corner or by clicking on the icon in the upper right corner of the help window.)

(*i*) Press the CapsLock key on your keyboard. Note what happens in the Status Bar as you turn the CapsLock on and off a few times. Repeat with the NumLock key and with function key F8 (if present).

(*j*) Close Excel by selecting File/Exit from the Menu Bar. (Note that this is an alternative to the method discussed in (*d*) above).

2.3 MOVING AROUND THE WORKSHEET

We now focus on the worksheet, which is the principal part of the spreadsheet. The worksheet is shown as an empty grid in Fig. 2.1. Each rectangle within the grid represents an empty cell. Cell A1 is the *currently active cell*, as indicated by the rectangle surrounding the cell. The graphical object shown in cell C6 is the *mouse pointer*. This indicates the current mouse position within the worksheet. In Excel the mouse pointer will take on different shapes, depending on the task being performed.

There are many ways to move around the worksheet and change the currently active cell. Cell movement is carried out in any of the ways described below.

Mouse

The easiest way to change the currently active cell is to set the mouse pointer over the desired cell and then click the left button.

If your mouse has a wheel between the buttons, you can turn the wheel to scroll vertically up or down within the worksheet. Or you can move through the worksheet by holding down the wheel and dragging the mouse in the desired direction.

Arrow Keys

The arrow keys are LeftArrow (←), RightArrow (→), UpArrow (↑), and DownArrow (↓). Pressing any of these keys once will result in movement in the indicated direction to an adjacent cell. Holding down one of these keys will result in a sustained movement in the indicated direction until the key is released. Once the key is released, whatever cell is selected will be the currently active cell.

PageUp, PageDown

Pressing the PgUp or PgDn keys results in vertical movement within the worksheet. Each move involves several lines. The movement will result in a new currently active cell.

Ctrl-LeftArrow, Ctrl-RightArrow, Ctrl-UpArrow, Ctrl-DownArrow

Holding down the Ctrl key (or the Command key) and pressing one of the arrow keys results in either horizontal or vertical movement to the opposite edge of the worksheet. As a result of this movement, a new cell will become the currently active cell.

Scroll Bars

Clicking on an arrow within a scroll bar results in movement one row at a time or one column at a time, depending on the scroll bar. Clicking at an arbitrary location within a scroll bar results in a larger movement. Dragging a scroll button results in a more controlled movement in the indicated direction. This type of movement does *not* result in a new currently active cell.

Home, End

Pressing Home or Ctrl-Home restores the upper left corner of the worksheet and causes cell A1 to be the active cell. (Under some conditions, pressing Home simply moves the active cell to the beginning of the current row.) Pressing Ctrl-End causes movement to the last cell containing data within the worksheet.

GoTo Key

Function key F5 on your keyboard is the GoTo key. Pressing this key generates a dialog box. Respond by entering the address of the cell that you wish to become active at the bottom of the dialog box. (This dialog box can also be generated by selecting Go To from the Edit menu.). The dialog box maintains a history of recent moves so that you can easily return to a previous location.

Problem

2.2 This problem provides practice in moving around the Excel worksheet.

(*a*) Using the mouse, move the pointer to some arbitrary location within the window and click left. Which cell is now currently active? Repeat several times.

(*b*)   Use the arrow keys to change the location of the currently active cell. Note what happens each time an arrow key is pressed.

(*c*)   Press the PageDown key a few times. What happens each time PageDown is pressed? Which cell is now currently active? Repeat using the PageUp key.

(*d*)   Hold down the Ctrl key (or the Command key) and press the RightArrow key ($\rightarrow$). What happens? Which cell becomes currently active? Repeat using Ctrl and each of the other arrow keys.

(*e*)   Using the mouse, drag the vertical scroll button up and down within the vertical scroll bar. Note what happens. Repeat with the horizontal scroll button. Is the currently active cell affected by this type of movement?

(*f*)   Using the mouse, click on the arrows at the ends of each scroll bar and note what happens. Also, try clicking the mouse in different locations within each scroll bar and note what happens.

(*g*)   Does your mouse have a wheel between the buttons? If so, turn the wheel and notice the resulting movement of the worksheet. Press the wheel down and release, then drag the mouse across the worksheet. Notice what happens.

(*h*)   Move the active cell to some arbitrary location within the worksheet. Then press Home and note what happens. Try the same thing with Ctrl-Home.

(*i*)   Press function key F5 and then specify some arbitrary cell address at the bottom of the dialog box. What happens when you press OK?

(*j*)   Select GoTo from the Edit menu. Then specify some arbitrary cell address at the bottom of the dialog box. Note what happens when you press OK. (This is an alternative to the use of function key F5.)

2.4 ENTERING DATA

You may enter either a numerical value (called a *number constant* in Excel) or a string (called a *text constant* in Excel) into any worksheet cell. Remember that errors can always be corrected using the techniques described in Sec. 2.5.

Numerical Values

Numerical values are entered as ordinary numbers, with or without a decimal point. Negative values should be preceded by a minus sign. Scientific notation is also permitted (e.g., to enter 4×10^8, type 4E+8 or simply 4E8). Thus, any of the following numerical values could be entered into the worksheet:

| 2 | −6 | 3.33 | 2.55E−12 | −7.08E+6 |
| 0.0 | 0.004 | −1.2e−7 | 4E8 | 1e10 |

Numerical values can also be entered in other ways; for example, with embedded commas, preceded by a currency sign ($), or followed by a percent sign (%). You may also enter a numerical date (e.g., 5/24/2005) or a time (e.g., 7:20 PM or 19:20:00). However, these formats are generally not used when carrying out technical calculations; hence, we will not discuss them further.

As you enter a numerical value in the currently active cell, you will see the value being entered in the Formula Bar as well as in the active cell. When you are finished typing the number, press the Enter key. You may also click on the ✓ symbol in the Formula Bar to accept the number, or click on the ✕ to reject the number. Figure 2.8 shows a number being entered in cell A1. Note the presence of the symbols ✕, ✓, and *fx* in the Formula Bar (provided the Formula Bar is the active element in the worksheet, as indicated by the blinking vertical cursor).

After the number has been entered, it will continue to be visible in the Formula Bar as well as in the active cell. However, the symbols ✕ and ✓ will no longer appear in the Formula Bar.

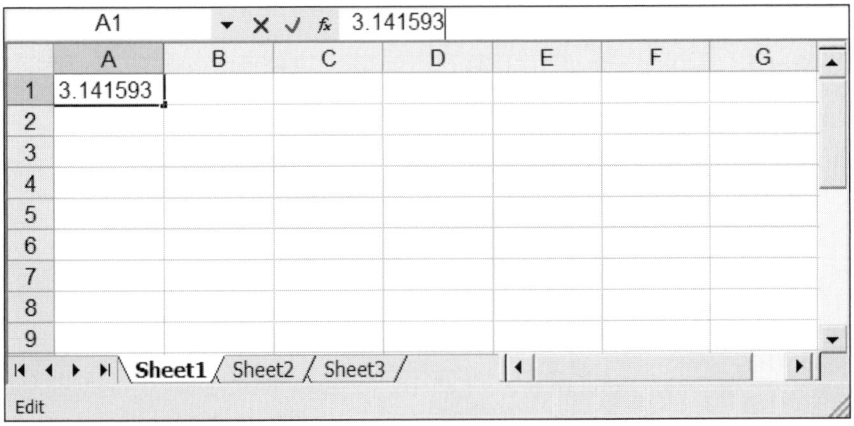

Figure 2.8 – Entering a numerical value in cell A1

The appearance of a numerical value (e.g., the number of decimal places) can be altered by selecting Cells... from the Format menu, then selecting the Number tab, and finally choosing Number from the Category listing. You may then make the appropriate changes in the resulting dialog box, as shown in Fig. 2.9. (You may also select a different numerical format from the Category listing if you wish, such as Currency or Scientific.) Figure 2.10 shows the same worksheet as Fig. 2.8 except that the numerical value has been rounded to one decimal place using this technique.

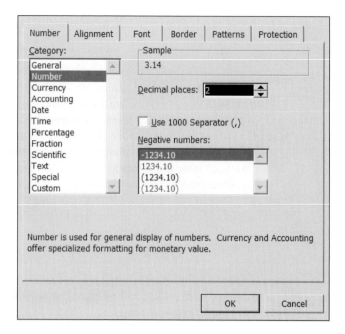

Figure 2.9 – The Format/Cells/Number menu

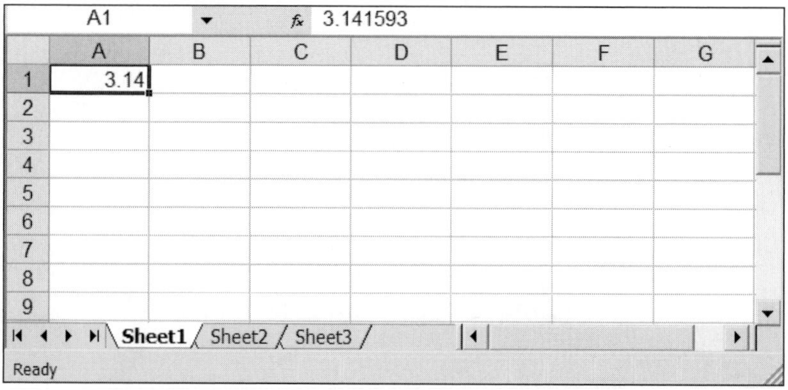

Figure 2.10 – Formatting a numerical value

Strings (Labels)

A *string* (also called a *label*) is entered as a text constant simply by typing the desired text into the active cell. The text will be visible in the Formula Bar as it is being entered and after it has been entered into the active cell. The ×, ✓, and *fx* symbols will appear in the Formula Bar while the string is being entered, just as with a numerical value. The string will extend to the right of the active cell if it is wider than the cell width.

Now suppose a long string has been entered into the currently active cell (for example, cell C5) and you wish to enter some other data in the cell to the right of the currently active cell (e.g., cell D5). When the new data is entered, the rightmost portion of the long string will be hidden behind the new data. Hence, only the leftmost portion of the string will appear within the original cell (C5). The entire string will still be stored within the computer's memory, however, and it will still be associated with the original cell (C5). If you again make this cell active, you will see the entire string in the Formula Bar, even though a portion of the string appears overwritten within the worksheet. You may also increase the width of the original cell (by dragging the mouse across the boundary separating the columns in the cell headings or by selecting Column/Width from the Format menu), thus displaying the entire string within the cell.

Problems

2.3 Enter the following grade report into an empty Excel worksheet. Place the heading (Grade Report) in cell A1. (The numerical grades correspond to letter grades, i.e., A = 4.0, B = 3.0, C = 2.0, D = 1.0, F = 0. Fractional values represent pluses and minuses, i.e., A– = 3.7, B+ = 3.3, B– = 2.7, etc.) Be sure that all of the strings are fully visible within the worksheet.

Grade Report
(Your full name)
(Your social security number)

Course	Credits	Grade
Chemistry I	4	3.0
Physics I	4	3.3
English Composition	3	2.0
Intro to Engineering	3	4.0
Calculus I	3	3.7
Seminar	1	2.7

2.4 Create an Excel worksheet that contains the following list of physical constants:

Physical Constants

Constant	Value	Units
Avogadro's number	6.022142×10^{23}	1/mol
Ideal gas constant	0.082054	liter atm/mol K
Pi	3.1415927	
Speed of light in a vacuum	2.997925×10^{8}	m/s
Speed of sound in air (20°C)	343	m/s

Place the title in row 1. Use three columns to represent the list, with the name in column A, the value in column B, and the units in column C. Include headings for each column. Show the title and the column headings in boldface. Show the numerical values to three-decimal accuracy. Use scientific notation to represent Avogadro's number and the speed of light. (From the Format menu, choose Cells... Then choose the Number tab, and choose Scientific from the Category listing.)

2.5 CORRECTING ERRORS

Errors that occur while entering data can easily be corrected. If you are still in the process of entering data (before pressing the Enter key or before clicking on the ✓ symbol in the Formula Bar), you can use the backspace key to remove unwanted characters, then proceed as usual. Or you can retype the entire data item after a number or a string has been entered incorrectly. The original (incorrect) data item will be replaced by the newer one once you press the Enter key. You can also clear out an active cell (i.e., erase its contents) by pressing the Delete key or by selecting Clear/All or Clear/Contents from the Edit menu.

If an erroneous data item is lengthy, you may prefer to edit a portion of the data item rather than retype the entire data item. To do so, make the cell active so that the cell content appears in the Formula Bar. Then edit the data item within the Formula Bar or within the cell once you have entered the edit mode (with the three special symbols visible in the Formula Bar). The editing is carried out in the usual manner, by positioning the cursor at the desired edit point and then deleting and/or inserting individual characters, as required. You may drag the cursor across several characters to highlight an area and then delete the entire highlighted area if you wish. You can also double-click on a number or a word within a string and then delete that number or word by pressing the Delete key.

When you are finished editing, you can accept the edited data item by pressing the Enter key or by clicking on the ✓ symbol. You can also cancel the edit by clicking on the ✗ symbol, thus restoring the data item to its original form. (Remember to save your work periodically to avoid unexpected data losses due to power failures, etc. Some methods for saving your data are explained in Chap. 3.)

Problem

2.5 Make the following changes to the worksheet described in Prob. 2.3.

(*a*) Change the heading from Grade Report to Semester Grade Report.

(*b*) Clear out your social security number.

(*c*) Change Intro to Engineering to Engineering Analysis.

(*d*)   Change the number of credits for Physics I from 4 to 3.

(*e*)   Change the grade for English Composition from 2.0 to 2.7.

When making these changes, try all of the various cell-editing features described above.

2.6 USING FORMULAS

We have already seen that formulas express interdependencies among the values in different cells within a worksheet. Formulas allow you to perform arithmetic operations on numerical values, combine strings, and compare the contents of one cell with another.

In Excel, *formulas begin with an equal sign* (=), *followed by a numerical expression* involving *constants*, *operators*, and *cell addresses*. Consider, for example, the formula =(B2+C3+5). In this formula, the numerical expression is (B2+ C3+5), the terms B2 and C3 are cell addresses, 5 is a numerical constant, and the plus sign (+) is an *arithmetic operator*. If this formula is entered into cell D7, then cell D7 will contain the sum of the numbers in cells B2 and C3, plus 5. Thus, if B2 contains the constant 2 and C3 contains 8, then D7 will contain 15 (2+8+5=15), as shown in Fig 2.11.

You should understand that *if either of the numbers in cells B2 and C3 is changed, the numerical value in cell D7 will change correspondingly.*

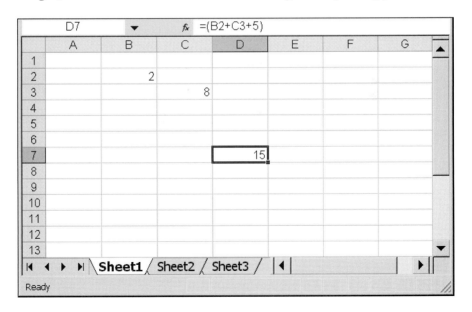

Figure 2.11 – Use of a formula

Operators

Excel includes *arithmetic operators*, a *string operator*, and *comparison operators*. The arithmetic operators combine numerical values (called *operands*) to produce a numerical value. The string operator is used to combine (i.e., to *concatenate*) two strings. The comparison operators compare one operand with another, resulting in conditions that are either *true* or *false*.

The permissible Excel operators are summarized below. (They are the same as the operators in Microsoft's Visual Basic programming language.)

To a beginner, the complete list of Excel operators may seem a bit overwhelming. However, most simple engineering applications use only the first five arithmetic operators (addition, subtraction, multiplication, division, and exponentiation).

Arithmetic Operator	*Purpose*	*Example*
+	Addition	A1+B1
−	Subtraction	A1−B1
*	Multiplication	A1*B1
/	Division	A1/B1
^	Exponentiation	A1^3
%	Percentage (divide by 100)	A1%

Note that the % operator requires only one operand (a cell reference or a constant). *Do not confuse this operator with the % icon in the Formatting Toolbar* (see Fig. 2.1).

String Operator	*Purpose*	*Example*
&	Concatenate	A1&B1

The individual operands can be either strings or numbers, though the result of the concatenation will always be a string.

Comparison Operators	*Meaning*	*Example*
>	Greater than	C1 > 100
>=	Greater than or equal to	C1 >= 100
<	Less than	C1 < 100
<=	Less than or equal to	C1 <= 100
=	Equal (equivalent)	C1 = C2
<>	Not equal (not equivalent)	C1 <> C2

Note that the result of a comparison operation is either *true* or *false*.

Example 2.1 Writing an Excel Formula

Suppose we wish to evaluate the algebraic equation

$$z = (5x^2 + 3y - 7) / (2x + 1)$$

within an Excel worksheet, where x and y are assigned the values $x = 10$ and $y = 12$. Suppose the value of x is entered in cell B1 as a numerical constant, y is entered in cell B3 as a constant, and the calculated value of z will be placed in cell B5. The appropriate Excel formula, which will appear in cell B5, is

=(5*B1^2+3*B2–7)/(2*B1+1)

Notice that the formula contains the *addresses* of the variables x and y, not the variables themselves. Also, note the multiplication operators preceding cell addresses B1 and B2. (Multiplication cannot be implied in Excel – it must be shown explicitly.)

Figure 2.12(*a*) shows a corresponding Excel worksheet. Notice that the calculated value is $z = 25.19048$, as shown in cell B5, as desired. (The value is automatically shown to five decimal accuracy.) The formula used to obtain this value is shown in the Formula Bar. Note that the formula is visible only when cell B5 is active.

Figure 2.12(*b*) shows the same worksheet except that the numerical result has been rounded to one decimal place, using the method presented in Sec. 2.4.

Figure 2.12(*a*) – Evaluating an algebraic expression in Excel

Most formulas include references to other cells. The cell references can, of course, be typed directly, or you can click on each referenced cell instead of typing it. Thus, to evaluate the formula =5*B1+3, you can type =5*, then click on cell B1, and then finish the formula by typing +3 at the end. This can simplify the process of typing long, tedious formulas and reduce the chances of typing the formula incorrectly.

Formulas that refer to other cells can be written two different ways – using *relative addressing* or *absolute addressing*. We will discuss this topic in Chap. 3.

B5	▼	f_x	=(5*B1^2+3*B3-7)/(2*B1+1)

	A	B	C	D	E	F	G
1	x =	10					
2							
3	y =	12					
4							
5	z =	25.2					
6							
7							
8							
9							

Sheet1 / Sheet2 / Sheet3 /

Ready

Figure 2.12(*b*) – Rounding the result to one decimal place

Circular References

A formula can refer to another formula, providing the second formula (i.e., the referenced formula) does not also refer to the first formula. If the formulas refer to each other, however, we have a *circular reference*, in which neither formula can be evaluated because each requires prior knowledge of the other. Thus, *circular references must be avoided when writing cell formulas*.

Suppose, for example, that cells A1 and A2 each contain a numerical constant, cell C1 contains the formula =A1+A2, and cell C2 contains the formula =2*C1. This is permissible, because the first formula (in cell C1) is independent of the second formula (in cell C2). When the second formula is evaluated, the value of the first formula will already be known. Hence, its evaluation is straightforward.

Now, however, suppose that cells A1 and A2 each contain a numerical constant, as before, but cell C1 contains the formula =C2+3. We assume that cell C2 still contains the formula =2*C1. Thus, the first formula (in cell C1) refers to the second formula (in cell C2), and the second formula refers to the first formula. This is a circular reference, because neither formula can be evaluated without first knowing the value of the other formula. Hence, neither formula can be evaluated.

When a circular formula is encountered, a message box appears, as shown in Fig 2.13. Clicking on OK automatically opens a help screen with instructions for correcting the problem, and the Circular Reference toolbar will appear on top of the worksheet, as shown in Fig 2.14. The features within the Circular Reference toolbar allow you to trace the dependency of a formula upon other cells, thus assisting in identifying the source of the circular reference. Note the marker connecting the cells with the conflicting formulas within the worksheet.

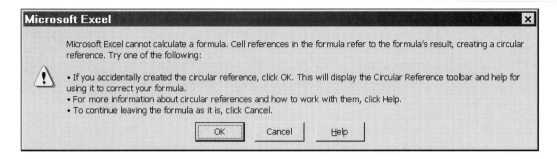

Figure 2.13 – Message box indicating a circular reference

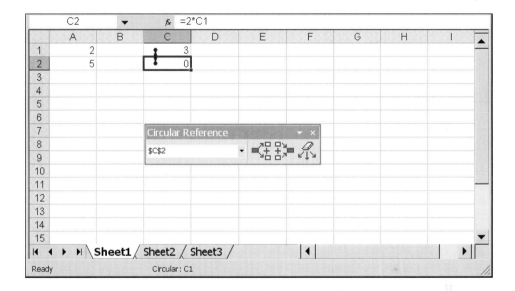

Figure 2.14 – A circular reference (C1 refers to cell C2, and C2 refers to C1)

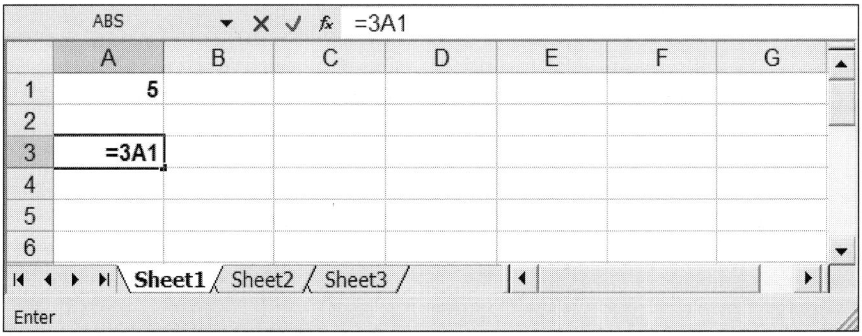

Figure 2.15 – Worksheet containing an incorrect formula in cell A3

Excel attempts to warn you if you enter a formula incorrectly. For example, Fig. 2.15 shows a worksheet in which the user has incorrectly entered the formula =3A1 (rather than =3*A1) into cell A3. The warning shown in Fig. 2.16 is generated when the user presses the Enter key. In addition, recent versions of Excel (e.g., Excel 2002) include a *Formula Auditing Toolbar*, as shown in Fig. 2.17. This toolbar allows you to evaluate formulas, check formulas for errors, and show the relationship between formulas and their dependent cells.

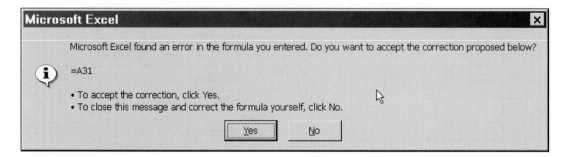

Figure 2.16 – Warning message indicating an incorrect formula

Figure 2.17 – The Auditing Toolbar

Example 2.2 A Simple Spreadsheet Application

A small machine shop has the following parts on hand:

Item	*Quantity*
Screws	8000
Nuts	7500
Bolts	6200

Create a worksheet that includes this information, plus the total number of parts on hand. Use a formula to calculate the total.

The desired worksheet is shown in Fig. 2.18. Note that the worksheet includes several strings (cells B2 through B5, cell B7, and cell C2), three numerical constants (cells C3, C4, and C5), and one formula (cell C7). The heavy border around cell C7 indicates that it is currently active. Notice that the formula =(C3+C4+C5) used to generate the numerical value in cell C7 is visible in the Formula Bar. Thus, you can distinguish between numerical constants and formula values by examining the content of the Formula Bar.

The sum in cell C7 could also have been obtained simply by selecting cell C7, clicking on the AutoSum icon (Σ) in the Standard Toolbar, and then pressing Enter. The AutoSum feature can be activated beneath a column of numbers or to the right of a row of numbers. Newer versions of Excel include other Auto select features for columns of numbers, such as AVERAGE, COUNT, MAX and MIN. To activate one of these features, click on the downward-pointing arrow next to the Σ symbol (see Fig. 2.1).

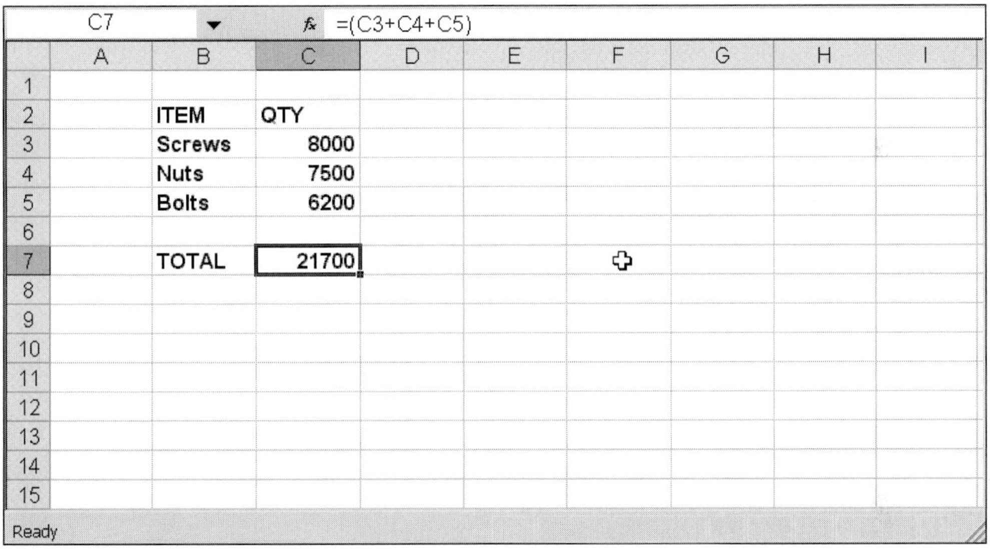

Figure 2.18 – Using a formula to calculate a sum

Operator Precedences

A formula may include several different operators. The question then arises as to which operation is carried out first, which is carried out second, and so on. To answer this question, Excel includes several groups of *operator precedences* that define the order in which the operations are carried out. These operator precedences are listed below, from highest to lowest. Thus, the percentage operation would be carried out first, followed by the exponentiation operation, and so on. The relational comparisons would be carried out last.

Operator Precedence	*Operators*
1	Percentage (%)
2	Exponentiation (^)
3	Multiplication and division (* and /)
4	Addition and subtraction (+ and −)
5	Concatenation (&)
6	Comparisons (>, >=, <, <=, =, <>)

If multiple operators within the same group appear consecutively, the operations will be carried out from left to right. For example, in the formula =(C1/D2*E3), both operators fall within the same group. Therefore, the division operation will be carried out first, and the resulting quotient will be multiplied by the contents of cell E3.

The normal precedences can be overridden by introducing pairs of parentheses. For example, the previous formula can be altered to read =(C1/(D2*E3)). Now the multiplication would be carried out first because this part of the formula is included within the innermost pair of parentheses. Then the contents of cell C1 would be divided by the resulting product. You may use these parentheses freely, in accordance with the logic of your particular problem.

We will say more about the use of cell formulas in Chapter 3, when we discuss copying and moving cells within a worksheet.

Cell Names

Excel allows you to assign names to individual cells so that you can use cell names rather than cell addresses in formulas. To do so, select a cell and highlight the cell address shown in the Name Box (the leftmost portion of the Formula Bar). Then type the desired cell name in place of the cell address (begin with a letter and avoid blank spaces), and press the Enter key. You may then rewrite any formulas that access these cells so that they use either the cell names or the cell addresses.

Example 2.3 Naming Cells

Alter the worksheet created in Example 2.2 so that cell C3 is called Part1, cell C4 is called Part2, cell C5 is called Part3, and cell C7 is called Part_Total. Then change the formula in cell C7 from =(C3+C4+C5) to =(Part1+Part2+Part3).

Figure 2.19 shows the worksheet with cell C3 active, after the cell has been named Part1. The cell name is shown in the Name Box, in the left portion of the Formula Bar. Note that the *value* within this cell (8000) is not affected by the change in the cell name.

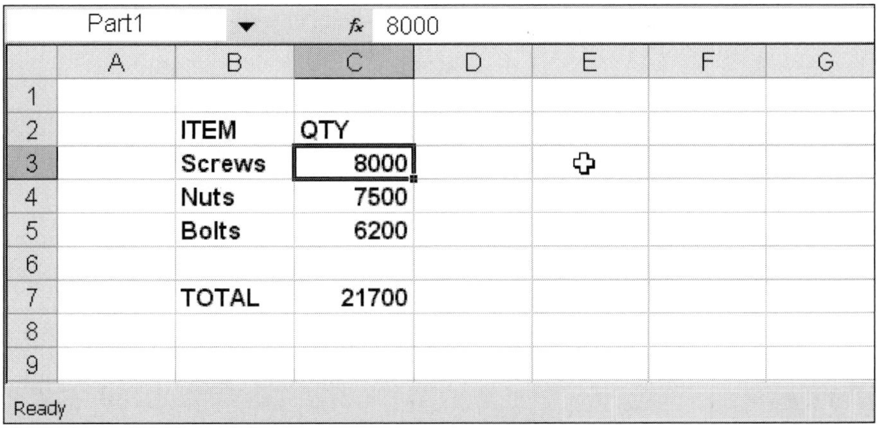

Part1	▼		*fx*	8000			
	A	B	C	D	E	F	G
1							
2		ITEM	QTY				
3		Screws	8000		✛		
4		Nuts	7500				
5		Bolts	6200				
6							
7		TOTAL	21700				
8							
9							
Ready							

Figure 2.19 – Cell C3 is named Part1

Figure 2.20 shows the same worksheet, with cell C7 active. Notice that this cell has been renamed Part_Total, as shown in the Name Box. The formula within this cell has also been changed to =(Part1+Part2+Part3), as shown in the right portion of the Formula Bar. Note that the *value* generated by the formula (21700) is the same as that shown in Example 2.2 (see Fig 2.18).

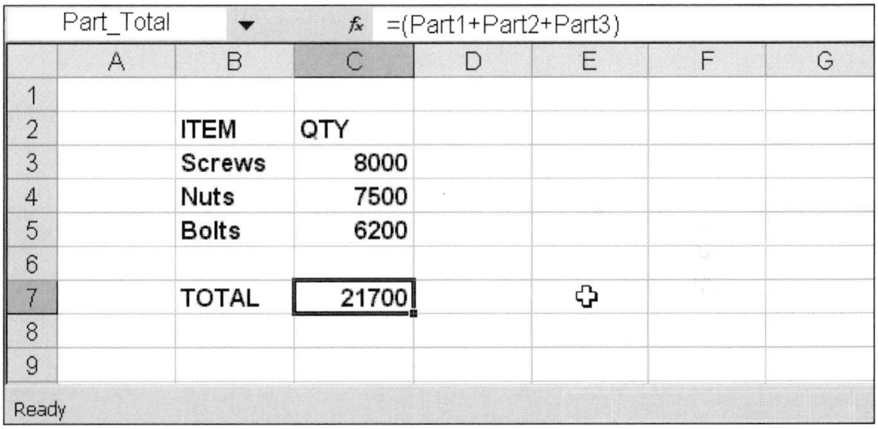

Part_Total	▼		*fx*	=(Part1+Part2+Part3)			
	A	B	C	D	E	F	G
1							
2		ITEM	QTY				
3		Screws	8000				
4		Nuts	7500				
5		Bolts	6200				
6							
7		TOTAL	21700		✛		
8							
9							
Ready							

Figure 2.20 – Cell C7 is named Part_Total

Problems

2.6   Change the numerical values in the worksheet shown in Fig 2.18, Example 2.2, as follows. Notice what happens to the total as each numerical value is changed.

Item	Quantity
Screws	6500
Nuts	9000
Bolts	5400

(a) Obtain the sum in cell C7 using a formula, as in Example 2.2.

(b) Use the AutoSum feature to obtain the total shown in cell C7.

(c) Rename the cells and then obtain the sum using a formula based upon the new cell names, as in Example 2.3.

2.7 Extend the worksheet shown in Figure 2.18, Example 2.2, as follows:

(a) Add a third column containing the following cost information:

Item	Cost
Screws	$0.02 each
Nuts	$0.03 each
Bolts	$0.05 each

(b) Add a fourth column containing the value of the screws, the nuts, and the bolts.

(c) At the bottom of this fourth column, add a cell containing the total value of all of the parts. Use a formula to obtain the total. Then try using AutoSum to obtain the total.

(d) Change the quantity of screws to 6500. What is the effect of this change elsewhere in the worksheet?

(e) Change the cost of nuts to $0.04 each. What is the effect of this change elsewhere in the worksheet?

(f) Based upon the original values, what is the average cost per part? (Do not distinguish between part types when answering this question.)

2.8 At the end of every term, each student's grade-point average (GPA) is calculated as

$$\text{GPA} = \frac{\sum_i C_i G_i}{\sum_i C_i}$$

where C_i refers to the number of credits for the ith course and G_i refers to the numerical grade for that course.

Enter the grade report described in Problem 2.3 and add a row at the bottom of the worksheet indicating the student's GPA for the given semester. Use an Excel formula based upon the above equation to determine the GPA.

2.9   Four hundred grams of nitrogen gas (molecular weight = 28 g/mole) are stored in a 10-liter container. Set up an Excel worksheet to determine the pressure of the gas (in atmospheres) as a function of temperature and volume, assuming nitrogen is an ideal gas. Place the value of each variable in its own cell. Label all of the quantities clearly.

Recall that the ideal gas law is

$$PV = nRT$$

where P = pressure (atmospheres)
V = volume (liters)
n = number of moles (mass of gas/molecular weight)
R = ideal gas constant (0.082054 liter atm/mole °K)
T = absolute temperature (°K = °C + 273)

(*a*)   What pressure corresponds to a temperature of 22°C?

(*b*)   What is the pressure if the temperature is changed to 37°C?

(*c*)   What is the pressure if the original temperature (22°C) is restored but the volume of the container is reduced to 7 liters?

(*d*)   What is the pressure if the original 10-liter container contains 300 grams of oxygen (molecular weight = 32 g/mole) at 18°C?

2.10   Set up an Excel worksheet to determine the future amount of money (*F*) that will accumulate over *n* years if *P* dollars is deposited in a bank account earning *r* percent interest each year, compounded annually. Place each quantity in its own cell. Label all of the quantities clearly.

The formula for computing compound interest is

$$F = P(1+i)^n$$

where i = annual interest rate expressed as a decimal = $r/100$.

(*a*)   How much money will accumulate if $1000 is deposited for 20 years earning 5.5 percent per year, compounded annually?

(*b*)   How much *additional* money will accumulate if $1000 is deposited for 20 years earning 6 percent per year, compounded annually?

(*c*)   How much money will accumulate if $1000 is deposited for 30 years earning 5.5 percent per year, compounded annually?

(*d*)   How much money will accumulate if $5000 is deposited for 25 years earning 5.75 percent per year, compounded annually?

2.11   Figure 2.21 shows an electrical circuit consisting of a DC source and three resistors connected in series. Suppose the source is a 12-volt battery, and the resistances have values of $R_1 = 150\ \Omega$, $R_2 = 250\ \Omega$ and $R_3 = 50\ \Omega$ (note that the symbol Ω represents resistance, in ohms).

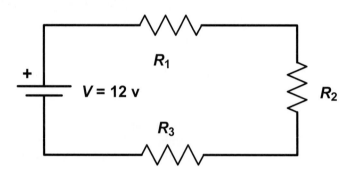

Figure 2.21 – Three resistors in series

Create an Excel worksheet to determine the current (i) flowing through the circuit, the voltage drop across each of the resistors, the power consumed by each resistor, and the total power consumption. Set the worksheet up in the following manner:

Source Voltage =

Current =

	Resistance	Voltage Drop	Power Consumed
R1 =			
R2 =			
R3 =			
Totals =			

Note that the current (in amperes) is determined as $i = V/(R_1+R_2+R_3)$, where V represents the source voltage. The voltage drop across each resistor (in volts) is determined as $V_1 = iR_1$, $V_2 = iR_2$, etc., and the power consumed by each resistor (in watts) is given by $P_1 = iV_1$, $P_2 = iV_2$, etc.

(*a*)   Determine all of the unknown quantities for this circuit.

(*b*)   What happens if the source voltage is increased to 15 V?

(*c*)   What happens if the source voltage is restored to 12 V, but the values of the resistors change to $R_1 = 175\ \Omega$, $R_2 = 100\ \Omega$, and $R_3 = 225\ \Omega$?

(*d*)   What happens if all of the quantities are restored to their original values and a fourth resistor is added whose value is 200 Ω?

2.12 Re-create the Excel worksheet shown in Fig. 1.8 to solve the parallel electrical circuit problem described in Example 1.2.

(a) Verify that the solution given in Example 1.2 is correct.

(b) Change the value of R_3 from 20 to 10 Ω. What effect does this have on the other variables in the problem?

2.13 The volumetric flow through a pipe is determined as

$$V = vA$$

where V = volumetric flow, in cu ft/sec
  v = mean fluid velocity, ft/sec
  A = cross sectional flow area, sq ft

Create an Excel worksheet that can be used to answer the following questions:

(a) What will be the volumetric flow through the pipe if the mean fluid velocity is 10 ft/sec and the pipe has an outer diameter of 4 inches and a thickness of 0.25 inch?

(b) What pipe diameter will be required to support a volumetric flow rate of 1.5 cu ft/sec, assuming the mean fluid velocity remains 10 ft/sec?

2.14 Heat exchangers often utilize concentric tubes, with a hot fluid flowing through the inner tube and a coolant (which itself will heat up) in the *annulus*; i.e., the space surrounding the inner tube (see Fig. 2.22).

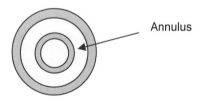

Annulus

Figure 2.22 – Cross-section of an annulus formed by two concentric tubes

Create an Excel worksheet to determine the outer pipe diameter that will support the same volumetric flow as the inner pipe, assuming the fluid velocities will be the same in the annulus and the inner pipe. *Hint*: Since the mean fluid velocities are the same, the cross-sectional flow area of the annulus must be the same as the cross-sectional flow area of the inner pipe: i.e., $A_2 = A_1$. Hence,

$$\frac{\pi}{4}(d_2 - 2t_2)^2 - \frac{\pi}{4}(d_1)^2 = \frac{\pi}{4}(d_1 - 2t_1)^2$$

or $d_2 = 2t_2 + \sqrt{d_1^2 + (d_1 - 2t_1)^2}$

where d_2 = the outside diameter of the outer pipe, in
 d_1 = the outside diameter of the inner pipe, in
 t_2 = the thickness of the outer pipe, in
 t_1 = the thickness of the inner pipe, in

Determine the outer pipe diameter for a pipe whose thickness is 0.25 inch, if the inner pipe has an outside diameter of 3 inches and a thickness of 1/8 inch.

2.15 A *manometer* is a U-shaped tube that is used to measure pressure differences. The tube contains a fluid, often mercury. If each end of the tube is at a different pressure, the pressure difference will cause the fluid to rise in one leg of the manometer (the low pressure side). The greater the pressure difference, the greater the height of the column (i.e., change in fluid level). The appropriate equation is

$$(p_2 - p_1) = \rho g h$$

where p_2 = the higher pressure, Pa (*note*: 1 Pa = 1 N/m^2)
 p_1 = the lower pressure, Pa
 ρ = the density of the fluid, kg/m^3
 g = the acceleration due to gravity, 9.80 m/s^2
 h = the increase in the level of the fluid in the low pressure side, m

Create an Excel worksheet to determine the pressure difference as a function of the height of the fluid column and the fluid density.

(*a*) What will be the pressure difference if the manometer contains mercury and the height of the column is 10 cm? The density of mercury is 13.6×10^3 kg/m^3.

(*b*) What column height corresponds to a pressure difference of 20 kPa (i.e., 20×10^3 Pa)?

2.7 USING FUNCTIONS

Excel includes many different *functions* that can be used to carry out a wide variety of operations. For example, there are groups of functions that carry out mathematical and statistical operations, process financial data, process text, and return information about the worksheet. Table 2.1 lists several commonly used functions that are used elsewhere in this book. A more extensive summary appears in the Appendix. For now, however, we will concentrate on general techniques for using functions.

Table 2.1   Commonly Used Excel Functions

Function	Purpose
ABS(x)	Returns the absolute value of x.
ACOS(x)	Returns the angle (in radians) whose cosine is x.
ASIN(x)	Returns the angle (in radians) whose sine is x.
ATAN(x)	Returns the angle (in radians) whose tangent is x.
AVERAGE($x1, x2, \ldots$)	Returns the average of $x1, x2, \ldots$.
COS(x)	Returns the cosine of x.
COSH(x)	Returns the hyperbolic cosine of x.
COUNT($x1, x2, \ldots$)	Determines how many numbers are in the list of arguments.
DEGREES(x)	Converts x radians to degrees.
EXP(x)	Returns e^x, where e is the base of the natural (Naperian) system of logarithms.
INT(x)	Rounds x down to the closest integer.
LN(x)	Returns the natural logarithm of x ($x > 0$).
LOG10(x)	Returns the base-10 logarithm of x ($x > 0$).
MAX($x1, x2, \ldots$)	Returns the largest of $x1, x2, \ldots$.
MEDIAN($x1, x2, \ldots$)	Returns the median of $x1, x2, \ldots$.
MIN($x1, x2, \ldots$)	Returns the smallest of $x1, x2, \ldots$.
MODE($x1, x2, \ldots$)	Returns the mode of $x1, x2, \ldots$.
PI( )	Returns the value of π. (The empty parentheses are required.)
RADIANS(x)	Converts x degrees to radians.
RAND( )	Returns a random value between 0 and 1. (The empty parentheses are required.)
ROUND(x, n)	Rounds x to n decimal places.
SIGN(x)	Returns the sign of x. (Returns $+1$ if $x > 0$, -1 if $x < 0$.)
SIN(x)	Returns the sine of x.
SINH(x)	Returns the hyperbolic sine of x.
SQRT(x)	Returns the square root of x ($x > 0$).
STDEV($x1, x2, \ldots$)	Returns the standard deviation of $x1, x2, \ldots$.
SUM($x1, x2, \ldots$)	Returns the sum of $x1, x2, \ldots$.
TAN(x)	Returns the tangent of x.
TANH(x)	Returns the hyperbolic tangent of x.
TRUNC(x, n)	Truncates x to n decimals.
VAR($x1, x2, \ldots$)	Returns the variance of $x1, x2, \ldots$.

A function consists of a *function name*, followed by one or more *arguments*. The arguments are enclosed in parentheses and separated by commas. Consider, for example, the function specification SUM(C1,C2,C3). This function calculates the sum of the quantities in cells C1, C2, and C3. The function name is SUM, and the arguments are the cell references C1, C2, and C3.

The previous function could also have been written as SUM(C1:C3). The term C1:C3 is called a *range*. It refers to all of the cells between the given extremities (i.e., between cells C1 and C3). The range designation is very convenient when a function requires many contiguous cells as arguments.

Function arguments need not be restricted to cell references. Any formula can be used as a function argument, as long as it is of the proper type (e.g., a numerical formula cannot be used if a string-type argument is required, and so on). A function argument can even include a reference to another function. Consider, for example, the function specification SUM(A1, SQRT(A2/2), 2*B3+5, D7:D12). This function has four arguments: the cell address A1, the function specification SQRT(A2/2), the formula 2*B3+5, and the range D7:D12. The SQRT function returns the square root of its own argument, which in this example is the formula A2/2. (That is, the SQRT function in this example will return the square root of one-half the quantity in cell A2.)

Example 2.4 Student Exam Scores

A group of students obtained the following exam scores in their Introduction to Engineering class.

Student	*Exam 1*	*Exam 2*	*Final Exam*
Davis	82	77	94
Graham	66	80	75
Jones	95	100	97
Meyers	47	62	78
Richardson	80	58	73
Thomas	74	81	85
Williams	57	62	67

Enter the information into an Excel worksheet. Then determine an overall score for each student, assuming that each exam carries equal weight. Also, determine the class average for each of the exams and determine a class average for the students' overall scores.

An Excel worksheet containing all of the student names and exam scores is shown in Fig 2.23. The class average for exam 1 (71.6) is shown in active cell B10, beneath the individual scores in cells B2 through B8. Notice the corresponding formula =ROUND(AVERAGE(B2:B8),1) in the Formula Bar. Within this formula, the AVERAGE function is used to obtain the actual average. This function is an argument within the ROUND function, which is used to round the calculated average to one decimal place. Similar formulas are used to generate the averages shown in cells C10 and D10.

Cells E2 through E8 contain the overall scores for the individual students. Each of these rounded averages was obtained using a similar formula. For example, the formula used to obtain the value shown in cell E2 is =ROUND(AVERAGE(B2:D2),1); the formula for cell E3 is =ROUND(AVERAGE(B3:D3),1); and so on. The value in cell E10, which represents the average of the students' overall scores, was obtained in a similar manner (see Problem 2.16 below).

B10	▼	f_x =ROUND(AVERAGE(B2:B8),1)						
	A	B	C	D	E	F	G	H
1	Student	Exam 1	Exam 2	Final Exam	Overall Score			
2	Davis	82	77	94	84.3			
3	Graham	66	80	75	73.7			
4	Jones	95	100	97	97.3			
5	Meyers	47	62	78	62.3			
6	Richardson	80	58	73	70.3			
7	Thomas	74	81	85	80			
8	Williams	57	62	67	62			
9								
10	AVERAGE	71.6	74.3	81.3	75.7			
11								
12								
13								
14								
15								
Ready								

Figure 2.23 – Student exam scores

Example 2.5 Evaluating Trigonometric Functions

Suppose we wish to evaluate the following algebraic formulas, taken from Example 1.1, using an Excel worksheet:

$$v_x = v_0 \cos(\theta)$$

$$v_y = v_0 \sin(\theta)$$

where the angle θ is expressed in radians. (*Note*: If θ is originally expressed in degrees, it must be converted to radians before the trigonometric functions can be evaluated.)

Figure 2.24(*a*) shows the required Excel worksheet. The corresponding cell formulas are shown in Fig. 2.24(*b*). Notice the use of embedded functions; e.g.,

=B3*COS(RADIANS(B5)) and =B3*SIN(RADIANS(B5))

We could have carried out the conversion to radians separately from the formula evaluations. For example, the formula

=RADIANS(B5)

might appear in cell B6, and the formulas

    =COS(B6) and =SIN(B6)

in cells B7 and B9, respectively. However, the method we have chosen in the worksheets is more concise.

	B7 ▼		*fx* =B3*COS(RADIANS(B5))		
	A	B	C	D	E
1	Evaluation of Algebraic Formulas				
2					
3	Initial velocity =	100	ft/sec		
4					
5	Theta =	35	degrees		
6					
7	v_x =	81.9	ft/sec		
8					
9	v_y =	57.4	ft/sec		
10					
11					
Ready					

Figure 2.24(*a*) – Evaluating trigonometric formulas

	B1 ▼	*fx*		
	A	B	Formula Bar	C
1	Evaluation of Algebraic Formulas			
2				
3	Initial velocity =	100		ft/sec
4				
5	Theta =	35		degrees
6				
7	v_x =	=B3*COS(RADIANS(B5))		ft/sec
8				
9	v_y =	=B3*SIN(RADIANS(B5))		ft/sec
10				
11				
Ready				

Figure 2.24(*b*) – Corresponding cell formulas

Selecting Functions

There are several different ways to select a function. You may type it in directly, as a part of a formula. This works well if you happen to remember the exact function name. But you may also search for a function by clicking on the Function button (*fx*) in the Formula Toolbar (the Standard Toolbar in older versions of Excel). Or, you may select Function... from the Insert menu. The last two methods display a list of functions and categories in a separate dialog box, as shown in Fig. 2.25(*a*). You may then select a category and a function from the accompanying list, as shown in Fig 2.25(*b*).

Once a function is selected from the list of functions, the proper way to write the function is outlined under the list. This search feature is very helpful if you do not remember the details of the function you wish to use.

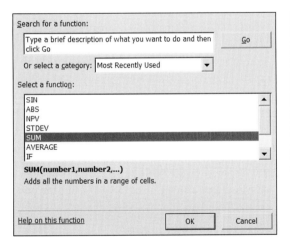

Figure 2.25(*a*) – Searching for a function

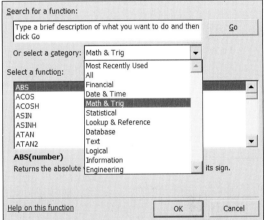

Figure 2.25(*b*) – Selecting a function

If you are typing a formula that requires a function, then clicking on the function within the list of functions shown in Fig. 2.25(*a*) or Fig. 2.25(*b*) will place the function directly into the formula being typed, ready to accept any required arguments. Thus, if the formula =ABS(A1) is being typed into cell B3, then you can simply type the equal sign, then select the ABS function from the list of functions as shown in Fig. 2.25(*b*). The formula will then appear as =ABS(). The cursor will appear within the empty parentheses and a dialog box will appear, as shown in Fig. 2.26. You may then click on cell A1 to add the required argument. The desired formula, =ABS(A1), will then appear in cell B3.

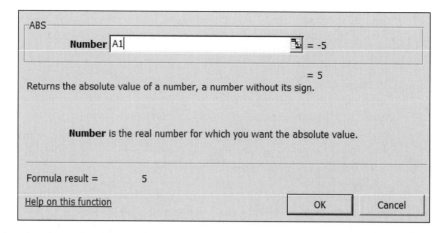

Figure 2.26 – Dialog box requesting an argument for the selected function

Problems

2.16 Answer the following questions for the worksheet shown in Fig 2.23, Example 2.4:

(*a*) Write the formula used to generate the value shown in cell C10.

(*b*) Write the formula used to generate the value shown in cell E4.

(*c*) Write two different formulas that can be used to generate the value shown in cell E10.

2.17 For the student exam scores given in Example 2.4, suppose that each of the first two exams contributes 30 percent to the student's overall score and the final exam contributes 40 percent.

(*a*) Construct a worksheet similar to that shown in Fig 2.22 for this situation. Be sure to include the new weighting factors within the formulas, as required.

(*b*) In column G, calculate the difference between each student's overall score and the class average (i.e., the average of the overall scores, shown in cell E10).

(*c*) In cell G10, calculate the average of the differences in cells G2 through G8. What value do you expect to see here, and why?

2.18 Modify your solution to Prob. 2.14 so that the formula used to calculate d_2 makes use of the SQRT function.

2.19 Create an Excel worksheet that will evaluate each of the following algebraic expressions. Use appropriate Excel functions to evaluate each expression.

(a) $y = 4e^{-3x}$ where $x = 0.3$

(b) $z = \sqrt{\left| x^2 - y^2 \right|}$ where $x = 2.5$ and $y = 3.7$

> *Note*: This expression will require two different Excel functions – one to obtain the absolute value of $x^2 - y^2$, and the other to determine the square root of the absolute value.

(c) $y = 2\sin(3\theta)$ where $\theta = 27$ degrees.

> *Note*: This expression will require two different Excel functions – one to convert θ from degrees to radians, and the other to determine the sin of θ (see Example 2.5).

2.20 An Excel worksheet contains the formulas shown in Fig. 2.27(*a*). These formulas produce the numerical results shown in Fig. 2.27(*b*). Whenever you press F9 the calculated values change.

What is the purpose of this worksheet?

E4	▼	*fx*

	A	B	C
1	One	Two	Sum
2	=RAND()	=RAND()	
3	=INT(6*A2)+1	=INT(6*B2)+1	=A3+B3
4			

Figure 2.27(a)

E4	▼	*fx*

	A	B	C
1	One	Two	Sum
2	0.330244	0.542368	
3	2	4	6
4			

Figure 2.27(b)

2.8 SAVING AND RETRIEVING A WORKSHEET

There are several ways to save a worksheet in Excel. The most common is to click on the Save icon in the Standard Toolbar (see Fig. 2.28) or to select Save from the File menu. Either of these methods will cause the current version of the worksheet to be saved under its current file name, thus replacing any earlier version saved under the same name.

You may also save a worksheet under a different name or to a different location by selecting Save As... from the File menu. This will cause the Save As dialog box to appear (see Fig. 2.29). The Save As dialog box will appear automatically if you attempt to save a worksheet that has not been saved previously, prompting you for a file name and a location.

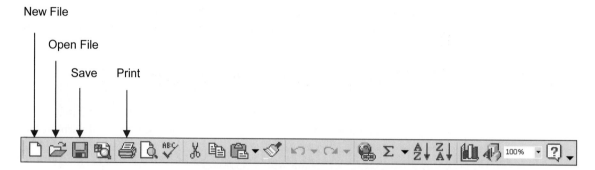

Figure 2.28 – The Standard Toolbar (note the Save and Print buttons)

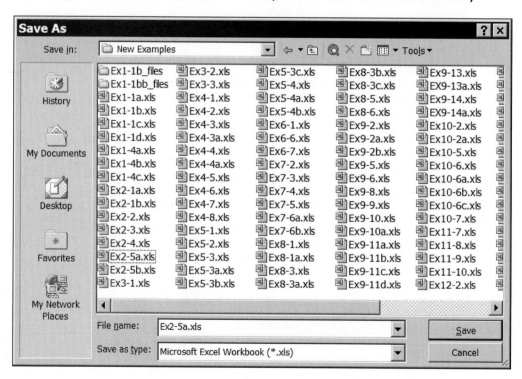

Figure 2.29 – The Save As dialog box

Example 2.6 Saving a Worksheet

Suppose you wish to save a worksheet using the name Sample Bridge Design in the folder called My Designs. Figure 2.30 shows the correct entries in the Save As dialog box Note that the location (My Designs) is shown in the Save in: data entry area at the top of the dialog box. The worksheet name is entered in the File name: entry area near the botton. The extension (.xls) need not be included – it will be added automatically, since the worksheet is being saved as a Microsoft Excel Worksheet (*.xls) file, as indicated at the bottom of the dialog box.

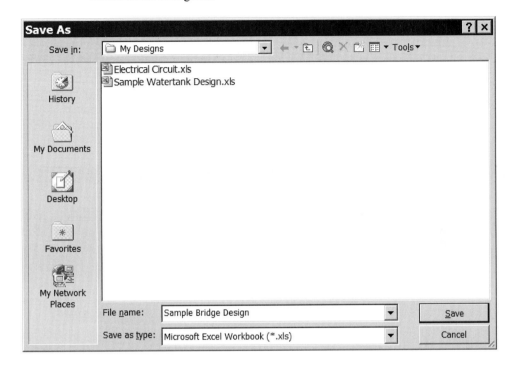

Figure 2.30 – Saving a new worksheet

If a worksheet has been changed in any manner since the last save and you attempt to exit from Excel, you will first see a dialog box asking you whether or not to save the current version of the worksheet before exiting. If you answer Yes, the Save As dialog box will appear, prompting you for a file name and storage location. This feature is intended to prevent you from leaving Excel without saving your most recent work, thereby losing whatever changes you may have made since the last save.

You should save your work frequently when building or editing a worksheet. This will minimize the damage caused by an unexpected power failure or a careless reply to a Save prompt (e.g., answering No rather than Yes).

To *retrieve* a worksheet that has been saved previously, either click on the Open icon in the Standard Toolbar (see Fig. 2.28), or select Open... from the File menu. In either case a dialog box will appear, prompting you for a file name within the current location (i.e., the current folder). You may also change folders if you wish.

2.9 PRINTING A WORKSHEET

Once you have built and edited a worksheet, you may want to print it out on paper. In Excel you can print either the entire workbook, selected worksheets within the workbook, or a selected portion of one worksheet. To print an *entire* worksheet, select Print... from the File menu. Then choose Active sheet(s) from the resulting dialog box, as shown in Fig. 2.31. (Note that there are several options available in this dialog box, most of which are self-explanatory.) The entire worksheet will then be printed out onto one or more pages, as required. (This assumes, of course, that Excel has been set up to recognize your particular printer, that the printer has been turned on and loaded with paper, etc.) You can also print the worksheet by simply clicking on the Print button in the Standard Toolbar (see Fig. 2.28).

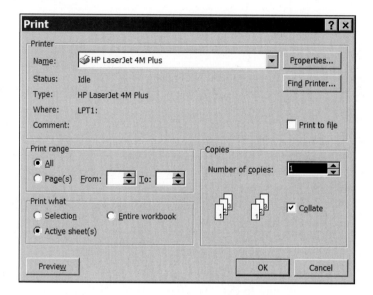

Figure 2.31 – The Print dialog box

When you print an entire worksheet in this manner, you may find that several printed pages are required to display the entire worksheet. This may be true even if the worksheet is relatively small. You may be able to get around this problem by printing only a portion of the worksheet, or by reducing the size of the worksheet image so that it fits on one page.

To reduce the size of the worksheet image so that it fits on one printed page, select File/Page Setup, then select the Page tab. Check the Fit to: box and choose 1 page wide and 1 tall, as shown in Fig. 2.32(*a*). Then click on the Print button.

To print a *part* of a worksheet, you must first select the block of cells to be printed, choose Print... from the File menu, and then choose Selection from the resulting dialog box (see Fig. 2.31). The selected cells will then be printed.

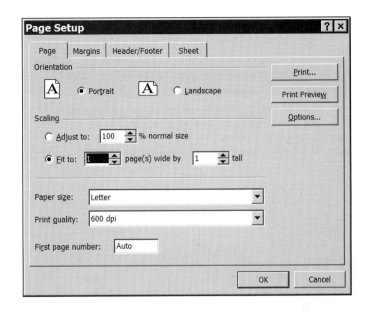

Figure 2.32(*a*) – The Page tab of the Page Setup dialog box

To display gridlines on the printed sheet, select File/Page Setup and choose the Sheet tab. Then check the Gridlines box click and click on the Print button. To display row headings and column numbers, click on the Row and column headings box. Figure 2.32(*b*) illustrates these optional settings.

Figure 2.32(*b*) – The Sheet tab of the Page Setup dialog box

You can also preview the appearance of the printed sheet on your screen by selecting Print Preview from the File menu (see Fig. 2.33). It is a good idea to do this before actually printing, to verify that you like the appearance of the fonts, column widths, page breaks, and so on.

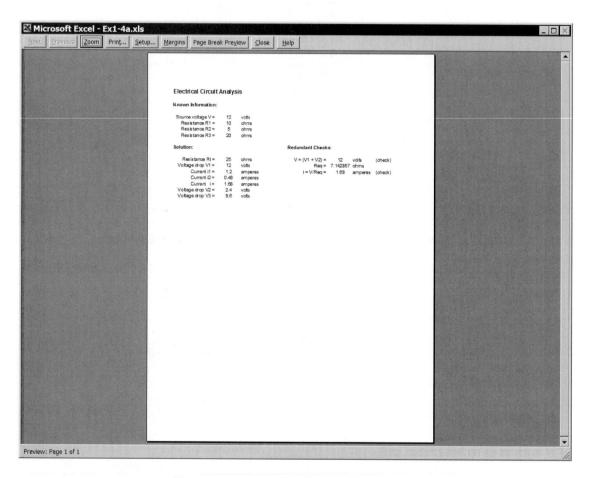

Figure 2.33 – Previewing a printed page

Example 2.7 Printing a Worksheet

Print the Excel worksheet shown in Fig. 1.1 two different ways:

(*a*)   As it would normally appear, without gridlines and row/column headings.

(*b*)   With gridlines and row/column headings.

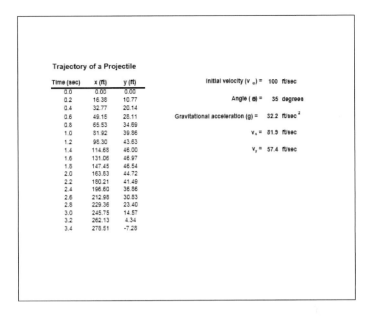

Trajectory of a Projectile

Time (sec)	x (ft)	y (ft)
0.0	0.00	0.00
0.2	16.38	10.77
0.4	32.77	20.14
0.6	49.15	28.11
0.8	65.53	34.69
1.0	81.92	39.86
1.2	98.30	43.63
1.4	114.68	46.00
1.6	131.06	46.97
1.8	147.45	46.54
2.0	163.83	44.72
2.2	180.21	41.49
2.4	196.60	36.86
2.6	212.98	30.83
2.8	229.36	23.40
3.0	245.75	14.57
3.2	262.13	4.34
3.4	278.51	-7.28

Initial velocity (v_o) = 100 ft/sec

Angle (θ) = 35 degrees

Gravitational acceleration (g) = 32.2 ft/sec^2

v_x = 81.9 ft/sec

v_y = 57.4 ft/sec

Figure 2.34(*a*) – An image of a printed worksheet

	A	B	C	D	E	F	G	H	I
1	Trajectory of a Projectile								
2									
3	Time (sec)	x (ft)	y (ft)			Initial velocity (v_o) =		100	ft/sec
4	0.0	0.00	0.00						
5	0.2	16.38	10.77			Angle (θ) =		35	degrees
6	0.4	32.77	20.14						
7	0.6	49.15	28.11			Gravitational acceleration (g) =		32.2	ft/sec^2
8	0.8	65.53	34.69						
9	1.0	81.92	39.86				v_x =	81.9	ft/sec
10	1.2	98.30	43.63						
11	1.4	114.68	46.00				v_y =	57.4	ft/sec
12	1.6	131.06	46.97						
13	1.8	147.45	46.54						
14	2.0	163.83	44.72						
15	2.2	180.21	41.49						
16	2.4	196.60	36.86						
17	2.6	212.98	30.83						
18	2.8	229.36	23.40						
19	3.0	245.75	14.57						
20	3.2	262.13	4.34						
21	3.4	278.51	-7.28						

Figure 2.34(*b*) – Adding gridlines and row/column headings

Figure 2.34(*a*) shows the most common form of the printed worksheet, without gridlines and row/column headings. This was accomplished with the Page Setup/Page options as shown in Fig. 2.32(*a*). Note the selection of Fit to: 1 page(s) wide by 1 tall. This assures that the entire worksheet will appear on one page.

To add gridlines and row/column headings to the printed page, we select these features on the Page Setup/Sheet tab, as shown in Fig. 2.32(*b*). Note, in particular, that the Gridlines and the Row and column headings boxes have been checked. An image of the printed worksheet is shown in Fig. 2.34(*b*).

Problems

2.21 Reconstruct the worksheet shown in Fig. 2.18. Print the worksheet in the following ways:

(*a*) By clicking on the Print button in the Standard Toolbar.

(*b*) By selecting Print/Active sheet(s) from the File menu.

(*c*) By selecting the Page Setup/Page dialog box and then choosing Fit to: 1 page(s) wide by 1 tall, as shown in Fig. 2.32(*a*).

(*d*) By selecting the Page Setup/Sheet dialog box and then clicking on the Gridlines and Row and column headings boxes, as shown in Fig. 2.32(*b*).

2.22 Reconstruct the worksheet shown in Fig. 2.23 and print it in the following ways:

(*a*) On one printed page, in the normal manner (without gridlines, and without row/column headings).

(*b*) Including gridlines and row/column headings.

(*c*) Displaying only cells B1 through E8.

CHAPTER 3

EDITING AN
EXCEL WORKSHEET

In Chapter 2 we learned the basics of creating Excel worksheets. We now focus on the fundamentals of editing an existing worksheet. This includes copying cells; moving cells; moving cell formulas; inserting and deleting individual cells, rows and columns; and formatting (i.e., changing the appearance of) individual cells or groups of cells. We will also learn about smart tags and hyperlinks, and we will see how cell formulas can be displayed within a worksheet.

3.1 EDITING THE WORKSHEET

Many spreadsheet operations are carried out on a *block* of cells (i.e., a group of contiguous cells included within a row, a column, or several adjacent rows or columns). For example, you might want to create a graph from a list of numbers, sort a list of names, print a block of data, or move a block of data from one part of a worksheet to another. Excel includes a number of editing commands that allow you to carry out these block-type operations.

Selecting a Block of Cells

To process a block of cells in Excel, you must *first select the cells and then carry out the desired operation*. Multiple-cell selection is best carried out with a mouse. To select a block of cells, move the cursor to one corner of the cell block, hold down the left mouse button, and drag the mouse across the worksheet to the opposite corner. The entire block of cells will then be highlighted.

Newer versions of Excel allow you to select groups of *nonadjacent* (i.e., physically separated) cells. To do so, select the first cell in the regular manner. Then hold down the Ctrl key while the remaining (nonadjacent) cells are selected. Nonadjacent *blocks* of cells can be selected in the same manner, by selecting the first block of cells in the regular manner and then holding down the Ctrl key while selecting the remaining blocks.

You can also select the *entire worksheet* by clicking on the Select All Button (the blank button in the upper left-hand corner, directly to the left of the column headings and directly above the row headings). Or you can hold down the Control and Shift keys and then press the space bar (or use the Command – Shift buttons and the space bar on a Macintosh).

Excel includes an AutoCalculate feature that automatically shows the sum of all numerical values within a selection. This sum appears within the Status Bar. If you place the mouse cursor over the sum and click the right mouse button, a drop-down menu will appear, allowing you to display other types of numerical values (e.g., the average value, the max, the min, and so on).

Example 3.1 Selecting a Block of Cells

In the worksheet containing student exam scores created for Example 2.4, select the block of cells extending diagonally from cell B2 to cell E10.

The worksheet originally shown in Fig. 2.23 is again seen in Fig. 3.1, with the desired block of cells selected. The selection was obtained by first making cell B2 active and then dragging the mouse diagonally down to cell E10 while holding the left mouse button.

	B2	▼	fx	82			
	A	B	C	D	E	F	G	H
1	Student	Exam 1	Exam 2	Final Exam	Overall Score			
2	Davis	82	77	94	84.3			
3	Graham	66	80	75	73.7			
4	Jones	95	100	97	97.3			
5	Meyers	47	62	78	62.3			
6	Richardson	80	58	73	70.3			
7	Thomas	74	81	85	80			
8	Williams	57	62	67	62			
9								
10	AVERAGE	71.6	74.3	81.3	75.7			
11								
12								
13								
14								
15								
Ready					Sum=2422.8			

Figure 3.1 – Selecting a block of cells

Note that the entire selection is highlighted (the colors are reversed). Cell B2 is currently active, which is why its appearance is different than the remaining cells within the block. Its value (a numerical constant) appears in both the Formula Bar and the cell.

Notice the message Sum=2422.8 within the Status Bar. This is the sum of all numerical values within cells B2 through E10. The value is calculated automatically and displayed in the Status Bar.

Clearing a Block of Cells

Blocks of cells can be cleared, just as single cells are cleared. (Remember that *any information placed in the cells will be lost when the cells are cleared.*) To clear a block of cells, select the block of cells and then press the Delete key. Or, you can select the block of cells and then choose Clear/All, Clear/Contents or Clear/Formulas from the Edit menu.

Copying a Cell to an Adjacent Block of Cells by Dragging

A constant or formula can easily be copied from one cell to an adjacent block of cells by dragging. To do so, proceed as follows:

1. Select the cell to be copied.

2. Place the mouse pointer over (or close to) the small square in the lower right corner of the cell. (The appearance of the mouse pointer will change to a + sign when it is correctly positioned.)

3. Hold down the left mouse button, and drag the selection across all of the desired target cells.

4. Release the mouse button. The content of the original cell will then be copied to the target cells, overwriting any information previously in the target cells.

Now suppose a numerical constant is being copied to one or more cells in an adjacent row or an adjacent column. Dragging the original constant across the desired target cells will cause the constant to be copied to the target cells, as described above. But if you hold down the Control key while dragging, the target cells will automatically fill with successively increasing values. Thus, if cell A1 contains the constant 5, then dragging cell A1 across cells A2, A3, and A4 while holding the Control key will place the values 6, 7, and 8 in those cells.

If the object being copied is a formula rather than a constant, *the cell addresses referred to within the formula will automatically be adjusted to accommodate each new cell location* (unless *absolute addressing* is used, as discussed in the section on copying formulas below). This provides a very convenient way to fill a worksheet with a self-adjusting repeated formula, as illustrated in the following example.

Example 3.2 Preparing a Table by Copying Cells

Prepare an Excel worksheet containing a table of y versus x, where

$$y = 2x^2 + 5$$

and $x = 0, 1, 2, \ldots, 10$

Figure 3.2 shows the worksheet in an early stage of construction. Note that the constant 0 has been entered in cell A4. In addition, notice that the mouse marker is positioned over the lower right corner of cell A4, and its shape has changed to a + sign.

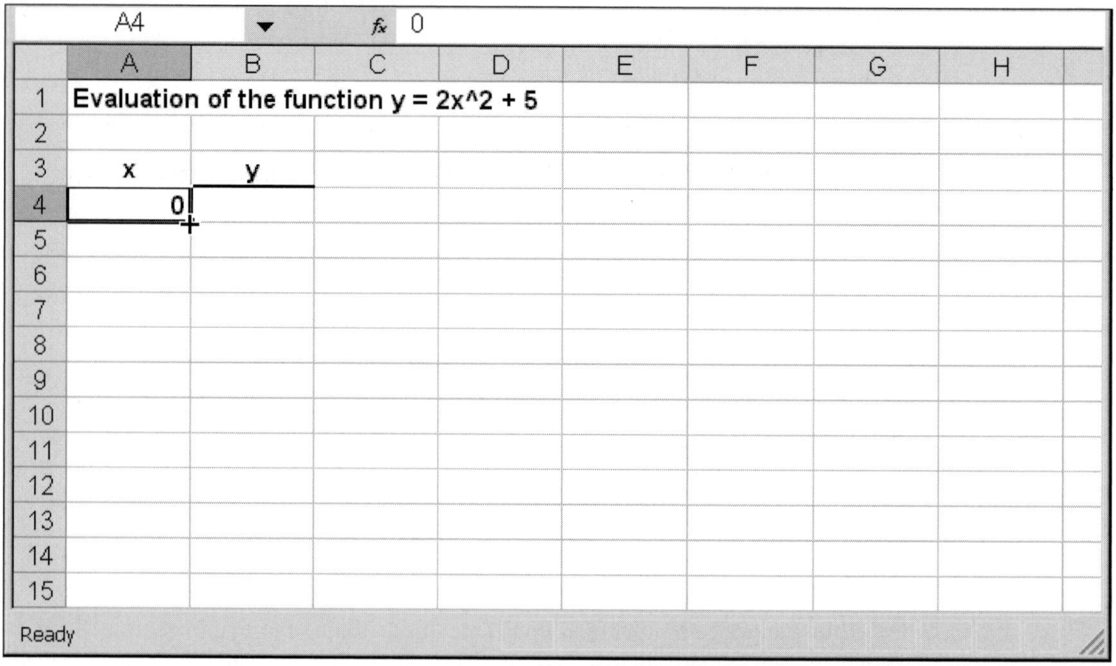

	A	B	C	D	E	F	G	H
	A4	▼	f_x	0				
1	Evaluation of the function y = 2x^2 + 5							
2								
3	x	y						
4	0							
5								
6								
7								
8								
9								
10								
11								
12								
13								
14								
15								

Ready

Figure 3.2 – First step when copying a list of consecutive digits

We now hold down both the Ctrl key and the left mouse button, and drag the mouse over cells A5 through A14. Figure 3.3 shows the resulting worksheet. Note that the values 1 through 10 are copied into cells A5 through A14, respectively.

Next we enter the formula =2*A4^2+5 into cell B4, resulting in the value 5 within this cell (see Fig. 3.4). We then copy this formula to cells B5 through B14. The copying is accomplished by holding down the left mouse button and dragging cell B4 across cells B5 through B14. The values 7, 13, 23, . . . , 205 then appear, as shown in Fig. 3.4. (Note that the Ctrl key is *not* held down when copying the formula.)

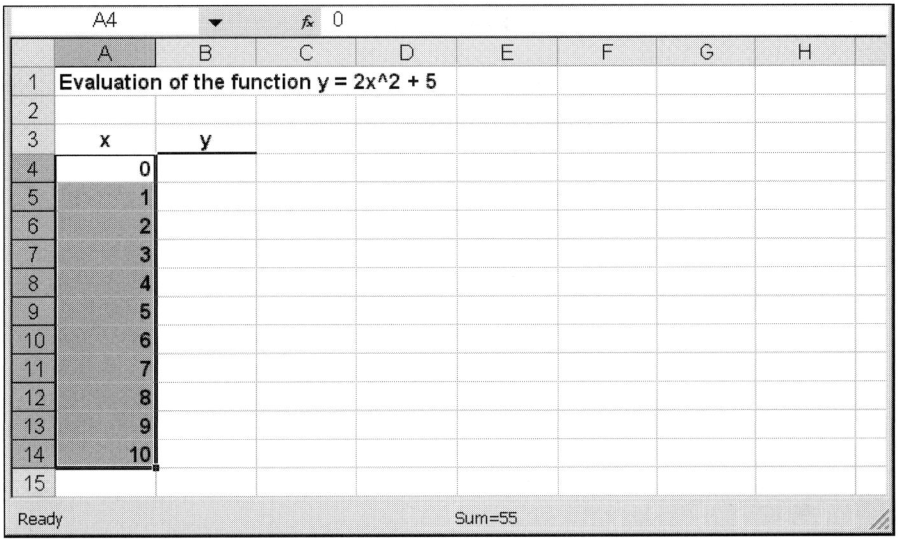

Figure 3.3 – Consecutive digits copied into cells A5 through A14

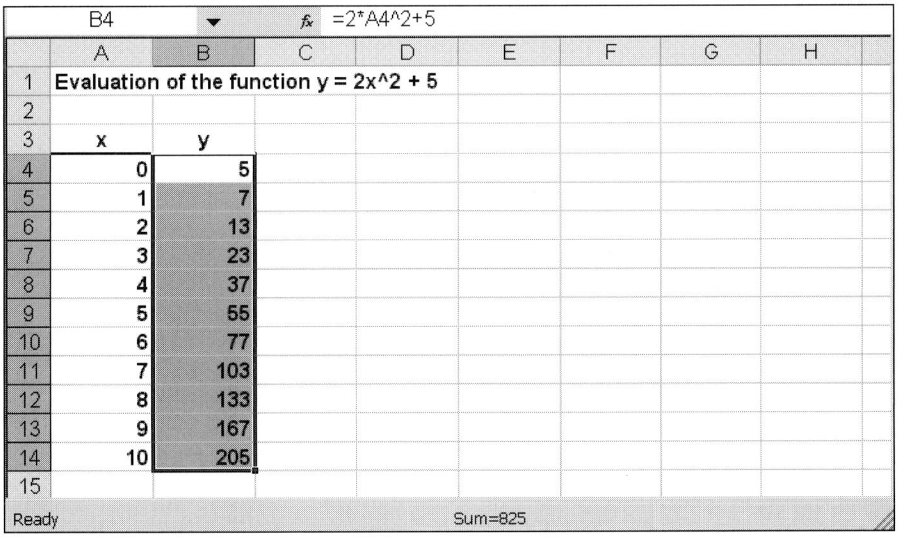

Figure 3.4 – Formula copied from cell B4 into cells B5 through B14

This completes the desired table. Observe that we entered only one numerical constant (in cell A4) and one formula (in cell B4). The remainder of the table was obtained by copying the appropriate information into the remaining cells, thus simplifying the entire process considerably.

Copying a Block of Cells to Nonadjacent Cells

You can copy a block of cells from one part of the worksheet to another by carrying out the following four steps:

1. Select the block of cells to be copied.
2. Select Copy from the Edit menu. The border around the selected cells will then appear to move, indicating that the block is ready to be copied.
3. Move the mouse pointer to the upper left cell of the new location.
4. Press Enter, or select Paste from the Edit menu. The original block of cells will then be copied to the new location. Note that the copied block of cells will overwrite whatever was originally in the new location.

Here is a quicker way to copy a block of cells from one location to another:

1. Select the block of cells to be copied.
2. Click on the Copy icon in the Standard Toolbar (see Fig. 3.5). The border around the selected cells will again appear to move, indicating that the block is ready to be copied.
3. Move the mouse pointer to the upper left cell of the new location.
4. Click on the Paste icon in the Standard Toolbar. The original block of cells will then be copied to the new location. Remember that the copied block of cells will overwrite whatever was originally present.

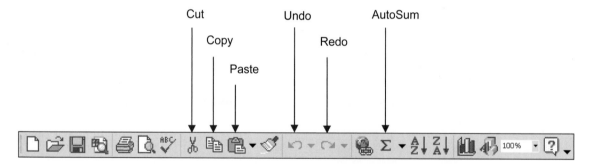

Figure 3.5 – The Standard Toolbar

You may also copy a block of cells from one location to another by dragging the block of cells from its original location to the new location. To do so, proceed as follows:

1. Select the block of cells to be copied.
2. Move the mouse pointer to any point on the border of the selected cells. (Note that shape of the pointer will change to an arrow.)
3. Hold down the Control key and the left mouse button, and drag the selected block to the new location.

4. Release the mouse button. The original block of cells will then be copied to the new location. Again, note that the copied block of cells will overwrite whatever was originally in the new location.

Moving a Block of Cells

Moving a block of cells from one location to another is similar to copying. Thus, to move a block of cells from one location to another, carry out the following four steps:

1. Select the block of cells to be moved.
2. Select Cut from the Edit menu. The border around the selected cells will then appear to move, indicating that the block is ready to be moved.
3. Move the mouse pointer to the upper left cell of the new location.
4. Press Enter, or select Paste from the Edit menu. The original block of cells will then be moved to the new location, overwriting whatever was there previously. The original location will then be empty.

Here is another way to move a block of cells from one location to another:

1. Select the block of cells to be moved.
2. Click on the Cut icon in the Standard Toolbar, as shown in Fig. 3.5. The border around the selected cells will appear to move, indicating that the block is ready to be moved.
3. Move the mouse pointer to the upper left cell of the new location.
4. Click on the Paste icon in the Standard Toolbar. The original block of cells will then be moved to the new location, overwriting whatever was there before. The original location will then be empty.

You can also move a block of cells by dragging. To do so, proceed as follows:

1. Select the block of cells to be copied.
2. Move the mouse pointer to any point on the border of the selected cells. (Again, note that shape of the pointer will change to an arrow.)
3. Hold down the left mouse button and drag the selected block to the new location.
4. Release the mouse button. The original block of cells will then be copied to the new location, overwriting whatever was originally in that location.

3.2 UNDOING CHANGES

If you change your mind after copying or moving a block of cells from one location to another, you may "undo" the change (i.e., restore the worksheet to its previous state) by selecting Undo from the Edit menu, or by clicking on the counterclockwise arrow (the Undo icon 🔄) in the Standard Toolbar (see Fig. 3.5). In fact, Undo is not restricted to undoing cell movements – it can be used to negate many different editing changes. Hence, Undo will always refer to the last editing change (e.g., Undo Paste, Undo Delete, and so on). The Redo arrow (the clockwise arrow in the Standard Toolbar next to Undo) can be used to negate the effect of the previous Undo.

3.3 COPYING AND MOVING FORMULAS

Special care must be taken when copying or moving formulas. Suppose, for example, that cell C1 contains the formula =A1+B1. This indicates that cell C1 will contain a value that is obtained by adding the values in the two cells to its left. Now suppose that the contents of cell C1 (the formula) is *copied* to cell C2. The formula will automatically change to =A2+B2, so that cell C2 will contain a value that is the sum of the values in the two cells to *its* left. A cell address that is written in this manner is called a *relative address* because it changes automatically when a formula containing the address is copied to another cell.

Relative vs. Absolute Cell Addresses

The formula in cell C1 can also be written differently, as =$A$1+$B$1. A cell address written in this manner, with dollar signs preceding both the row number and column heading, is called an *absolute address*. An absolute address does *not* change when a formula containing the address is copied or moved to another location. Thus, if the contents of cell C1 is copied to cell C2, the formula in cell C2 will be the same as the formula in cell C1; i.e., =$A$1+$B$1.

A single cell formula can include both relative and absolute addresses, such as =$A$1+B1 ($A$1 is an absolute address, but B1 is a relative address), or $A1 + $B1 (the column headings, $A and $B, are absolute but the row number is relative). If such formulas are copied to another cell, *the relative addresses will change automatically, but the absolute addresses will remain unchanged.* Thus, if the formula =$A$1+B1 is copied from cell C1 to cell E1, the formula in cell E1 will be =$A$1+D1. Such formulas containing mixed address types may be desirable in certain types of situations.

Toggling between Relative and Absolute Cell Addresses Using F4

When writing a cell address within a formula, function key F4 can be used to toggle between relative, absolute, and mixed relative/absolute addressing. This is very convenient, as it avoids the need to type the $ sign, which can be awkward. For example, suppose you type the formula =A1+B1. If you highlight A1 within the formula and press F4, the formula will change to =$A$1+B1. (Note that A1 has changed to $A$1 – an absolute address.) Pressing F4 again will change the formula to =A$1+B1. And pressing F4 once more will change the formula to =$A1+B1. Finally, pressing F4 one more time will restore the original relative address; i.e., =A1+B1.

You need not highlight an entire cell address to use F4 in this manner. You may simply place the mouse cursor ahead of the address (in front of the A), within the address (between A and 1), or behind the address (behind the 1). Or, you may highlight multiple cell addresses within a formula and toggle all of them simultaneously using F4. Thus, if you highlight the entire formula =A1+B1 and press F4, the formula will change to =$A$1+$B$1. Multiple toggling is also possible, as described above.

Moving Formulas

If a formula is *moved* rather than copied, *all of the cell addresses within the formula will remain unchanged*. Thus, when moving a formula, it does not matter whether the cell addresses are written as relative addresses or absolute addresses.

On the other hand, if an *object* cell (i.e., a cell that is *referred to* in a formula) is moved, *the formula will automatically change to accommodate that move, whether the cell reference is relative or absolute*. For example, suppose that cell C1 contains the formula =A1+B1. Now suppose that the content of cell A1 is moved to cell B5. Then the formula in cell C1 will automatically change to =B5+B1. Moreover, if the original formula had been =$A$1+B1, the formula would become $B$5+B1. Thus, absolute addresses and relative addresses *both* change automatically if a cell *reference* changes within a formula.

Working with cell formulas is not as complicated as it may appear. Remember that relative cell addressing is *usually* (but not always) appropriate in most elementary applications. Be especially careful when copying formulas from one cell location to another.

3.4 INSERTING AND DELETING ROWS AND COLUMNS

Sometimes a need arises to insert one or more rows or columns into an existing worksheet without destroying any of the information currently in the worksheet. To insert a single row into a worksheet, select any cell in the area where the new row is to be inserted. Then choose Rows from the Insert menu. The new row will then be inserted in the location of the selected cell. Or, you may right-click on the row number (at the left of the worksheet) that is to contain the new row. Then select Insert from the resulting menu. In either case, the information that initially occupied this row and all of the rows beneath it will be "pushed down," so that none of the information originally in the worksheet will be lost. A new empty row will then be inserted into the worksheet. The cell addresses appearing in formulas will automatically be adjusted for the new location of any rows that were moved.

A *block* of rows is inserted in the same way. First select several contiguous cells within a *column*, indicating where the new rows will appear. Then choose Rows from the Insert menu, causing the new rows to be inserted into the worksheet. Or, select the row numbers (at the left of the worksheet) that will contain the new rows. Then right-click on any one of the selected row numbers and choose Insert from the resulting menu. The information originally in these rows and all rows beneath them will be "pushed down" under the insertion. Cell addresses appearing in formulas will automatically be adjusted for the new location of the rows that were moved.

Figures 3.6(*a*) and (*b*) illustrate the appearance of a worksheet before and after inserting two new rows. Notice the manner in which room has been made for the new rows, by moving the displaced rows (and all rows beneath them) below the insertion point.

Column insertions are made in the same manner as row insertions. To insert a single column, select any cell in the area where the column is to be inserted. Then choose Columns from the Insert menu, causing the new column to be inserted in the location of the selected cell. Or, right-click on the column letter (at the top of the worksheet) that is to contain the new column. Then select Insert from the resulting menu. In either case, the information that initially occupied the selected column and all of the columns to its right will be "pushed further to the right," so that all of the information originally in the worksheet will be retained. The cell addresses appearing in all existing formulas will automatically be adjusted, as required.

To insert a *block* of columns, the procedure is much the same as the insertion of a single column. First select several contiguous cells within a *row*, indicating where the new columns will appear. Then choose Columns from the Insert menu. Or, select the column letters (at the top of the worksheet) that will contain the new columns. Then right-click on any one of the selected column letters and choose Insert from the resulting menu. The new columns will then be inserted into the worksheet. The information originally in these columns and all columns to their right will be "pushed to the right" of the insertion.

Rows and columns can be *deleted* from a worksheet in much the same manner as they are inserted. Thus, to delete a row, either select any cell within the row and then choose Delete/Entire Row from the Edit menu, or right-click on the row number and select Delete from the resulting menu. The worksheet will "close up" following the deletion. To delete a *block* of rows, either select a group of adjacent cells within the unwanted rows and choose Delete/Entire Row from the Edit menu, or select the row numbers and choose Delete.

Column deletions work the same way as row deletions. A single column is deleted by selecting any cell within the column and then choosing Delete/Entire Column from the Edit menu. Similarly, a *block* of columns is deleted by selecting a group of adjacent cells within the unwanted columns and then choosing Delete/Entire Column from the Edit menu. Or, by selecting the column letters, right-clicking on any one of the selected column letters, and then choosing Delete from the resulting menu.

A2	▼		fx	Davis			
	A	B	C	D	E	F	G	H
1	Student	Exam 1	Exam 2	Final Exam	Overall Score			
2	Davis	82	77	94	84.3			
3	Graham	66	80	75	73.7			
4	Jones	95	100	97	97.3			
5	Meyers	47	62	78	62.3			
6	Richardson	80	58	73	70.3			
7	Thomas	74	81	85	80			
8	Williams	57	62	67	62			
9								
10	AVERAGE	71.6	74.3	81.3	75.7			
11								
12								
13								
14								
15								
Ready					Sum=632			

Figure 3.6(*a*) – Preparing to insert two new rows above row 2

A2	▼		fx				
	A	B	C	D	E	F	G	H
1	Student	Exam 1	Exam 2	Final Exam	Overall Score			
2								
3								
4	Davis	82	77	94	84.3			
5	Graham	66	80	75	73.7			
6	Jones	95	100	97	97.3			
7	Meyers	47	62	78	62.3			
8	Richardson	80	58	73	70.3			
9	Thomas	74	81	85	80			
10	Williams	57	62	67	62			
11								
12	AVERAGE	71.6	74.3	81.3	75.7			
13								
14								
15								
Ready								

Figure 3.6(*b*) – The worksheet after inserting two new rows

Remember that *the deletion of a row or column will result in a loss of information* within the worksheet. If you inadvertently make an unwanted deletion, however, you can "undo" the deletion, thus recovering the deleted information. To do so, select Undo Delete from the Edit menu or click on the Undo icon (the counterclockwise arrow in the Standard Toolbar) shown in Fig. 3.5. Note that the undo must be carried out *immediately* after the deletion.

Any remaining cell formulas that refer to a deleted row or a deleted column will result in an error message.

3.5 INSERTING AND DELETING INDIVIDUAL CELLS

Individual cells can also be inserted and deleted within a worksheet. Cell insertions cause existing cells to be shifted either to the right or down. To insert one or more adjacent cells, first select the location of the new cells. Then choose Cells/Shift cells right or Cells/Shift cells down from the Insert menu, or right-click and choose either Insert/Shift cells right or Insert/Shift cells down from the resulting menu. The cells displaced by the insertion will then be shifted to the right or down, as requested. Note that Shift cells right causes the insertion to occur within existing *rows*, whereas Shift cells down causes the insertion within existing *columns*. The cell insertion process is illustrated in Figs. 3.7(*a*) through 3.7(*c*). Note that the contents of cells C4:E5 have been shifted to cells D4:F5.

Cell *deletions* work the same way. Simply select the block of cells to be deleted and choose Delete/Shift cells left or Delete/Shift cells up from the Edit menu, or right-click and choose either Delete/Shift cells left or Delete/Shift cells up from the resulting menu. The first menu selection (Delete/Shift cells left) will cause the cells to the right of the deleted cells (in the same *rows*) to be shifted to the left, thus filling the "hole" formed by the deleted cells. Similarly, the second menu selection (Delete/Shift cells up) will result in the cells beneath the deleted cells (in the same *columns*) being shifted up, filling the vacancy left by the deleted cells.

	C4		f_x 100					
	A	B	C	D	E	F	G	H
1	Student	Exam 1	Exam 2	Final Exam	Overall Score			
2	Davis	82	77	94	84.3			
3	Graham	66	80	75	73.7			
4	Jones	95	100	97	97.3			
5	Meyers	47	62	78	62.3			
6	Richardson	80	58	73	70.3			
7	Thomas	74	81	85	80			
8	Williams	57	62	67	62			
9								
10	AVERAGE	71.6	74.3	81.3	75.7			
11								
12								
13								
14								
15								
Ready					Sum=162			

Figure 3.7(*a*) – Preparing to insert two cells in locations C4 and C5

You should realize that *deleting a cell is not the same as clearing a cell*. The distinction is important when the deleted cell is referenced in a formula that resides elsewhere. Suppose, for example, that cells A1 and B1 contain the numerical constants 10 and 20, and cell C1 contains the formula =A1+B1. The value 30 (10+20=30) will appear in cell C1. Now suppose cell A1 is *cleared*. Its numerical value will then become zero so that the value displayed in cell C1 will be 20 (0+20=20). If cell A1 is *deleted*, however, the formula in cell C1 will indicate an *error* because the reference to cell A1 will no longer be valid.

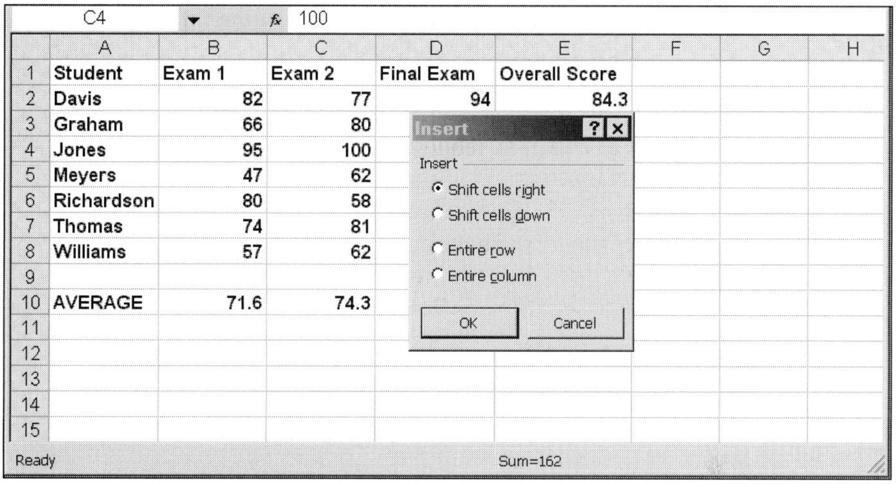

Figure 3.7(*b*) – The Insert Cells menu

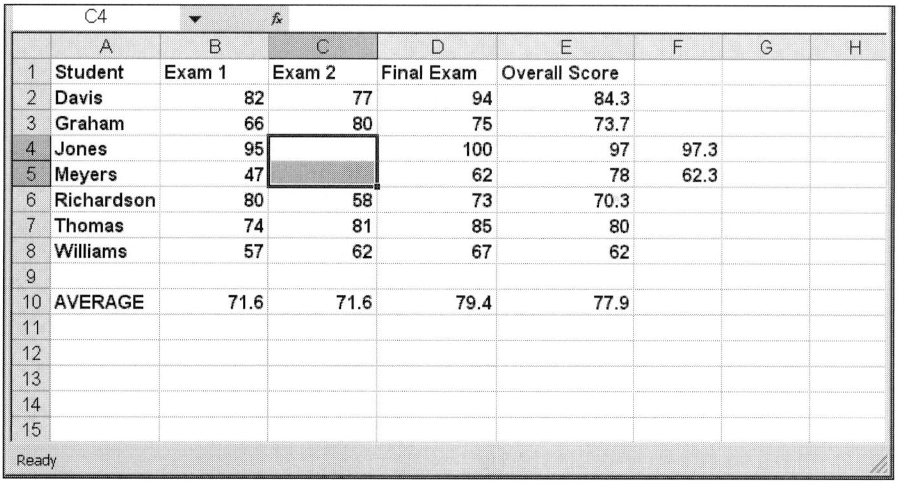

Figure 3.7(c) –The worksheet after inserting two new cells

When inserting or deleting cells, remember that all remaining cell formulas respond in the same manner as insertions or deletions of entire rows or columns. Thus, the formulas will automatically be adjusted for any existing cells that are displaced as a result of an insertion. An error message will result from any cell formula that refers to a deleted cell.

3.6 SMART TAGS

Newer versions of Excel include *smart tags* – small buttons offering choices that facilitate various editing and error-correcting tasks. Smart tags appear automatically, in response to various actions taken by the user. For example, Fig. 3.8(*a*) shows a worksheet in which the information in cell A1 (simply the constant 1) is copied into cells A2 through A4. A smart tag (specifically, the Auto Fill Options button) is shown adjacent to cell A4, in response to the copy operation. By clicking on the smart tag, the choices shown in Fig. 3.8(*b*) are displayed. Selecting Fill Series (rather than Copy Cells, which is the default selection) causes cells A2 through A4 to fill automatically with the values 2 through 4 (each cell increasing the value from the previous cell by 1).

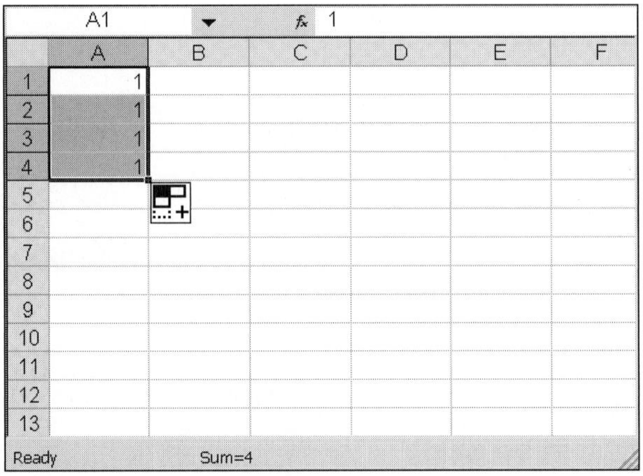

Figure 3.8(*a*) – A smart tag

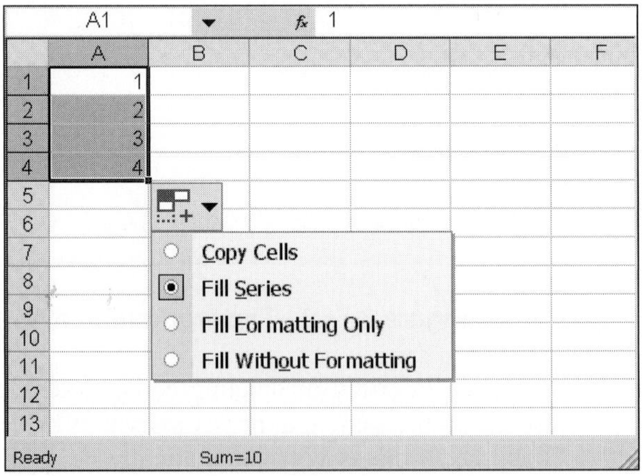

Figure 3.8(*b*) – Available choices when clicking on a smart tag

Smart tags are convenient but they can also be intrusive, since they tend to pop up frequently and are difficult to remove. Beginners, in particular, may find them annoying. Smart tags can be disabled by following the steps below. (There are several different types of smart tags. Hence, several different actions must be taken in order to disable them.)

- Select AutoCorrect Options... from the Tools menu. Then select the AutoCorrect tab, and deselect the Show AutoCorrect Options buttons box, as shown in Fig. 3.9(*a*). (*Note*: *Deselecting* a box means the box must not be checked.)
- Select AutoCorrect Options... from the Tools menu. Then select the Smart Tags tab, and deselect the Label data with smart tags box, as in Fig.3.9(*b*).
- Select Options... from the Tools menu. Then select the Edit tab, and deselect the Show Paste Options buttons box and the Show Insert Options buttons box, as shown in Fig. 3.9(*c*).

 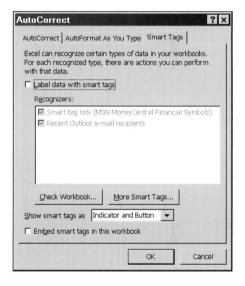

Figure 3.9(*a*)   **Figure 3.9(*b*)**

The AutoCorrect Menu, showing (*a*) the AutoCorrect Tab and (*b*) the Smart Tags tab

3.7 ADJUSTING COLUMN WIDTHS

You can improve the appearance of a worksheet by changing the width of one or more columns, particularly if a column contains very short numbers (e.g., two or three digits), very long numbers, or lengthy strings. You can change the width of the columns individually, or you can change the width of a block of columns collectively.

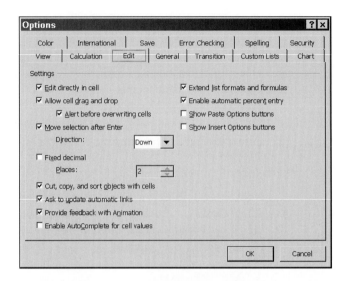

Figure 3.9(*c*) – The Edit tab in the Options menu

The easiest way to change the width of a single column is by dragging. To do so, position the mouse pointer on the column heading (the row containing A, B, C, etc.) at the right edge of the desired column. The pointer will then change its shape to a plus sign with horizontal arrows. You may then change the width of the column by holding down the left mouse button and dragging the column edge in the desired direction.

If a cell is too narrow to display a number, several *pound signs* (e.g., ###) will appear in place of the number. To remedy this situation, simply drag the column heading so that the column is wider. (Some versions of Excel will display the hidden number within a separate small box if you position the mouse pointer over the ###s.)

The width of a column can also be changed by selecting Column/Width... from the Format menu. This will result in a dialog box requesting a numerical value for the column width. Or, you may choose Column/AutoFit Selection from the Format menu. This customizes the column width to the data item within the currently active cell. (In some versions of Excel, AutoFit can also be activated by double-clicking on the right edge of a column heading.) To restore the column width to its original (default) size, choose Column/Standard Width... from the Format menu.

You can change the width of a *block* of columns by first selecting the appropriate cells within any row and then choosing Column/Width... from the Format menu. (Or select Column/Standard Width... or Column/AutoFit Selection from the Format menu.) Then proceed in the same manner as when resizing a single column. Note, however, that the AutoFit Selection will size the columns separately, to accommodate each individual data item.

3.8 FORMATTING DATA ITEMS

Another way to improve the appearance of a worksheet is to *format* the data items within their individual cells. Formatting refers to the appearance of the numerical values, the alignment of the data items within the cells, the choice of fonts, etc. To format a data item, you must first select the data item and then select Cells... from the Format menu. Then choose the appropriate tab from the resulting dialog box (Number, Alignment, Font, etc.) and respond to the items that are of interest.

We have already discussed some different ways to represent numerical values (see Sec. 2.4). The various selections can be seen by choosing the Number tab from the Format Cells dialog box, as shown in Fig. 3.10(*a*). This results in a display of several different categories of numbers, most of which have multiple format codes. A detailed discussion of all of these choices is beyond the scope of this chapter. For most technical applications, however, the Number and Scientific categories are the most frequently used. The Scientific selection causes the numerical value to be displayed in scientific notation. Both of these selections allow you to specify the number of decimal places that are displayed. (Remember that *the total number of significant figures in a calculated result should not exceed the number of significant figures in the associated input data.* This usually limits the number of decimal places.) Try formatting some numbers in a worksheet using several of these format codes.

Cell alignment is another useful formatting feature. Here we are concerned primarily with the horizontal and/or vertical alignment of data items within their cells. The cell alignment can be altered by first selecting one or more cells and then choosing the Alignment tab from the Format Cells dialog box, as shown in Fig. 3.10(*b*). The desired horizontal and vertical alignment features can then be selected from the choices that are displayed. The orientation of the text can also be selected from this dialog box. (Note that the horizontal cell alignment can also be controlled from the Formatting Toolbar.)

Excel permits other types of formatting, including the choice of several *fonts* and *font sizes*, the use of *italic* and *boldface* fonts, the placement of *borders* around various cells, the use of *subscripts* and *superscripts*, and the use of *cell patterns*. These features can be accessed by selecting Cells... from the Format menu. Some (e.g., italics, bold) can also be accessed from the icons on the Formatting Toolbar. The details are straightforward and can be determined by simple experimentation.

3.9 EDITING SHORTCUTS

Many of the features in the Edit and Format menus can be accessed more rapidly by selecting one or more cells and then pressing the right mouse button (or by pressing the Command+Options keys and then pressing the mouse button on a Macintosh). Thus, you can easily access frequently used Edit features such as Cut, Copy, Paste..., Insert..., Delete..., and Clear Contents. You can also access the

Paste Special feature, which allows you to paste selectively (e.g., formulas only, values only, etc.), or the multitab Format Cells dialog box in this manner.

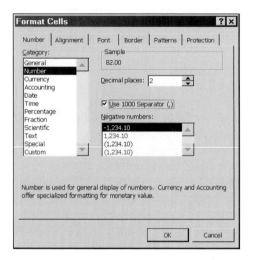

Figure 3.10(*a*)

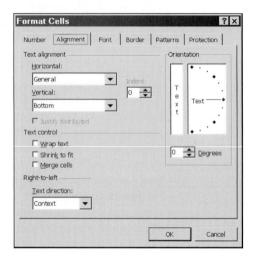

Figure 3.10(*b*)

The Format Cells dialog box, showing two different groups of formatting options

3.10 HYPERLINKS

Recent versions of Excel are able to automatically create *hyperlinks* to Internet web pages and e-mail addresses. Thus, if a cell entry appears to be a World Wide Web address (e.g., www.mcgraw-hill.com), Excel may automatically convert it to a web hyperlink. Once this hyperlink has been established, clicking on it will automatically open your web browser. Similarly, if a cell entry appears to be an e-mail address (e.g., author@mcgraw-hill.com), Excel may automatically convert it to an e-mail hyperlink. Clicking on this hyperlink will automatically open your e-mail program. You can also insert hyperlinks to other objects or other places in the current spreadsheet, though the details of this topic are beyond the scope of our current discussion.

In recent versions of Excel (e.g., Excel 2002 and 2003), an unwanted hyperlink can be removed by right-clicking on it and selecting Remove Hyperlink from the resulting menu. Moreover, automatic hyperlink creation can be disabled in recent versions of Excel by selecting AutoCorrect Options.../AutoFormat As You Type from the Tools menu, and then *deselecting* the Internet and network paths with hyperlinks box (i.e., remove the check from the box).

In Excel 2000, an unwanted hyperlink can be removed by right-clicking on it and selecting Hyperlink/Remove Hyperlink from the resulting menu. Unfortunately, the automatic hyperlink creation feature cannot be disabled in Excel 2000.

Example 3.3 Editing a Worksheet

Enhance the appearance of the worksheet shown in Fig. 2.18, which was originally developed in Example 2.2.

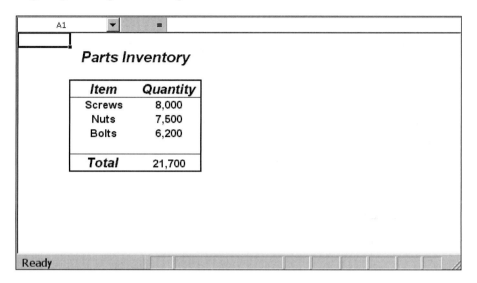

Figure 3.11 – A visually enhanced worksheet

The edited worksheet is shown in Fig. 3.11. Notice that a title and several blank rows have been added, and columns B and C have been widened (by selecting cells within the columns and then choosing Column/Width from the Format menu). In addition, some of the fonts have been changed, the data items have been centered within their cells, a border has been placed around the original block of cells, horizontal lines have been added beneath the headings and above the total, and the numbers are displayed differently (note the commas). These changes were all made from the Format Cells dialog box, obtained by selecting Cells... from the Format menu. Finally, note that the grid lines and row/column headings have been removed (by selecting Options.../View from the Tools menu).

Problems

3.1   Reconstruct the worksheet shown in Fig. 1.1, Example 1.1. Enter the appropriate formulas into cells F9, F11, A5, B4 and C4 (see Fig. 1.3). Then copy the remaining formulas into cells A6 through A21, B5 through B21, and C5 through C21 using the formula copying technique illustrated in Example 3.2. Verify that the results are correct.

3.2   Reconstruct the worksheet containing student exam scores as shown in Fig. 2.23. Then enhance its appearance by editing it in the following ways:

(*a*) Add some empty space at the top of the worksheet. Place a worksheet title (e.g., **Engineering Analysis 100 Semester Grade Report**) at some convenient location within this space. Try a single-line title and a two-line title, and use the one you like the best.

(*b*) Format the numerical values so that they all show two decimal places.

(*c*) Adjust the column widths as required to make the worksheet more readable.

(*d*) Align the numerical values and their headings so that they are centered horizontally within their respective cells.

When you are finished editing, print the edited version of the worksheet. Fill the printed page with the worksheet, using a landscape layout.

3.3 In Prob. 2.8 you were asked to prepare a worksheet containing a student's first-semester grade report including a calculation of the student's grade-point average (GPA) for that semester. The data were given in Prob. 2.3.

This problem requires you to expand that worksheet by adding a second-semester grade report, thus showing a full academic year (two semesters).

To expand the worksheet, proceed as follows

(*a*) Add two blank lines between the social security number and the heading line (**Course Credits Grade**). Then add the title **Semester 1** within this space.

(*b*) To the right of the first-semester grade report, add the following second-semester grade report. Remember to include a grade-point average for the second semester beneath the grade report, as described in Prob. 2.8. (You can enter this information easily by copying the first-semester grade report and then editing as required.)

Semester 2

Course	Credits	Grade
Chemistry II	4	3.7
Physics II	3	3.0
Economics	3	3.3
Engineering Analysis	3	4.0
Calculus II	3	3.3
Seminar	1	3.0

(*c*) Add a calculation of the overall (first-year) grade point average at the bottom of the worksheet.

(*d*) Enhance the appearance of the spreadsheet by adding blank rows and columns, adjusting column widths, formatting numerical values, underlining headings, and so on.

(e)   Try changing the text by selecting different typefaces, selecting larger fonts for the headings, and using boldface fonts selectively.

When you are finished adding information, print the new version of the worksheet. Fill the printed page with the worksheet, using a landscape layout.

3.4   Rearrange the worksheet created in Prob. 3.3 in the following ways:

(a)   Place the second-semester grade report beneath rather than to the right of the first-semester grade report.

(b)   Place the grade-point average for each semester to the right of the individual course listings and grades.

(c)   Place the overall (first-year) grade-point average beneath the semester grade-point averages.

(d)   Adjust the overall appearance of the worksheet by adding or removing blank rows and columns, readjusting column widths, repositioning the headings, and so on.

3.5   Mechanical engineers often study the movement of damped oscillating objects. Studies of this type play an important role in the design of automobile suspensions, among other things.

   The horizontal displacement of an object is given as a function of time by the equation

$$x = x_0 e^{-\beta t} \left[\cos(\omega t) + (\beta / \omega) \sin(\omega t) \right]$$

The parameters β and ω depend upon the mass of the object and the dynamic characteristics of the system.

   Prepare a table of x versus t over the interval $0 \leq t \leq 30$ seconds, using the formula-copying technique illustrated in Example 3.2. When entering formulas, be careful to use relative and absolute cell addressing correctly, as required. To construct the table, assume $x_0 = 8$ in, $\beta = 0.1$ sec^{-1}, and $\omega = 0.5$ sec^{-1}. Use 1-second intervals (i.e., let $t = 0, 1, 2, \ldots, 30$ seconds). Estimate the point where x first crosses the t-axis (i.e., estimate the value of t corresponding to $x = 0$).

3.6   Electrical engineers often study the transient behavior of RLC circuits (containing resistors, inductors, and capacitors). Such studies play an important role in the design of various electronic components.

   Suppose an open RLC circuit has an initial charge of q_0 stored within a capacitor. When the circuit is closed (by throwing a switch), the charge flows out of the capacitor and through the circuit in accordance with the equation

$$q = q_0 e^{-Rt/(2L)} \cos\left[\sqrt{\frac{1}{LC} - \left(\frac{R}{2L}\right)^2}\, t\right]$$

In this equation, R is the resistive value of the circuit, L is the value of the inductance, and C is the value of the capacitance.

Prepare a table of q versus t over the interval $0 \le t \le 0.5$ second, using the formula copying technique illustrated in Example 3.2. When entering formulas, be careful to use relative and absolute cell addressing correctly, as required. To construct the table, assume that $q_0 = 10^3$ coulombs, $R = 0.5 \times 10^3$ ohms, $L = 10$ henries, and $C = 10^{-4}$ farads. Let t vary by 0.01 second intervals (in other words, let $t = 0, 0.01, 0.02, \ldots, 0.5$ second). Estimate the point where q first crosses the t-axis (i.e., estimate the value of t corresponding to $q = 0$).

Notice the similarity between this problem and the vibrating mass described in Problem 3.5.

3.11 DISPLAYING CELL FORMULAS

When a formula is being entered into a cell, it is displayed both within the cell and within the Formula Bar. Once the formula has been completed and the Enter key has been pressed, however, the formula is no longer visible within the cell. Instead, the *value* generated by the formula is displayed within the cell (though the formula may still be seen within the Formula Bar of the *currently active* cell).

There are times when it may be desirable to see the actual cell formulas within their respective cells, rather than the values generated by the formulas. If you hand in a homework assignment, for example, your instructor may want to verify that you have been using cell formulas rather than typing in numbers. Or you may wish to examine the cell formulas within a worksheet yourself, simply to verify that they are correct.

To view the cell formulas within their respective cells, choose Options... from the Tools menu. When the Options dialog box appears, select the View tab and then select Formulas under the Window options heading, as shown in Fig. 3.12. (This will cause the cells to increase in width somewhat.) You can remove the formula display at a later time by deactivating the Formulas selection from the Options dialog box.

Figure 3.13 shows a simple worksheet in which the cell formulas are displayed (notice the formula displayed in cell C7). This is the same worksheet that was originally shown in Fig. 2.20. Note that the cell widths are now wider, to accommodate the cell formulas.

When printing cell formulas it is convenient to include gridlines as well as row and column headings so that the cell addresses can be clearly identified. To do so, select Page Setup... from the File menu, choose the Sheet tab, and then select Gridlines and Row and column headings, as shown in Fig. 3.14.

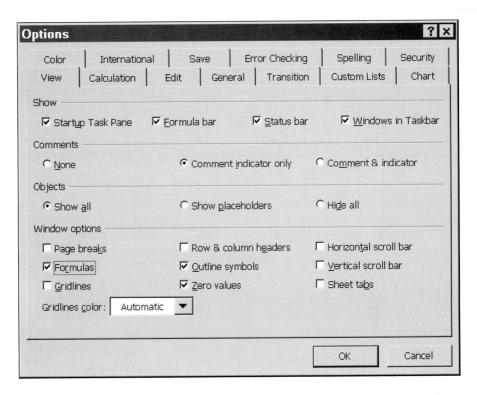

Figure 3.12 – The Options menu, showing how to display cell formulas

	A	B	C	D
1				
2		ITEM	QTY	
3		Screws	8000	
4		Nuts	7500	
5		Bolts	6200	
6				
7		TOTAL	=(C3+C4+C5)	
8				
9				
10				
11				
12				

Figure 3.13 – Displaying a formula in cell C7

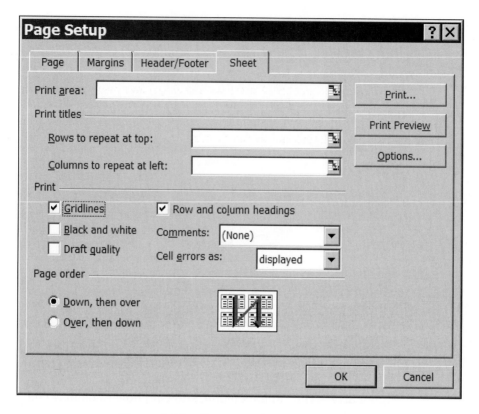

Figure 3.14 – Some useful Print settings

Problem

3.7   Display the cell formulas for the following worksheets. Then print each worksheet showing the cell formulas.

(*a*)   The gas law problem described in Prob. 2.9.

(*b*)   The compound interest problem described in Prob. 2.10.

(*c*)   The electrical circuit described in Prob. 2.11.

(*d*)   The heat exchanger problem described in Prob. 2.14.

(*e*)   The student exam grades described in Prob. 2.17.

(*f*)   The table of displacement versus time, described in Prob. 3.5.

(*g*)   The table of electrical charge versus time, described in Prob. 3.6.

3.12 CLOSING REMARKS

In this chapter and the last we have focused on some of the most frequently used features of Excel, in preparation for the various engineering applications to be discussed in the chapters that follow. Excel includes many other features, most of which are simple to use but beyond the scope of our present discussion. You are encouraged to learn more about Excel from the excellent online help and tutorials included with Excel and the many detailed Excel texts that are currently on the market.

CHAPTER 4

GRAPHING DATA

Have you ever tried to make sense out of a list or table of numbers? It can be done, but it is certainly tedious. A list of numbers does not readily reveal trends, nor does it show interdependencies between variables. If you plot the data as a graph, however, such trends and interdependencies become obvious and easy to understand. Surely, if a picture is worth a thousand words, then a graph is worth a thousand numbers.

Graphing data is one of the most common tasks carried out with a spreadsheet program. Excel includes the capability to generate a wide variety of graphs. (In Excel, graphs are referred to as *charts*.) All graphs are relatively easy to create in Excel, though some types of graphs are more easily generated than others. Our interest will be directed primarily toward the use of *x-y graphs* (called *Scatter Charts* or *XY Charts* in Excel), as this is the type of graph that is most commonly used by engineers. We will, however, also discuss other types of graphs.

Engineers generally refer to the horizontal axis (the *x*-axis) as the *abscissa* and the vertical axis (the *y*-axis) as the *ordinate*. In Excel, however, the horizontal axis is called either the *Category Axis* or the *Value Axis*, depending on the chart type. (*Category Axis* is used in most charts, though *Value Axis* is used for XY charts.) The vertical axis is always called the *Value Axis*.

4.1 CHARACTERISTICS OF A GOOD GRAPH

Constructing a graph involves more than simply plotting the data points. The graph must convey information in a manner that is clear and unambiguous, and it should be attractive. For example, Fig. 4.1 shows a graph of the height of a projectile as a function of time. This graph was generated by a mathematical formula. Notice that the graph includes a title, and the axes are clearly labeled, including the applicable units. The grid is readable but uncluttered. In this graph the plotted points are shown explicitly against a light clear background, with a smooth curve running through them. Note, however, that graphs created in this manner (by a mathematical formula) need not necessarily show the plotted points.

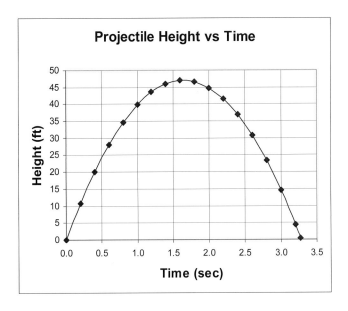

Figure 4.1 – A clear, attractive graph created mathematically

Figure 4.2(*a*) shows a graph in which the data points represent individual measurements. In such situations, the data points may not define a smooth curve because of small errors or variations in the measurements. This is called *data scatter*. Measured data points should always be shown individually, using a symbol for each data point, so the data scatter can be seen clearly, as shown in Fig. 4.2(*a*).

Figure 4.2(*b*) shows the same set of measured data with *trendline* passed through the data. We will say much more about trendlines in Chapter 8, when we discuss fitting equations to data.

Finally, in Fig. 4.3 we see a graph showing the area and volume of a sphere as a function of the sphere's radius. Hence, we have two different data sets within the same graph, each shown in a different color. In this case, the data are again generated by mathematical formulas, though the data points are shown explicitly. Notice that the graph includes a *legend*, which allows the viewer to identify each data set. The data sets could also have been plotted using separate value axes (ordinates), one on the left and the other on the right.

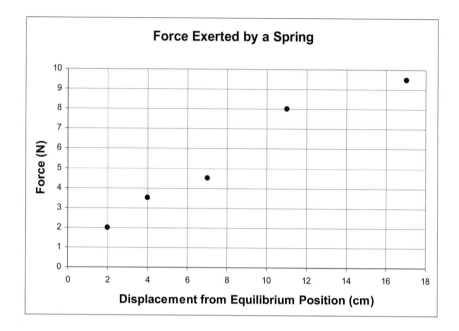

Figure 4.2(*a*) – A graph showing measured data points

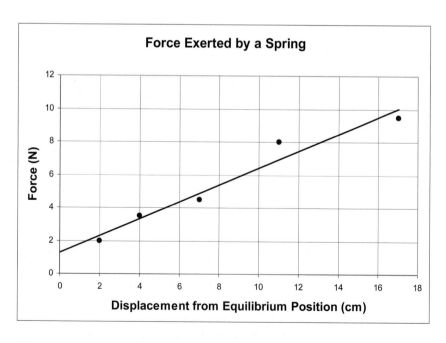

Figure 4.2(*b*) – A trendline representing a set of measured data points

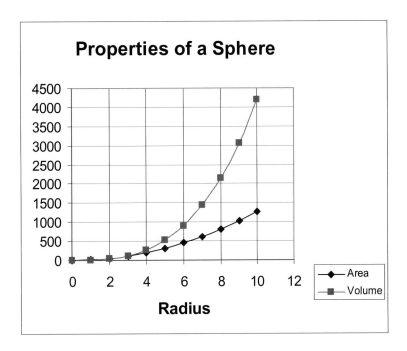

Figure 4.3 – A graph showing two different data sets

4.2 CREATING A GRAPH IN EXCEL

The easiest way to generate a graph in Excel is to use the *Chart Wizard*, which is found in the Standard Toolbar (see Fig. 4.4). Or, you can select Chart from the Insert menu.

In either case, the general procedure is as follows:

1. Select a block of contiguous cells containing the data to be plotted (the *source data*). The selection may include column headings or row labels.

Chart Wizard

Figure 4.4 – The Standard toolbar

2. Click on the *Chart Wizard* icon within the Standard Toolbar or select Chart from the Insert menu. A dialog box will appear, as shown in Fig. 4.5(*a*). (This is the first of four successive dialog boxes.)

3. Select a chart type from either the Standard Types tab shown in Fig. 4.5(*a*) or the Custom Types tab. If one of the standard types has been selected (which will usually be the case), choose a chart subtype from the selections appearing to the right. Then press either Next or Finish. (Pressing Next allows you to specify additional information about the final appearance of the chart.) You can also press Cancel if you wish to start over again.

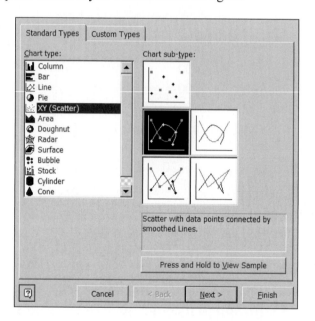

Figure 4.5(*a*) – Chart Wizard, first dialog box

4. If you pressed Next in step 3 a second dialog box will appear, as shown in Fig. 4.5(*b*). The default Data Range tab allows you to specify or verify the block of source data. The Series tab allows you to change the data that represent either the *x*- or *y*-values. When finished, press either Next or Finish. (You can also press Back, to change something entered earlier, or Cancel, to terminate the entire process.)

5. If you pressed Next in step 4 a third dialog box will appear, as shown in Fig. 4.5(*c*). This dialog box allows you to verify or change other information, such as a title, the labels along the axes, the gridlines, the location of the data legend (if you want to display a legend), and the presence of data labels (i.e., labels along the axes). Press either Next or Finish when you are ready.

6. If you pressed Next in step 5 a fourth and final dialog box will appear, as shown in Fig. 4.5(*d*). This dialog box asks you whether to display the graph within an existing worksheet or a separate worksheet. (The worksheet containing the data is the default.) Press Finish once this information has been supplied.

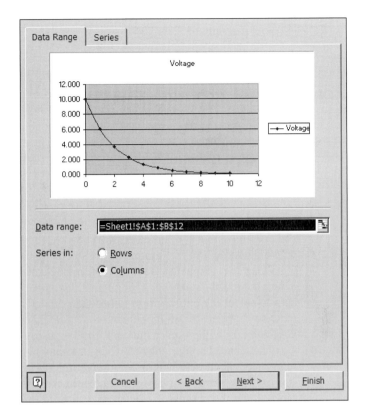

Figure 4.5(*b*) – Chart Wizard, second dialog box

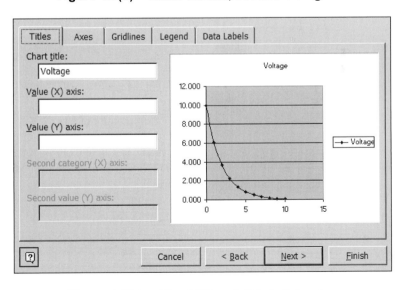

Figure 4.5(*c*) – Chart Wizard, third dialog box

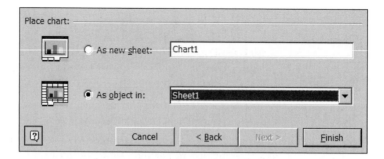

Figure 4.5(*d*) – Chart Wizard, fourth dialog box

When these steps have been completed, the desired graph will appear somewhere within the indicated worksheet. In more recent versions of Excel, the data included within the graph will be surrounded by an outline.

Once the graph has been completed, it can be activated by clicking on it. When the graph is activated, eight small squares will appear along its edges. (Be sure to select the entire graph rather than one of its components.) The graph can then be dragged to another location within the worksheet, or its shape can be altered by dragging one of the eight small squares located along its edges.

The Chart Toolbar shown in Fig. 4.6 may also appear somewhere within the worksheet. (You may access it by selecting Toolbars/Chart from the View menu.) This toolbar allows you to select and format various parts of a graph, change the type of graph, or change the appearance of the graph. You can relocate this toolbar by dragging it to a new location, and you can change its shape by pulling one of its edges either toward or away from the center.

Figure 4.6 – The Chart Toolbar

Before you create a graph, remember that the data selected must be appropriate to the type of graph desired. Thus, an *x-y* graph should be used to represent *paired data*; i.e., a set of *x*-values and a corresponding set of *y*-values. On the other hand, a line graph, bar graph, or pie chart is appropriate for *single valued* data; i.e., *y*-values only.

When constructing an *x-y* graph, the data are typically entered in adjacent cells, in either columns or rows. The data selection can include adjoining labels (e.g., column headings or row labels) but not empty cells.

A graph embedded within a worksheet can easily be moved to its own worksheet by clicking on the graph and then selecting Location/As new sheet from the Chart menu. Or the following procedure can be used, providing more control over the size and location of the graph:

1. Click on the graph so that it becomes the active object within the worksheet. Then choose Cut from the Edit menu.

2. Select another worksheet from the tabs displayed at the bottom of the screen. (The Sheet Tabs feature found in the View tab under Tools/Options must be enabled so that the tabs are visible, as in Fig. 2.1.)

3. Select a cell whose location will define the upper left corner of the chart in the new worksheet. Then choose Paste from the Edit menu.

4. If the graph is to be *copied* rather than *moved* from the original worksheet, select Edit/Copy rather than Edit/Cut in step 1.

4.3 *X-Y* GRAPHS (EXCEL *XY CHARTS*, OR *SCATTER CHARTS*)

An *x-y graph* is created by passing a line or a curve through a series of *paired data points*. Each paired data point consists of an *x*-value and a *y*-value; hence, the name *x-y* graph. Most engineering and scientific data are recorded as paired data points and displayed in the form of *x-y* graphs. The data points need not be uniformly spaced along the *x*-axis.

Do not confuse x-y graphs (called *XY Charts* or *Scatter Charts* in Excel) with *line graphs* (called *Line Charts* in Excel). If the independent variable varies continuously (e.g., time or distance), you should always use an *x-y* graph rather than a line graph to represent the data. (We will discuss line graphs later in this chapter, in Sec. 4.7.)

The most common type of *x-y* graph is one in which the *x*-axis is subdivided into a series of equally spaced intervals, and the *y*-axis is subdivided into another set of equally spaced intervals. When the axes are subdivided in this manner, we refer to the graph as having *arithmetic coordinates* (or *cartesian coordinates*).

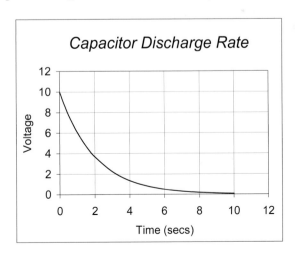

Figure 4.7 – An *x-y* graph with arithmetic (cartesian) coordinates

Figure 4.7 shows an x-y graph with arithmetic coordinates. This type of graph is normally created in Excel is by entering the x- and y-values in adjacent rows or columns, with the x-values in the top row or the leftmost column. The x- and y-values are then selected as a single block of cells. You can, however, also enter the data in *nonadjacent* rows or columns, selecting the x-data in the usual manner and then holding down the Control key (or Command key) while selecting the y-data.

After selecting the data, click on the *Chart Wizard* icon or select Chart from the Insert menu, as explained in Sec. 4.2. Choose an *XY (Scatter) Chart* from the first dialog box (*do not* select a *Line Chart*!), and select the desired type of x-y graph from the available subtypes. The selections allow you to display or hide the individual data points, pass line segments or a curve through the data points, etc.

Once the chart type has been selected, finish the graph by providing the information requested in the remaining dialog boxes, as described earlier. (Note that both axes are called *Value Axes* in an XY chart.)

Example 4.1 Creating an *X-Y* Graph in Excel

The voltage drop across a capacitor varies with time in accordance with the formula

$$V = 10e^{-0.5t}$$

where V represents voltage drop, in volts, and t represents time, in seconds. Prepare an x-y graph of the voltage as the time varies from 0 to 10 seconds. Display the data to three-decimal precision using arithmetic coordinates. Label the graph so that it is legible and attractive.

B2 = =10*EXP(-0.5*A2)

	A	B	C	D	E	F	G	H	I
1	Time	Voltage							
2	0	10.000							
3	1	6.065							
4	2	3.679							
5	3	2.231							
6	4	1.353							
7	5	0.821							
8	6	0.498							
9	7	0.302							
10	8	0.183							
11	9	0.111							
12	10	0.067							
13									
14									
15									

Ready

Figure 4.8 – An Excel worksheet containing paired data

Figure 4.8 shows a worksheet containing several values of *V* versus *t* spanning the 10-second time interval. The independent variable (time) is tabulated in the first column (column A), and the dependent variable (voltage) is tabulated in the second column. The dependent variables are generated by a formula, as seen in the formula bar for cell B2.

To create an *x-y* graph, we first select the data in columns A and B (cells A1 through B12). We then click on the *Chart Wizard* icon in the Standard Toolbar, resulting in the dialog box shown in Fig. 4.9. Notice that an XY (Scatter) Chart has been selected from the left column, and the first subtype, showing the individual data points without any interconnection, was selected from the right display area.

Figure 4.10 shows the Series tab dialog box in step 2 of the Chart Wizard. This dialog box allows you to specify the *x*- and *y*-values. The default values are used in this example; i.e., the data in column A are plotted along the *x*-axis, and the data in column B along the *y*-axis. You may change these, if you wish, by entering the appropriate ranges in the areas labeled X Values: and Y Values: within this dialog box.

We then select Finish, resulting in the graph shown within Fig. 4.11. At this point, you may change the data legend within the graph by clicking on it and retyping the new legend. You may also remove the legend by clicking on it and pressing the Delete key.

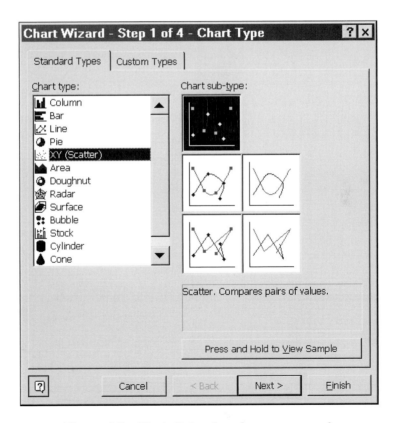

Figure 4.9 – First dialog box for an *x-y* graph

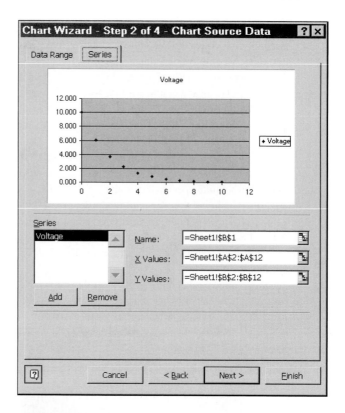

Figure 4.10 – Second dialog box for an *x-y* graph

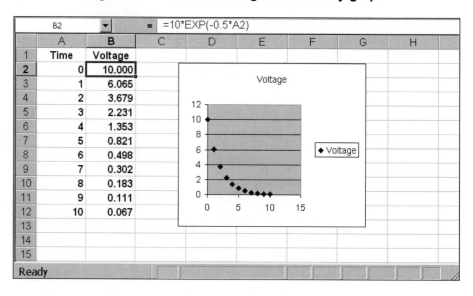

Figure 4.11 – The final, unedited *x-y* graph within the worksheet

The graph was then edited to improve its appearance and legibility. The editing was initiated by clicking on the graph, causing it to be activated for editing. Figure 4.12 shows the worksheet after the editing was completed.

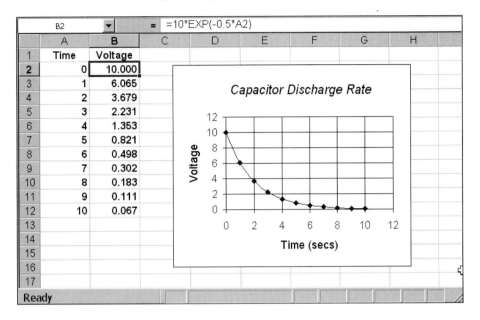

Figure 4.12 – The edited *x-y* graph within the worksheet

To create the graph shown in Fig. 4.12, the following changes were made to the original graph:

1.   The graph was relocated by dragging it to a new location within the worksheet and was then enlarged by dragging its edges in the desired directions.

2.   The chart subtype was changed by clicking on the chart, selecting Chart Type from the Chart menu, and then choosing a different subtype from the available selections.

3.   The data legend was removed by clicking on it and then pressing the Delete key.

4.   The chart title was changed from Voltage to Capacitor Discharge Rate. This was accomplished by selecting Chart Options/Titles from the Chart menu and then entering the new title into the Chart title area near the top of the dialog box.

5.   The new title was then made larger and formatted in bold italic, by selecting the font size from the Formatting Toolbar and then clicking on the *Bold* and *Italic* icons.

6.   Labels were added along the axes. To do so, we again selected Chart Options/Titles from the Chart menu and then entered the labels into the indicated areas for the *x*- and *y*-axes. (Note that each axis is referred to as a Value Axis in an *x-y* chart.)

7.   The background color within the plot area was changed by clicking on the plot background and choosing Selected Plot Area from the Format menu. A new color (white) was then selected from the resulting dialog box.

8. The scale along the x-axis was changed and vertical grid lines added by clicking on the chart and then choosing Selected Axis/Scale from the Format menu. Within the resulting dialog box, the Auto check marks were removed from the Maximum, Major unit, and Minor unit selections, and the values were changed to Maximum 12, Major unit 2, and Minor unit 2.

Note that there are other ways to perform many of these editing changes. For example, we could have used the shortcuts resulting from clicking the right mouse button (on an Apple computer, by holding down the Command+Options keys and pressing the mouse button). Or we could have clicked directly on certain objects (such as the chart title) and then changed them as desired.

4.4 ADDING DATA TO AN EXISTING DATA SET

Once a graph has been created, additional data points can be added at any time. To do so, proceed as follows:

1. Add the new data points to the worksheet, with the new x-values in their correct sequential order. The new data points should automatically be included within the graph if the x-values fall within the existing range of values.

2. If the new x-values fall outside of the existing range of values, activate the graph by clicking on it. An outline will appear around the existing x- and y-values.

3. Locate the small black squares at the edges of the outline surrounding the data points. Drag these squares so that the new data ponts are included within the outline. The new data points will then show up within the graph.

Example 4.2 Adding Data to an *X-Y* Graph

Suppose we wish to add two additional data points to the existing data that we have plotted in Example 4.1. The voltage will be calculated as before, using the expression

$$V = 10e^{-0.5t}$$

for the values $t = 0.5$ and $t = 12$ seconds.

Figure 4.13 shows the worksheet with the original data and the corresponding graph. Note that the graph has been activated by clicking on it, as evidenced by the outlines surrounding the graph, the values of the independent variable (column A), and the values of the dependent variable (column B).

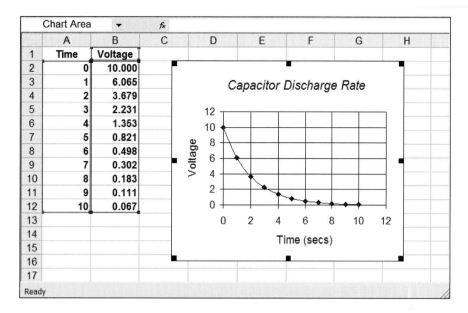

Figure 4.13 – The original voltage vs time data

We then insert a new row above row 3, and enter the value 0.5 in (the new, now empty) cell A3. Copying the formula in cell B2 to cell B3 then produces the corresponding value of the voltage. Since this new data point falls within the existing range of time values, it automatically shows up in the graph, as seen in Fig. 4.14(*a*).

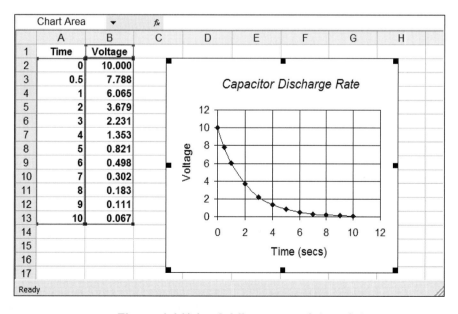

Figure 4.14(*a*) – Adding a new data point

We now enter the value 12 in cell A14, and copy the formula in cell B13 to cell 14. We then drag the outlines surrounding the data in columns A and B so that they include the new data in row 14. Once this is complete, the new data point will appear in the graph, as shown in Fig. 4.14(*b*).

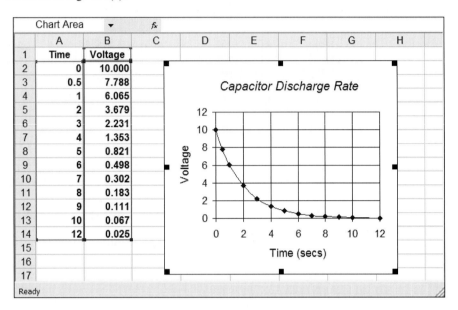

Figure 4.14(*b*) – Adding a second data point

Adding a New Set of Dependent Variables

Now suppose you want to enter and plot an additional set of dependent variables. This assumes, of course, that the new data and the existing data both correspond to the same set of independent variables. The procedure for doing this is similar to that for adding individual data points, except that the new dependent variables are entered into a separate column (or a separate row, if the worksheet uses a row-by-row layout.) In particular:

1.   Add the new dependent variables in a separate row or column. It is probably easiest if this new row or column is located alongside the existing dependent variables.

2.   Activate the graph by clicking on it. An outline will appear around the existing data set.

3.   Locate the small black squares at the edges of the outline surrounding the data set. Drag these squares so that the new dependent variables are included within the outline. The new data set will then show up within the graph automatically.

As an alternative, you may, of course, start from scratch by highlighting the entire data set and then proceeding through the Chart Wizard.

Example 4.3 Adding Dependent Variables to an *X-Y* Graph

Now suppose we wish to add an additional set of voltages to the data that we have plotted in Example 4.1. The new voltage data will apply to a second device. These voltages will be calculated using the expression

$$V = 8e^{-0.3t}$$

where *V* represents voltage and *t* represents time, in seconds.

Figure 4.15 shows the worksheet with the original data in columns A and B, the new data tabulated in column C, and a graph containing only the original data. The size of the graph has been increased somewhat for clarity. Note that the graph has been activated by clicking on it, as shown by the outlines surrounding the graph, the time values (column A), and the original voltages (column B).

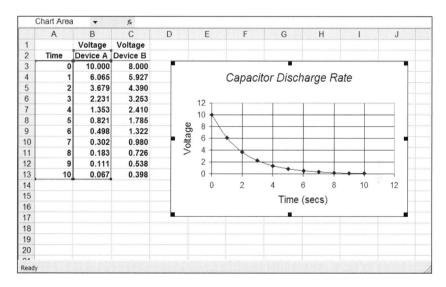

Figure 4.15 – Two sets of voltage vs time data, only one set is plotted

To add the new voltage data to the graph, simply drag the corner of the outline so that it includes column C. The new data will then be included in the graph automatically, as shown in Fig. 4.16(*a*).

Unfortunately, we cannot easily distinguish between the two data sets within the graph, even though the data sets are plotted in two different colors. We can correct this problem by adding a *legend* to the graph. To do so, right-click within the graph area and select Chart Options... Then select the Legend tab, and click on the box labeled Show Legend. The final result is shown in Fig. 4.16(*b*). Note that the legend is taken from the headings shown in cells B2 and C2.

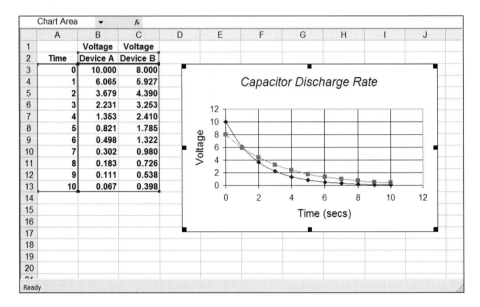

Figure 4.16(*a*) – Adding the new data to the graph

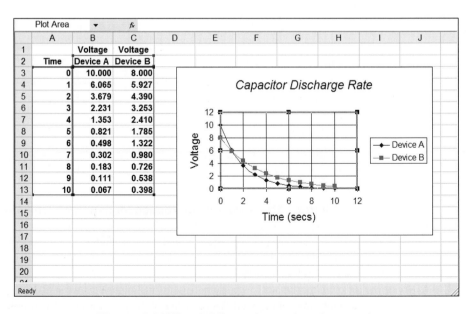

Figure 4.16(*b*) – Adding a legend to the graph

Nonadjacent Data

In all of the examples we have seen so far the data have been *contiguous*; i.e., the values of the independent variables and the dependent variables have been adjacent to one another. Thus, the data can easily be selected because it falls into one large block. This is not a requirement, however. The values of the independent variables and the dependent variables need not be adjacent to one another – they can be separated by one or more rows or columns. Similarly, if several sets of dependent variables are included within one worksheet, their values need not be adjacent to one another. In such situations, simply hold down the Ctrl key (or Command key) when selecting the nonadjacent independent and dependent variables.

Problems

4.1 The following measurements of voltage versus time were obtained for the capacitor described in Example 4.1. (Note that there is some scatter in the data since these are *measured* rather than *calculated* values.)

Time (sec)	Voltage (V)	Time (sec)	Voltage(V)
0	9.8	6	0.6
1	5.9	7	0.4
2	3.9	8	0.3
3	2.1	9	0.2
4	1.0	10	0.1
5	0.8		

(*a*) Place the data in a worksheet using a row-oriented layout. (Place the values of the independent variables in the top row and the values of the dependent variables in the bottom row.)

(*b*) Construct an *x-y* graph of the data. *Do not* interconnect the data points. Add appropriate titles and edit the overall appearance of the graph, as in Example 4.1.

(*c*) Change the location of the *x-y* graph from the original worksheet to a separate worksheet. Test the various editing features of the graph after it has been relocated.

4.2 A polymeric material contains a solvent that dissolves as a function of time. The concentration of the solvent, expressed as a percentage of the total weight of the polymer, is shown in the following table as a function of time.

Solvent Concentration (weight percent)	Time (sec)
55.5	0
44.7	2
38.0	4
34.7	6
30.6	8
27.2	10
22.0	12
15.9	14
8.1	16
2.9	18
1.5	20

Enter the data into an Excel worksheet and plot the data as an *x-y* graph with time as the independent variable. Show the individual data points.

4.3   The relationship between pressure, volume, and temperature for many gases can be approximated by the ideal gas law, which is written as

$$PV = RT$$

where P is the absolute pressure, V is the volume per mole, R is the ideal gas constant (0.082054 liter atm/mole °K), and T is the absolute temperature.

(*a*)   Construct an Excel worksheet containing a table of pressure versus absolute temperature for absolute temperatures ranging from 0 to 800 °K, and specific volumes of 20, 35, and 50 liters/mole.

Note that the worksheet should contain four columns. The values of the absolute temperature should be placed in the first column, and the second column should contain the corresponding pressure values for a specific volume of 20 liters/mole. The third and fourth columns should contain the pressure values for specific volumes of 35 and 50 liters/mole, respectively.

(*b*)   Plot all of the data (i.e., all three pressure versus temperature curves) on the same *x-y* graph. Edit the graph so that it is legible and attractive. Include a legend, indicating the specific volume associated with each curve.

4.4   The force exerted by a spring is given by

$$F = -kx^2$$

where F is the force, in newtons; x is the displacement of the spring from the equilibrium position, in centimeters; and k is a spring constant. Prepare an Excel worksheet, including an x-y graph, showing force as a function of distance for two springs whose spring constants are 0.1 newtons/cm^2 and 0.5 newtons/cm^2, respectively. Plot both curves on the same graph. For each spring, consider displacements ranging from 0 to 20 cm.

4.5 Several engineering students have built a wind-driven device that generates electricity. The following data have been obtained with the device:

Wind velocity (mph)	Power (watts)
0	0
5	0.26
10	2.8
15	7.0
20	15.8
25	28.2
30	46.7
35	64.5
40	80.2
45	86.8
50	88.0
55	89.2
60	90.3

Enter the data into an Excel worksheet and plot the electrical power as a function of wind velocity. Show the individual data points.

4.6 The following data describe the current, in milliamps, passing through an electronic device as a function of time.

Time (sec)	Current (mA)	Time (sec)	Current (mA)
0	0	9	0.77
1	1.06	10	0.64
2	1.51	12	0.44
3	1.63	14	0.30
4	1.57	16	0.20
5	1.43	18	0.14
6	1.26	20	0.091
7	1.08	25	0.034
8	0.92	30	0.012

Enter the data into an Excel worksheet and plot the current as a function of time. Show the individual data points.

4.7   The following data describe the sequence of chemical reactions A→B→C. The concentrations of A, B, and C, in moles per liter, are tabulated as a function of time.

Time (sec)	Conc A	Conc B	Conc C
0	5.0	0.0	0.0
1	4.5	0.46	0.02
2	4.1	0.84	0.06
3	3.7	1.2	0.13
4	3.4	1.4	0.22
5	3.0	1.6	0.33
6	2.7	1.8	0.45
7	2.5	1.9	0.58
8	2.3	2.0	0.72
9	2.0	2.1	0.87
10	1.8	2.2	1.0
12	1.5	2.2	1.3
14	1.2	2.2	1.6
16	1.0	2.1	1.9
18	0.83	2.0	2.2
20	0.68	1.85	2.5
25	0.41	1.5	3.1
30	0.25	1.2	3.5
35	0.15	0.93	3.9
40	0.09	0.71	4.2

Enter the data into an Excel worksheet. Plot all three concentrations as a function of time on the same graph. Show the individual data points. Display each data set in a different color.

4.5 SEMI-LOG GRAPHS

Sometimes it is advantageous to use *logarithmic coordinates* when plotting y against x. To understand logarithmic coordinates, consider Fig. 4.17, which shows a logarithmic scale directly beneath an arithmetic scale. Note the nonuniform spacing of the units along the logarithmic scale. Also, note that the location of 1 on the logarithmic scale is equivalent to the location of log 1 on the arithmetic scale (because log 1 = 0, as shown on the arithmetic scale). Similarly, the location of 2 on the logarithmic scale is equivalent to the location of log 2 on the arithmetic scale (because log 2 = 0.30), and so on. Thus, we see that *plotting* x *(or* y*) on a logarithmic scale is equivalent to plotting log* x *(or log* y*) on an arithmetic scale.*

Note that you simply plot the value of x *(or* y*) directly when plotting data on a logarithmic scale. You need not calculate the log of* x *(or the log of* y*) – the logarithmic scale automatically does this for you.*

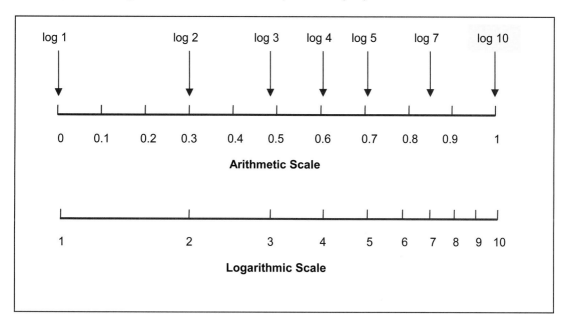

Figure 4.17 – Comparing arithmetic and logarithmic scales

A graph in which the *y*-axis (the ordinate) has a logarithmic scale and the *x*-axis (the abscissa) has an arithmetic scale is known as a *semi-log* graph. Semi-log graphs are commonly used in many diverse fields, including engineering, chemistry, physics, biology, and economics. Figure 4.18(*a*) shows an arithmetic graph of the voltage versus time data originally presented in Example 4.3. The accompanying graph shown in Fig. 4.18(*b*) uses semi-log coordinates and is shown for comparative purposes. Both plots are based upon the exponential equation

$$V = 10e^{-0.5t}$$

over the interval $0 \leq t \leq 10$. Notice that the curve appears as a straight line in Fig. 4.18(*b*). Also, note that the lower limit of the *y*-axis in Fig. 4.18(*b*) is 0.01 rather than 0, since the log of 0 is undefined.

Engineers plot data on a semi-log graph for two reasons: First, the range of *y*-values can be much greater, often spanning several orders of magnitude (i.e., several powers of 10). For example, the *y*-axis in Fig. 4.18(*b*) ranges from 0.01 to 10, which is three orders of magnitude. Accordingly, Fig. 4.18(*b*) is known as a *three-cycle* semi-log graph.

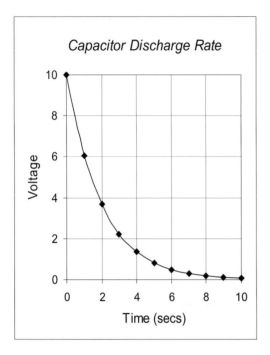

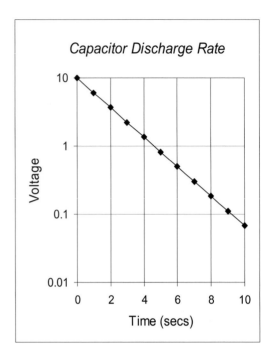

Figure 4.18(*a*) – An arithmetic graph   **Figure 4.18(*b*) – A semi-log graph**

Second, the *exponential equation*

$$y = ae^{bx} \tag{4.1}$$

appears as a *straight line* when it is plotted on a semi-log graph (i.e., when the log of y is plotted against x). Many phenomena in science and engineering are governed by this equation or some variation, such as

$$y = a(1 - e^{bx}) \tag{4.2}$$

Hence, if a data set shows up as a straight line when plotted on a semi-log graph, we can conclude that an exponential-type equation will best represent the data (or, more significantly, we can conclude that the data were generated by a *process* governed by an exponential-type equation).

To see why Equation (4.1) plots as a straight line on a semi-log graph, let us take the natural log of each side of the equation. We obtain

$$\ln y = \ln a + bx \tag{4.3}$$

In general terms, we can write the equation for a straight line as

$$ordinate = constant + slope \times abscissa \tag{4.4}$$

Thus, we see that Equation (4.3) results in a straight line if we plot ln y against x on an *arithmetic x-y* graph; that is,

$$\ln y = \ln a + b\,x \tag{4.5}$$

abscissa

slope

constant (*y*-intercept)

ordinate

Since plotting ln y against x on an arithmetic graph is equivalent to plotting y against x on a semi-log graph, we conclude that Equation (4.1) plots as a straight line on a semi-log graph.

The value of the y-intercept, a, is simply the value of y corresponding to $x = 0$. This value can be read directly from either the arithmetic graph or the semi-log graph. The value of the slope, b, can be determined by selecting two points, (x_1, y_1) and (x_2, y_2) on the semi-log graph (since this is a straight line). Thus,

$$b = (\ln y_2 - \ln y_1)/(x_2 - x_1) = \ln(y_2/y_1)/(x_2 - x_1) \tag{4.6}$$

The linearity of exponential-type equations on semi-log graphs is not restricted to the use of natural logarithms as a base. It is a general result that is valid for all logarithmic bases.

In Excel, a semi-log graph can easily be generated from an ordinary (arithmetic) *x-y* graph, provided the *y*-values are all positive. To do so, click on the *y*-axis and choose Selected Axis/Scale from the Format menu (or double-click on the *y*-axis). Then check the box labeled Logarithmic scale in the resulting dialog box.

A semi-log graph can also be generated directly by selecting Custom Types/Logarithmic within the Chart Type dialog box. (Recall that the Chart Type dialog box appears in step 1 of the *Chart Wizard* when generating a graph from scratch. It can also be obtained by clicking on an existing graph and then selecting Chart Type from the Chart menu.) Remember, however, that *the* Logarithmic *chart selection in Excel refers to a semi-log graph, not a log-log graph* (see below).

In either case you may want to relocate the labels along the *x*-axis so that they appear at the bottom of the graph. To do so, click on the *y*-axis and choose Selected Axis/Scale from the Format menu. Then enter the lowest *y*-value in the area labeled Value (X) axis Crosses at:. Be sure that the Auto box in front of this selection is not checked.

Example 4.4 Creating a Semi-Log Graph in Excel

Convert the arithmetic *x-y* graph developed in Example 4.1 and shown in Fig. 4.18(*a*) into the semi-log graph shown in Fig. 4.18(*b*).

We begin with the Excel worksheet shown in Fig. 4.12. To convert the graph into a semi-log graph, we click on the *y*-axis and choose **Selected Axis/Scale** from the **Format** menu. This results in the Format Axis dialog box shown in Fig. 4.19. Within this dialog box, we check the first four **Auto** boxes in the **Value (Y) axis scale** area and the **Logarithmic scale** box near the bottom. We also enter the value 0.01 (the minimum *y*-value) in the area labeled **Value (X) axis Crosses at:**. Figure 4.19 shows all of the new selections. Finally, we move to the **Number** tab in the Format Axis dialog box and select the **General** category, to automatically display the correct number of decimal places along the *y*-axis.

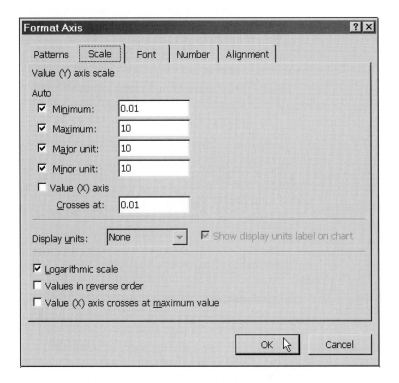

Figure 4.19 – The Format Axis dialog box

The resulting graph is shown within the worksheet in Fig. 4.20. Notice that the chart now appears as a semi-log graph, and the exponential equation appears as a straight line.

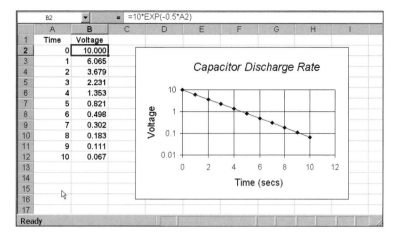

Figure 4.20 – Plotting data as a semi-log graph

Problems

4.8   Using Excel, generate an arithmetic *x-y* plot of ln *V* versus *t*, based upon the equation that appeared in Example 4.1; that is,

$$V = 10e^{-0.5t}$$

Compare the resulting graph with the semi-log graph shown in Fig. 4.20. What can you conclude about the relationship between the two graphs?

4.9   A chemical reaction is being carried out in a well-stirred tank. The concentration of the substance being created can be calculated as a function of time using the formula

$$C = a(1 - e^{-bt})$$

where *C* is the concentration in moles per liter and *t* is the time in seconds.

(*a*)   Construct a table of concentration versus time within an Excel worksheet for the case where $a = 8$ mole/L and $b = 0.25$ sec^{-1}. Include individual cells that contain the current values of *a* and *b*. Select a long enough time period so that the concentration approaches its equilibrium value.

(*b*)   Create an *x-y* graph of concentration versus time from the tabulated values. Connect the individual data points with line segments. Add an appropriate title and label the axes.

(*c*)   Change the values of *a* and *b* to 12 mole/L and 0.5 sec^{-1}, respectively. Notice what happens to the tabulated values and the graph. Note the comparative ease with which these changes are carried out.

(*d*) Restore the original values of *a* and *b*. Then alter the type of graph by replacing the arithmetic coordinates with semi-logarithmic coordinates. Explain why the resulting graph is not a straight line.

(*e*) Can you rearrange the equation so that it will appear as a straight line when plotted on semi-logarithmic coordinates?

4.10 Enter the data given in Prob. 4.2 into an Excel worksheet and plot the data as a semi-log graph with time as the independent variable. Show the individual data points. Compare the resulting graph with the graph obtained in Prob. 4.2.

4.6 LOG-LOG GRAPHS

A *log-log* graph has logarithmic coordinates along both axes. Thus, it is equivalent to a plot of log *y* against log *x* using arithmetic coordinates. Log-log graphs are useful for plotting scientific and technical data because they allow the data to span several orders of magnitude, and because a *power equation* of the form

$$y = ax^b \tag{4.7}$$

will appear as a straight line.

To see why Equation (4.7) plots as a straight line on a log-log graph, we take the log of each side of the equation, resulting in

$$\log y = \log a + b \log x \tag{4.8}$$

This is the equation for a straight line in which log *y* is the dependent variable, log *a* is a constant, *b* is the slope, and log *x* is the independent variable. (To obtain this equation, we may use natural logarithms, logs to the base 10, or any other logarithmic system; the results are the same.) In other words, on an *arithmetic* graph,

$$\log y = \log a + b \,(\log x) \tag{4.9}$$

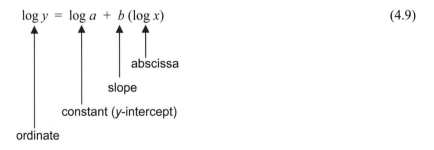

Since a plot of log y against log x on an arithmetic graph is equivalent to a plot of y against x on a log-log graph, we conclude that Equation (4.7) plots as a straight line on a log-log graph.

The value of the constant a is simply the value of y corresponding to $x = 1$ (because log 1 = 0). This value can be read directly from the log-log graph. The value of the slope, b, can then be determined by selecting two points, (x_1, y_1) and (x_2, y_2) on the log-log graph and making use of the expression

$$b = (\log y_2 - \log y_1)/(\log x_2 - \log x_1) = \log(y_2/y_1)/\log(x_2/x_1) \qquad (4.10)$$

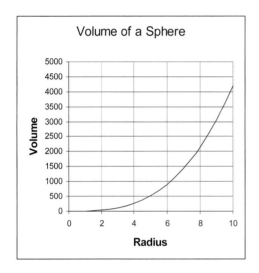

Figure 4.21(a)

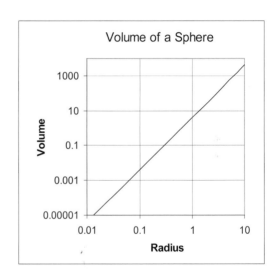

Figure 4.21(b)

For example, Figure 4.21(a) shows an arithmetic graph of the well-known equation for the volume of a sphere; that is,

$$V = \frac{4}{3}\pi r^3$$

within the interval $0 \le r \le 10$. Figure 4.21(b) shows a log-log graph of the same equation. Notice that the curve appears as a straight line on the log-log graph. In addition, note that the lower limit of each axis on the log-log graph is some small positive number rather than 0, since the log of 0 is undefined.

A log-log graph cannot be created directly in Excel, but it can easily be constructed from an ordinary (arithmetic) x-y graph or from a semi-log graph, provided all values of the independent and dependent variables are positive. (Remember that the log of zero is undefined, as is the log of a negative number. Hence, Excel will generate an error message if any of the x- or y-values are either zero or negative.)

To create a log-log graph from an ordinary x-y graph, simply click on the y-axis and choose Selected Axis/Scale from the Format menu. Then check the box labeled Logarithmic scale in the resulting dialog box. You may also want to relocate the labels along the x-axis by entering the lowest y-value in the area labeled Value (X) axis Crosses at: (be sure that the Auto box in front of this selection is not checked).

The process is then repeated for the x-axis; that is, click on the x-axis and choose Selected Axis/Scale from the Format menu and check the box labeled Logarithmic scale. You can also relocate the labels along the y-axis by entering the lowest x-value in the area labeled Value (Y) axis Crosses at: (again, be sure that the Auto box in front of this selection is not checked).

If you begin with a semi-log graph rather than an ordinary (arithmetic) x-y graph, then only the x-axis must be altered, using the method described in the preceding paragraph.

Example 4.5 Creating a Log-Log Graph in Excel

Create a log-log graph of the area and volume of a sphere as a function of the radius within the interval $0 \leq r \leq 10$, using the following two formulas.

$$A = 4\pi r^2 \qquad \text{and} \qquad V = \frac{4}{3}\pi r^3$$

Let us begin with a worksheet containing tabular values and an ordinary (arithmetic) x-y graph of A and V, as shown in Fig. 4.22. The x-y graph was created by first selecting the data in columns A, B, and C, and then following the steps described in Sec. 4.2. In order to convert this graph into a log-log graph, we first change the value of r shown in cell A2 from 0 to some small positive number, say 0.001. We then click on the x-axis and choose Selected Axis from the Format menu. If we then select the Scale tab, we obtain the dialog box shown in Fig. 4.23(a). Notice the values that have been entered for the minimum, the maximum, and the y-axis crossing point. Also note that the Logarithmic scale box has been checked.

We then click on the y-axis and repeat the entire process. This results in the dialog box shown in Fig. 4.23(b). Again, notice the values that have been entered for the minimum, the maximum, and the x-axis crossing point, and note that the Logarithmic scale box has been checked.

The resulting worksheet, containing the desired log-log graph, is shown in Fig. 4.24. Notice that the two curves shown in Fig. 4.22 are now represented as straight lines. In addition, notice that the y-axis has been subdivided into eight cycles and the x-axis into three cycles, where each cycle represents an order of magnitude (factor of 10) increase. Hence, this is an 8×3 log-log graph.

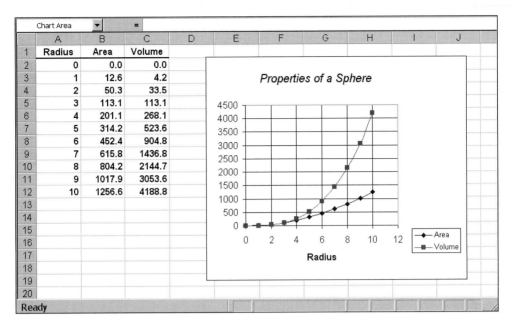

Figure 4.22 – An arithmetic graph showing the properties of a sphere

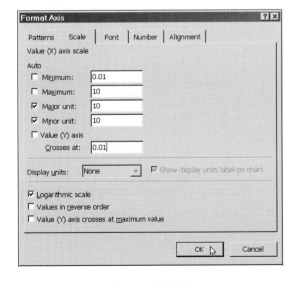

Figure 4.23(a)

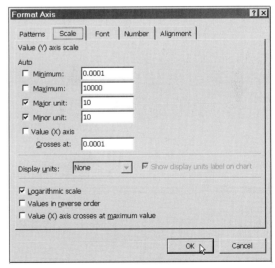

Figure 4.23(b)

Defining logarithmic coordinates along the x- and y-axes

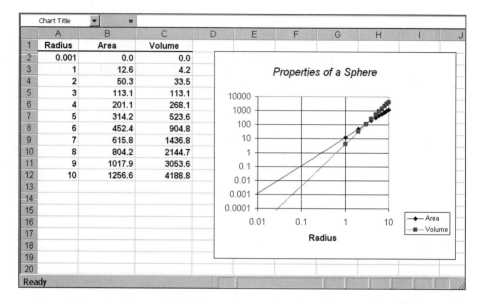

Figure 4.24 – A log-log graph showing the properties of a sphere

Problems

4.11 Construct a worksheet containing values of y versus x generated by the equation

$$y = 1.5\,x^{0.8}$$

over the interval $0 \le x \le 100$. Plot the data in the following three ways:

(*a*)   Plot y versus x using arithmetic coordinates.

(*b*)   Plot $\log_{10} y$ versus $\log_{10} x$ using arithmetic coordinates.

(*c*)   Plot y versus x using logarithmic (log-log) coordinates.

Compare the resulting graphs and explain any similarities in the shape of the curves.

4.12 The force exerted by a spring is given by

$$F = -kx^2$$

where F is the force, in newtons; x is the displacement of the spring from the equilibrium position, in centimeters; and k is a spring constant. Prepare an Excel worksheet, including a log-log graph, showing the magnitude of the force as a function of distance for two springs whose spring constants are 0.1 N/cm^2 and 0.5 N/cm^2, respectively. Plot both curves on the same log-log graph. For each spring, consider displacements ranging from 0.001 to 20 cm. Compare with the results obtained in Prob. 4.4.

4.13 A group of students have measured the quantity of water discharged from a tank as a function of time. The following data have been obtained:

Time (min)	Volume (gal)	Time (min)	Volume (gal)
0	0	35	70.2
5	17.2	40	77.2
10	27.8	45	83.7
15	38.4	50	89.9
20	46.5	55	96.9
25	54.3	60	103.4
30	62.5		

Enter the data into an Excel worksheet and plot the volume as a function of time using semi-log coordinates. Then plot the same data in a separate graph, using log-log coordinates. Show the individual data points in both graphs. What, if anything, can you conclude about the type of equation that might be used to represent the data?

4.14 An environmental engineer has obtained a bacteria culture from a municipal water sample and allowed the bacteria to grow within a petri dish. The following data were obtained:

Time (min)	Bacteria Concentration (ppm)
0	6
1	9
2	15
3	19
4	32
5	42
6	63
7	102
8	153
9	220
10	328

Enter the data into an Excel worksheet and plot the data several different ways. Use the resulting graphs to solve the following problem:

Suppose the growth of Type A bacteria is governed by a process described by the equation

$$C_A = ae^{bt}$$

and the growth of Type B bacteria is governed by a process described by the equation

$$C_B = at^b.$$

What type of bacteria is the engineer dealing with?

4.7 LINE GRAPHS (EXCEL *LINE CHARTS*)

Line graphs, unlike *x-y* graphs, are used to represent *single-valued* (categorical) data. Usually, the data points represent different values of the same entity, such as the average daily temperature for each of several consecutive days. Figure 4.12 shows a typical line graph. The data points are usually shown individually, as illustrated in Fig. 4.25.

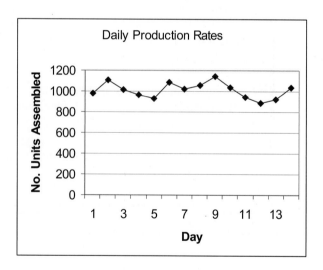

Figure 4.25 – A typical line graph

Example 4.6 Creating a Line Graph in Excel

A computer assembly plant operates seven days a week, 16 hours per day. The number of units assembled during a two-week period is given below.

Day	Units Assembled	Day	Units Assembled
1	980	8	1060
2	1108	9	1141
3	1016	10	1033
4	963	11	945
5	927	12	885
6	1088	13	918
7	1020	14	1039

Prepare a line graph showing the daily assembly rates.

Figure 4.26 shows a worksheet containing the data. Note that only the actual data (in column B) have been highlighted. The resulting graph will be somewhat simpler than if we had also highlighted the accompanying information in column A.

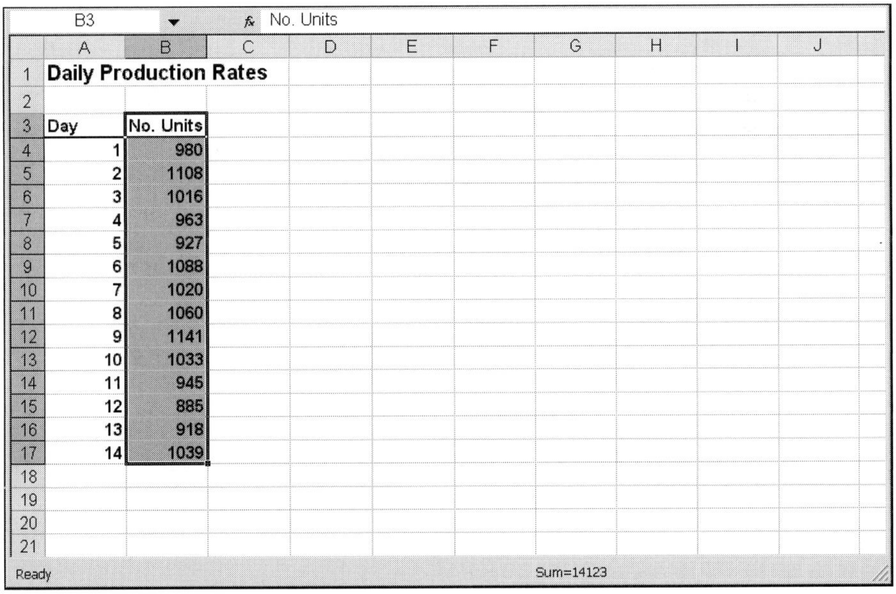

Figure 4.26 – A worksheet containing single-valued production data

The steps used to create the line graph follow the general procedure described in Sec. 4.2. Thus, we first select the data to be plotted and then click on the Chart Wizard button located in the standard toolbar. Figure 4.27 shows the first dialog box, in which a *Line Chart* (line graph) is selected. (Note that we have selected a line graph in which the data points are shown individually, since they represent measured data.) The data range, specified by highlighting the data in cells B3 through B17 of the original worksheet, is indicated in the second dialog box, shown in Fig. 4.28. The chart title and the axis labels are entered into the third dialog box, as shown in Fig. 4.29.

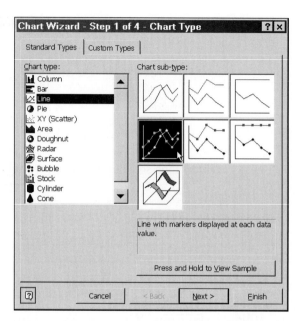

Figure 4.27 – First dialog box for a line graph

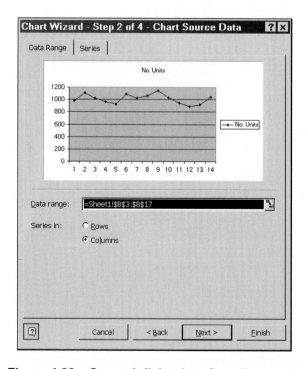

Figure 4.28 – Second dialog box for a line graph

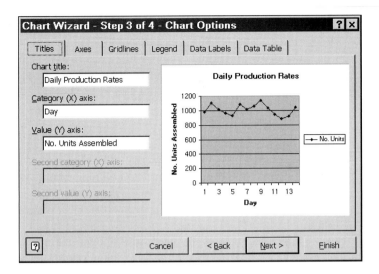

Figure 4.29 – Third dialog box for a line graph

The fourth (and last) dialog box is identical to that shown in Fig. 4.5(*d*). It allows us to specify that the graph will appear within the original worksheet. The result is the line graph embedded within the original worksheet, as shown in Fig. 4.30. Note, however, that the appearance of the line graph shown in Fig. 4.30 has been altered to improve its appearance. In particular, the size and location of the graph have been adjusted, the background color has been changed from gray to white, and the legend shown in Figs. 4.28 and 4.29 (No. Units) has been removed.

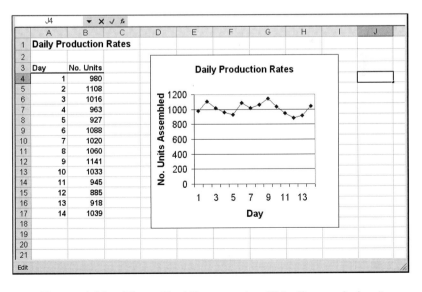

Figure 4.30 – The edited line graph within the worksheet

You are again reminded that line graphs are used only to represent *single-valued* data, resulting in data points that are uniformly spaced along the *x*-axis. Do not confuse line graphs with *x-y* graphs, which are used to represent *paired data points* (where each data point is defined by unique *x*- and *y*-values).

Problems

4.15 Reconstruct the worksheet shown in Fig. 4.26 on your own computer. Highlight the data in both columns A and B, and create a line graph of the highlighted data. Compare with the line graph shown in Fig. 4.30, which is based upon the data in column B only. Which graph is clearer?

4.16 Reconstruct the worksheet shown in Fig. 3.6(*a*) containing student exam scores. Construct a line graph of the overall scores for the individual students. Do not include the overall class average in the graph. Add a chart title and a title along each of the axes.

4.17 The following data show the repair times for two machines that have been experiencing frequent breakdowns:

Breakdown No.	Machine 1 (min)	Machine 2 (min)
1	12.1	22.5
2	27.8	15.1
3	18.5	8.2
4	6.5	11.9
5	24.6	7.7
6	33.7	19.4

(*a*) Prepare a line graph showing the repair times for machine 1.

(*b*) Prepare a separate line graph showing the repair times for machine 2.

(*c*) Prepare a single line graph showing the repair times for both machines.

4.8 BAR GRAPHS (EXCEL *COLUMN CHARTS*)

A *bar graph*, like a line graph, is used to represent *single-valued* (categorical) data. Unlike a line graph, however, a bar graph utilizes a series of vertical rectangles (bars) to represent the data. For example, the bar graph shown in Fig. 4.31 represents the inventory level of several different parts within a small machine shop. Each vertical bar represents the inventory level for a particular item.

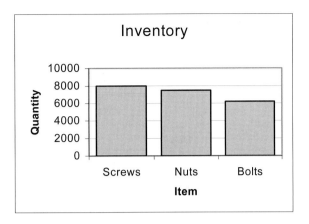

Figure 4.31 – A typical bar graph

Example 4.7 Creating a Bar Graph in Excel

Modify the worksheet developed in Example 2.2 by embedding a bar graph within the worksheet (see Fig. 2.18). Include the individual part quantities, but not the total, within the bar graph.

Figure 4.32 shows a highlighted portion of the worksheet (cells B2 through C5) that will be used to create the graph. Note that columns B and C are both highlighted. The information in column B will provide labels for the actual data in column C.

Again following the general procedure outlined in Sec. 4.2, we select a *column chart*, from the first Chart Wizard dialog box, as shown in Fig. 4.33. This is the graph type that will be used to produce the desired bar graph. The data range is specified in the second dialog box, as shown in Fig. 4.34. Note that a sample bar chart is included as a part of this dialog box. The third dialog box is shown in Fig. 4.35. It allows us to specify additional information, such as a title, labeling of the axes, appearance of gridlines, and so on.

	B2	▼	*fx* ITEM				
	A	B	C	D	E	F	G	H
1								
2		ITEM	QTY					
3		Screws	8000					
4		Nuts	7500					
5		Bolts	6200					
6								
7		TOTAL	21700					
8								
9								
Ready			Sum=21700				

Figure 4.32 – A worksheet containing single-valued inventory data

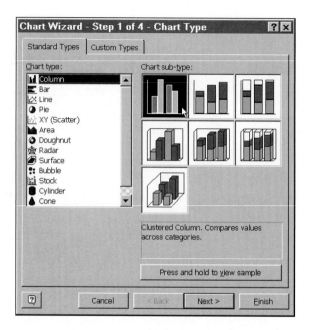

Figure 4.33 – First dialog box for a bar graph

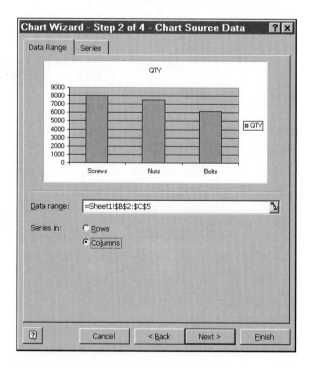

Figure 4.34 – Second dialog box for a bar graph

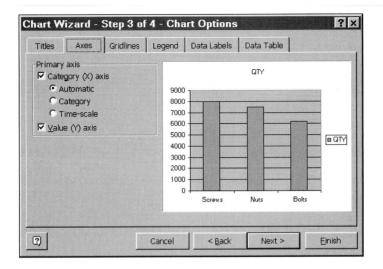

Figure 4.35 – Third dialog box for a bar graph

The resulting worksheet containing the bar chart is shown in Fig. 4.36. Notice that the numerical scale is automatically placed along the *y*-axis. Also note that the labels in column B are automatically copied to the bar graph, beneath their corresponding bars. These labels would not appear if the selection had been based solely on the numerical data in cells C3 through C5.

The legend QTY, shown to the right of the actual graph, and the title QTY, centered above the graph, were taken from cell C2 and automatically placed within the chart area. These items could have been edited (i.e., altered or deleted) while the chart was being created, in the third dialog box (see Fig. 4.35). Note that the data included in the graph are outlined, because the graph is active.

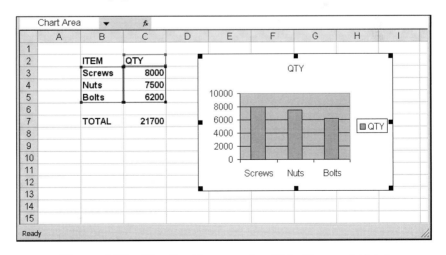

Figure 4.36 – The final bar graph within the worksheet

Figure 4.37 shows the same worksheet after the graph has been edited. In particular, note that the legend has been deleted, the chart title has been changed to Inventory, titles (Item and Quantity) have been added to the *x*- and *y*-axes, the colors (shown as different shades of gray) of the background and the vertical bars have been changed, and the bars have been widened.

The color of the vertical bars was changed by clicking on one of the bars and then selecting Selected Data Series/Patterns from the Format menu. The vertical bars were widened by clicking on one of the bars and then selecting Selected Data Series/Options from the Format menu. Note that we could have completely removed the space between the bars if we had wished.

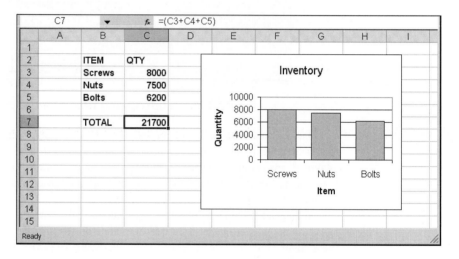

Figure 4.37 – The bar graph after editing

Problems

4.18   Reconstruct the worksheet shown in Fig. 4.36 on your own computer. Then modify the embedded bar graph in the following ways:

(*a*)   Add the following item to the list of parts on hand:

Item	*Quantity*
Washers	7200

(*b*)   Edit the new graph by adding the chart title *Parts in Stock*. In addition, add the title *Number* along the *y*-axis and the title *Part* along the *x*-axis. Then print the entire worksheet.

4.19 Reconstruct the worksheet shown in Fig. 3.6(*a*) containing student exam scores. Construct a bar graph of the overall scores for the individual students. Do not include the overall class average in the graph. Construct the graph in such a manner that the individual vertical bars touch one another. Add a chart title and a title along each of the axes. Compare with the line graph created for Prob. 4.16, using the same data.

4.20 Modify the worksheet shown in Fig. 3.6(*a*) to include the difference between each student's overall score and the class average. Enter these values in column F of the worksheet. Construct a bar graph showing the differences. Add a chart title and titles along the axes.

4.21 Problem 4.17 presented a list of repair times for two machines that have been experiencing frequent breakdowns. The data are reproduced below, for your convenience:

Breakdown No.	Machine 1 (min)	Machine 2 (min)
1	12.1	22.5
2	27.8	15.1
3	18.5	8.2
4	6.5	11.9
5	24.6	7.7
6	33.7	19.4

(*a*) Prepare a bar graph showing the repair times for machine 1.

(*b*) Prepare a separate bar graph showing the repair times for machine 2.

(*c*) Prepare a single bar graph showing the repair times for both machines.

Compare each of the resulting bar graphs with the corresponding line graph created for Prob. 4.17.

4.9 PIE CHARTS

A *pie chart* shows the distribution of individual data items within a data set. Like line graphs and bar graphs, pie charts represent *single-valued data*. For example, Fig. 4.38 shows a pie chart based upon the worksheet shown in Fig. 4.32. The pie chart shows how the individual items (screws, nuts, and bolts) are distributed among the total.

Pie charts can be labeled to show the individual data values, as in Fig 4.38, or they can show the *percentage* of each data item within the data set.

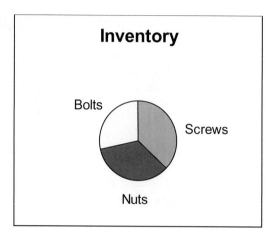

Figure 4.38 – A typical pie chart

Example 4.8 Creating a Pie Chart in Excel

Add a pie chart to the worksheet containing inventory data shown in Examples 2.2 and 4.7 (see Figs. 2.18 and 4.32). Include the individual part quantities, but not the total, within the pie chart. Label the individual segments by name, as shown in Fig. 4.37, and by percentage of the total part inventory.

We begin by highlighting cells B3 through C5 in the original worksheet, as shown in Fig. 4.39. (Note that we have not included the column headings shown in row 2, as we did in Example 4.1.) We then select a *Pie Chart* from the *Chart Wizard*, as shown in Fig. 4.40.

	B3	▼	*fx*	Screws				
	A	B	C	D	E	F	G	H
1								
2		ITEM	QTY					
3		Screws	8000					
4		Nuts	7500					
5		Bolts	6200					
6								
7		TOTAL	21700					
8								
9								
10								
11								
12								
Ready				Sum=21700				

Figure 4.39 – A worksheet containing single-valued inventory data

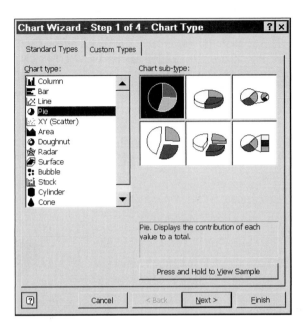

Figure 4.40 – First dialog box for a pie chart

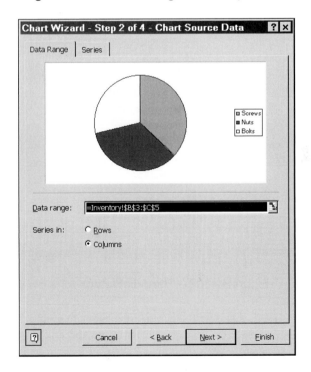

Figure 4.41 – Second dialog box for a pie chart

Figure 4.41 shows the second Chart Wizard dialog box, which identifies the data set. The chart title (Inventory) is provided in the Titles tab of the third dialog box, as shown in Fig. 4.42(*a*). The legend can be removed by selecting the Legend tab. In addition, we can select the data labels within the Data Labels tab of this dialog box, as shown in Fig. 4.42(*b*). Note that we have selected both the category names and the percentages.

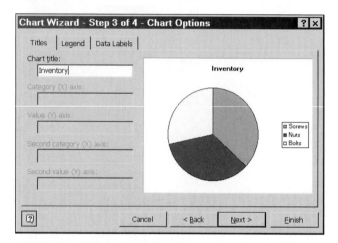

Figure 4.42(*a*) – Third dialog box for a pie chart showing the *Titles* tab

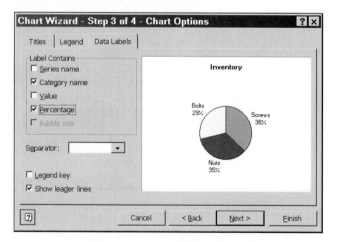

Figure 4.42(*b*) – Third dialog box for a pie chart showing the *Data Labels* tab

Now select either the Next button to specify the location of the final graph, or Finish to place the desired pie chart within the original worksheet (the default location). Figure 4.43(*a*) shows the resulting pie chart within the given worksheet. The objects within the pie chart, including the color of each segment, can be altered by right-clicking on them and then editing, as desired. Figure 4.43(*b*) shows an edited version of the pie chart.

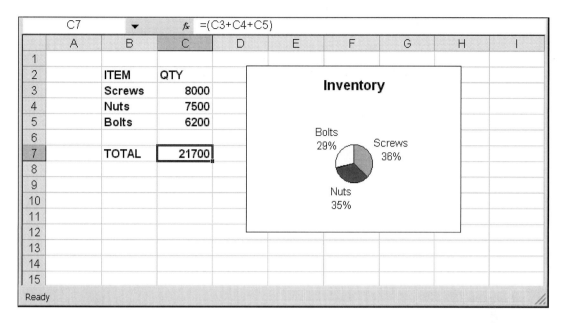

Figure 4.43(a) – The final pie chart within the worksheet

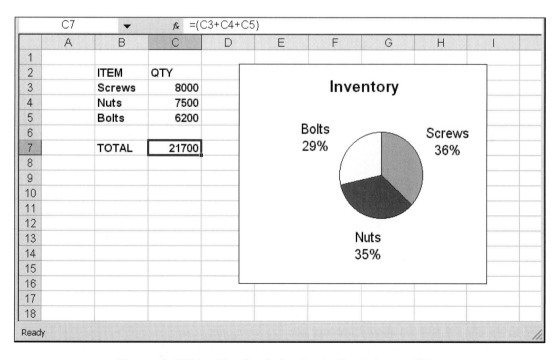

Figure 4.43(b) – The final pie chart after minor editing

Problems

4.22 The 40 students in Professor Boehring's *Introduction to Engineering* class received the following letter grades at the end of the semester:

Grade	Number of Students
A	5
B	11
C	14
D	7
F	3

Enter the data in an Excel worksheet and create a pie chart showing the grade distribution. Label each segment of the pie chart with the appropriate letter grade. Include a chart title. Select attractive sizes and locations for the circle, the labels, and the title, using Fig. 4.43(*b*) as a rough guide.

4.23 Modify the pie chart prepared for Prob. 4.22 so that the individual pie chart segments are separated from one another and the percentages are shown as labels. Include a legend that identifies the grade associated with each chart segment.

4.24 A commonly used over-the-counter cold medication contains the following types of active ingredients:

Ingredient	Weight
Pain reliever	500 mg
Cough suppressant	15 mg
Nasal decongestant	30 mg

Enter the data in an Excel worksheet and create a pie chart showing the distribution of active ingredients. Label each segment of the pie chart and include a chart title. Arrange the objects so that the graph is attractive, using Fig. 4.43(*b*) as a rough guide.

4.25 An aluminum alloy has the following chemical composition:

Element	Weight (percent)
Aluminum (Al)	93.25
Chromium (Cr)	0.5
Magnesium (Mg)	0.6
Titanium (Ti)	0.15
Zinc (Zn)	5.5

Enter the data in an Excel worksheet and create a pie chart showing the composition. Include appropriate labels and a title. Arrange the objects so that the graph is attractive, using Fig. 4.43(*b*) as a rough guide.

4.10 CLOSING REMARKS

In this chapter, we have focused primarily on x-y graphs (including semilog and log-log x-y graphs), since these are the types of graphs most commonly used in science and engineering. We also introduced line graphs, bar graphs, and pie charts. You should understand, however, that Excel is able to generate many other types of graphs. Some are used for business or financial applications, while others are intended simply to add pizzazz to displays of numerical data. You are encouraged to explore them on your own.

CHAPTER 5

ORGANIZING DATA

Engineers often store information in lists. Within the context of a spreadsheet, a *list* is a collection of one or more columns of data, usually arranged next to one another. The list may include both *numerical data* and *alphanumeric data* (i.e., words, names, and so on). Thus, a professor may maintain a class roster in the form of a list containing student names in one column, semester averages (expressed as percentages) in the second column, and corresponding letter grades in the third column. The columns may include *headings* in the first row. Each additional row will contain a complete set of data for one student; that is, a student name, semester average, and letter grade. The information within each of these rows is often referred to as a *record*.

A list can be created and edited in Excel using the standard worksheet editing procedures described in Chapters 2 and 3. Once a list has been created, we can *sort* the list; that is, rearrange the rows so that the information in a designated column increases or decreases. Or we can *filter* the data; that is, determine the rows for which the information in a designated column satisfies some particular criterion. For example, most professors create class rosters with their students listed alphabetically by their last names. At the end of the semester, the professor will usually rearrange (*sort*) the student records by exam scores, from highest to lowest. In addition, the professor may want to retrieve (*filter*) the names of those students whose exam scores exceed a certain value or fall within a certain range. Such procedures are generally referred to as *database operations*.

Excel includes a number of features that allow database operations to be carried out quickly and easily. In this chapter, we will consider the most common of these features, which allow the data within a list to be sorted and filtered. We will also consider the use of *pivot tables*, which allow data originally organized into "flat" lists to be recast into more more meaningful forms.

5.1 CREATING A LIST IN EXCEL

To create a list within an Excel worksheet, identify a block of cells that can be used to store the required rows and columns. Often, it is most convenient to place the list in the upper left portion of the worksheet so that there is room to expand the list into the worksheet (i.e., below and to the right). Each column should contain information of the same type. Each row should should contain one individual record; thus, each row will include one value from each of the columns.

The top row is often used for column headings; in fact, two or more rows can be used for this purpose if the headings are lengthy. Excel is usually able to distinguish automatically between the column headings and the actual information within the columns. This distinction may be based upon differences in the data types (e.g., alphanumeric headings and numerical data) or appearance (e.g., selection of a different font; larger type; or the use of boldface, italic, or underlining in the headings). If the headings are recognized, Excel will automatically omit the column headings from its various database operations.

Example 5.1 Creating a List in Excel

Consider the table shown below, containing information on thirteen states within the United States. Each row (each record) contains the name, capital, population (based upon the U.S. 1990 census), and size of a state.

State	Capital	Population	Size (square miles)
Alaska	Juneau	550,043	615,230
California	Sacramento	29,760,021	158,869
Colorado	Denver	3,294,394	104,100
Florida	Tallahassee	12,937,926	59,988
Missouri	Jefferson City	5,117,073	69,709
New York	Albany	17,990,455	53,989
North Carolina	Raleigh	6,628,637	52,672
North Dakota	Bismarck	638,800	70,704
Pennsylvania	Harrisburg	11,881,643	45,759
Rhode Island	Providence	1,003,464	1,231
Texas	Austin	16,986,510	267,277
Virginia	Richmond	6,187,358	42,326
Washington	Olympia	4,866,692	70,637

Enter this information as a list within an Excel worksheet. Determine the population density (population per square mile) for each state, based upon the given information. Include a heading for each column.

Figure 5.1 shows the Excel worksheet containing this information. Rows 1 and 2 contain various column headings. The remaining rows each contain the information for one state. Note that the states are listed alphabetically, as in the given table.

Also, note that the population densities are determined by a formula. For example, the population density for Alaska, shown in cell E3, is determined by the formula =C3/D3. Thus, the population density, in persons per square mile, is determined by dividing each state's population by its area.

Finally, note that the list includes column headings. Excel will be able to distinguish automatically between the headings and the information within columns A and B based upon format differences (the headings are centered, underlined, and shown in boldface). The headings in columns C, D, and E will be recognized based upon differences in data type (alphanumeric headings versus numerical data).

	E3		=	=C3/D3			
	A	B	C	D	E	F	G
1				Area	Population		
2	**State**	**Capital**	**Population**	**(sq. miles)**	**Density**		
3	Alaska	Juneau	550,043	615,230	0.9		
4	California	Sacramento	29,760,021	158,869	187.3		
5	Colorado	Denver	3,294,394	104,100	31.6		
6	Florida	Tallahassee	12,937,926	59,988	215.7		
7	Missouri	Jefferson City	5,117,073	69,709	73.4		
8	New York	Albany	17,990,455	53,989	333.2		
9	North Carolina	Raleigh	6,628,637	52,672	125.8		
10	North Dakota	Bismarck	638,800	70,704	9.0		
11	Pennsylvania	Harrisburg	11,881,643	45,759	259.7		
12	Rhode Island	Providence	1,003,464	1,231	815.2		
13	Texas	Austin	16,986,510	267,277	63.6		
14	Virginia	Richmond	6,187,358	42,326	146.2		
15	Washington	Olympia	4,866,692	70,637	68.9		
16							
17							
Ready							

Figure 5.1 – A list of states and their capitals, populations, and sizes

Once a list has been created, new records can be added by inserting new rows between existing rows or by adding new rows at the bottom of the list. Existing records can be modified simply by retyping the appropriate material, and they can be deleted by removing the appropriate rows.

Another way to add, delete, or modify a record, however, is to use a *form*. This is basically a dialog box that requests all of the information required to add, change, or delete a record, based upon the structure of the list. Forms can be accessed by selecting a cell within the record, choosing Form from the Data menu, and then supplying the information requested by the resulting dialog box, as shown in Fig. 5.2.

Figure 5.2 shows a *form* dialog box superimposed over the worksheet containing the list. The left portion of the dialog box shows four data entry areas, corresponding to columns A through D of the list. A *calculated* value, based upon the formula used in column E, is also shown. This value is displayed differently, however, because calculated values cannot be altered directly within a *form* dialog box.

The right portion of the dialog box contains a number of buttons that are used to create a new record (New), delete the existing record (Delete), move backward or forward (Find Prev, Find Next), or modify the contents of the current record (Criteria). The Restore button is used to restore the original data within a record after one or more changes have been entered into the data entry areas (using the Criteria button).

The use of forms does not provide any new capabilities within Excel. They are useful, however, for persons who may lack experience or confidence in working directly with the rows and columns in a list.

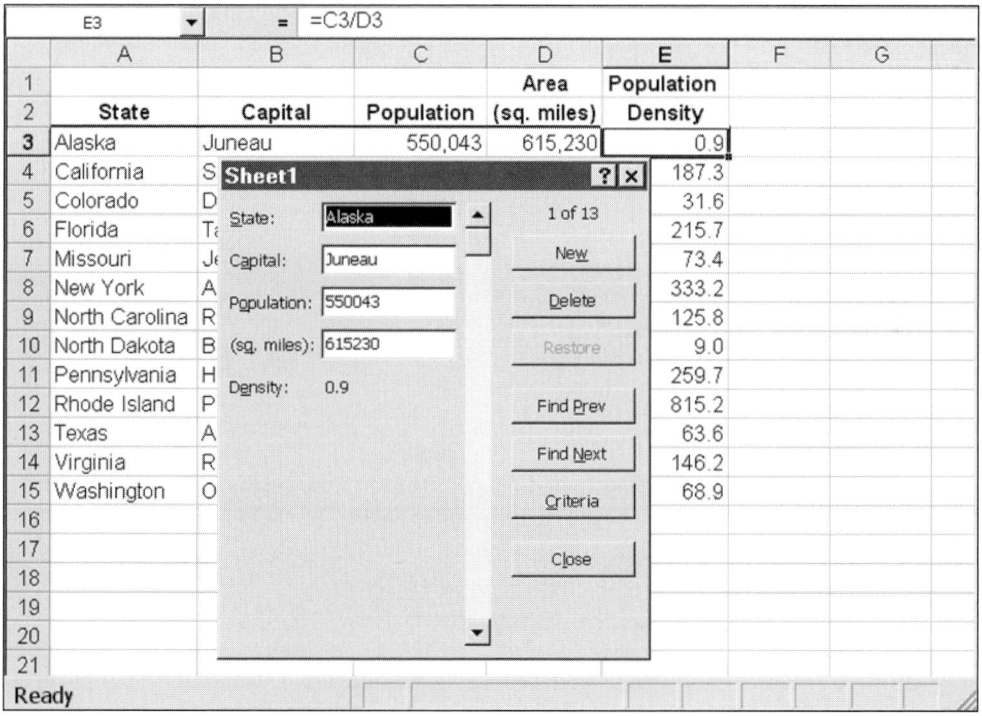

Figure 5.2 – A Form dialog box

5.2 SORTING DATA IN EXCEL

Once a data set has been correctly entered into a Worksheet as a list, the records can easily be sorted into ascending or descending order, based upon the contents of any column. To do so, simply select any cell within the desired column (including the column heading) and then click on either the *Ascending Sort Button* (lowest to highest, or A to Z) or the *Descending Sort Button* (highest to lowest, or Z to A) in the Standard Toolbar (see Fig. 5.3). Note that nothing needs to be highlighted.

Ascending Sort Button

Descending Sort Button

Figure 5.3 – The Standard Toolbar

Alternatively, you can sort a chart by selecting any cell within the desired column and then choosing Sort from the Data menu. Then provide the required values within the resulting dialog box. Multiple sorts (i.e., sorting with respect to a second column if the primary sort column contains identical entries) can be carried out in this manner. You can also rearrange the *columns* based upon the contents of a *row*, as an option. Again, note that nothing needs to be highlighted.

Example 5.2 Sorting a List in Excel

Sort the list of states created in Example 5.1 by population density, from highest to lowest.

Figure 5.4 shows the results of the sorting operation. These results were obtained by first selecting a cell within column E (in this case, cell E3) and then clicking on the Descending Sort Button. The results indicate that Rhode Island has the highest population density (815.2 persons per square mile), followed by New York (333.2), and so on, down to Alaska, which has the lowest population density (0.9).

We could have obtained these same results by selecting Sort from the Data menu and then responding to the Sort dialog box, as shown in Fig. 5.5. This method is a bit more complicated, but it offers the possibility of several options that are not available using the Sort buttons.

The original list shown in Fig. 5.1 can be restored by selecting any cell within column A and then rearranging the list in ascending order; i.e., in alphabetical order from A to Z.

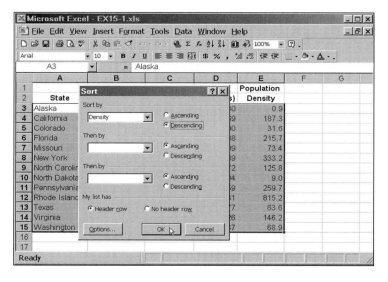

	E3	▼	=	=C3/D3			
	A	B	C	D	E	F	G
1				Area	Population		
2	State	Capital	Population	(sq. miles)	Density		
3	Rhode Island	Providence	1,003,464	1,231	815.2		
4	New York	Albany	17,990,455	53,989	333.2		
5	Pennsylvania	Harrisburg	11,881,643	45,759	259.7		
6	Florida	Tallahassee	12,937,926	59,988	215.7		
7	California	Sacramento	29,760,021	158,869	187.3		
8	Virginia	Richmond	6,187,358	42,326	146.2		
9	North Carolina	Raleigh	6,628,637	52,672	125.8		
10	Missouri	Jefferson City	5,117,073	69,709	73.4		
11	Washington	Olympia	4,866,692	70,637	68.9		
12	Texas	Austin	16,986,510	267,277	63.6		
13	Colorado	Denver	3,294,394	104,100	31.6		
14	North Dakota	Bismarck	638,800	70,704	9.0		
15	Alaska	Juneau	550,043	615,230	0.9		
16							
17							
Ready							

Figure 5.4 – The list of states sorted by population density in descending order

Figure 5.5 – The Sort dialog box

A sort need not apply to an entire list. You can sort an individual column, a block of adjacent columns, or a block of adjacent rows. To do so, simply select the block of cells that will be sorted, starting with the mouse pointer in the column that the sort will be based upon. Then use either of the sort procedures described above. (*Caution*: Be very careful when sorting only a part of a list. Your database can become hopelessly scrambled if you rearrange some of the data and leave other parts of the database in its original order.)

Problems

Note: Typing lengthy databases into Excel worksheets is tedious, and it may be unnecessary. Before entering a database for any of the problems in this chapter, check for the availability of downloads at the McGraw-Hill student website www.mhhe.com/gottfried3e.

5.1 Enter the data given in Example 5.1 into an Excel worksheet. Then carry out the following sorting operations:

(*a*) Sort the records in reverse order, beginning with Washington and ending with Alaska. Then restore the original list.

(*b*) Sort the records so that the state capitals are arranged alphabetically.

(*c*) Sort the records by population, from smallest to largest.

(*d*) Sort the records by size (area), from largest to smallest.

(*e*) Sort the column containing the state capitals alphabetically, leaving all other columns in their original order. Then sort the capitals in reverse order. Then restore the original list of capitals (so that they correspond to the correct states) by pressing the "undo" arrow (the counterclockwise arrow in the Standard toolbar) twice.

(*f*) Select the last six records (North Dakota through Washington). Sort these records by population, from largest to smallest. Then restore their original order.

5.2 The final scores for Professor Boehring's *Introduction to Engineering* course are shown below (all names are fictitious):

Student	Departmental Major	Final Score	Final Grade
Barnes	EE	87.2	B
Davidson	ChE	93.5	A
Edwards	ME	74.6	C
Graham	ME	86.2	B
Harris	ChE	63.9	D
Jones	IE	79.8	B
Martin	EE	99.2	A
O'Donnell	CE	80.0	B
Prince	ChE	69.2	C
Roberts	EE	48.3	F
Thomas	ME	77.5	C
Williams	IE	94.5	A
Young	CE	73.2	C

Enter the data into an Excel worksheet. Then carry out the following sorting operations:

(*a*)   Sort the student records by final score, from highest to lowest.

(*b*)   Sort the student records by departmental major. List the student names alphabetically within each departmental category. To do so, select Sort from the Data menu to access the Sort dialog box. Specify Sort by Major/Ascending, Then by Student/Ascending.

(*c*)   Sort the student records by departmental major. Now list the students by final scores, from highest to lowest, within each departmental category. Use the Sort dialog box, as in part (*b*). Consider each Ascending/Descending choice carefully.

(*d*)   Rearrange the student records so that the names are listed alphabetically, thus restoring the list to its original order.

5.3   A group of students has measured the time required for a series of droplets to vaporize after being dropped onto a very hot surface. The students have collected the following data:

Fluid	Initial Drop Size (mL)	Temperature Difference (°C)	Vaporization Time (sec)
Water	0.0335	200	90
		250	81
		300	74
		350	67
		400	61
		450	56
		500	50
		550	45
	0.0265	200	80
		250	72
		300	66
		350	59
		400	54
		450	49
		500	46
		550	45
Ethyl Alcohol	0.0156	200	25
		250	22
		300	20
		350	18
		400	17
		450	16
		500	15
		550	14

(*Listing of data continues on next page*)

Fluid	Initial Drop Size (mL)	Temperature Difference (°C)	Vaporization Time (sec)
	0.0121	200	22
		250	19
		300	18
		350	16
		400	15
		450	14
		500	13
		550	12
Benzene	0.0177	200	17
		250	15
		300	13
		350	12
		400	11
		450	10
		500	10
		550	9
	0.0141	200	15
		250	13
		300	12
		350	11
		400	10
		450	9
		500	9
		550	8

Note that the list contains six sets of data, where each data set corresponds to a particular fluid and specified initial drop size. The third column (temperature difference) represents the difference between the temperature of the hot surface and the boiling point temperatue of the fluid. (Can you figure out why it takes so long for the droplets to vaporize?)

Enter the data into an Excel worksheet. Be sure to fill all of the cells within the list, even if that requires repeating information from other cells. Then carry out the following sorting operations:

(*a*)   Sort the data alphabetically with respect to the fluid (i.e., rearrange the list so that the data for benzene is followed by the data for ethyl alcohol, and then water).

(*b*)   Restore the original list. Then sort each set of data in the order of descending temperature differences, maintaining the descending drop sizes. To do so, select Sort from the Data menu to access the Sort dialog box. Then specify the following: Sort by Fluid/Descending, Then by Size/Descending, and Then by Temperature/Descending.

(c) Restore the original list. Then sort the data sets in the order of ascending drop size, maintaining the ascending temperature differences for each data set (i.e., each drop size). Use the Sort dialog box, as in part (b).

(d) Restore the original list. Then sort the data alphabetically with respect to the fluid. For each fluid, sort the data sets in the order of ascending drop size. For each drop size, sort the data in the order of descending temperature differences. Carry out the entire sorting operation at once using the Sort dialog box, as in parts (b) and (c).

5.3 FILTERING DATA IN EXCEL

One of the most common database activities involves the retrieval of information that satisfies a certain condition. For example, a professor may seek the names of those students whose exam scores exceed 90 percent. Similarly, the professor may wish to determine which students are electrical engineering (EE) majors or, combining criteria, which EE majors have exam scores exceeding 90 percent.

Operations of this type can easily be carried out in Excel by *filtering* a list of data. Filtering shows only those records that satisfy the stated criteria. (The remaining records will be hidden but not deleted.) To filter the data, select any cell within the list and then choose Filter from the Data menu. You may then use either the AutoFilter or the Advanced Filter features. We will restrict our attention to AutoFilter since it can be used for most common data-retrieval operations.

When you choose AutoFilter, a downward-pointing arrow will appear at or near the top of each column. Clicking on one of these arrows will permit you to choose any of several different retrieval criteria based upon the data within that column. The criteria include selection of the top 10 records, the selection of those records that satisfy various equality conditions based upon the contents of the column, or the selection of all records. More complicated "customized" selections are also possible. The details are illustrated in the next example.

The hidden records can be restored by selecting All from a drop-down menu under a column heading, or by selecting Filter/Show All from the Data menu.

Example 5.3 Filtering a List in Excel

In this example, we will perform several filtering operations on the data within the list of states, originally created in Example 5.1. Using Excel's AutoFilter feature, determine:

1. The 10 states having the highest population density.
2. Which state has its capital in Richmond.
3. Which states have areas exceeding 100,000 square miles.
4. Which states have populations between 10 and 20 million people.

To answer part 1, we first select an arbitrary cell within the list (in this case, cell E3) and then choose Filter/AutoFilter from the Data menu. This results in a series of downward-pointing arrows within the column heading area, as shown in Fig. 5.6.

	E3 ▾	=	=C3/D3			
	A	B	C	D	E	F	G
1				Area	Population		
2	State ▾	Capital ▾	Populatio ▾	(sq. mile ▾	Density ▾		
3	Alaska	Juneau	550,043	615,230	0.9		
4	California	Sacramento	29,760,021	158,869	187.3		
5	Colorado	Denver	3,294,394	104,100	31.6		
6	Florida	Tallahassee	12,937,926	59,988	215.7		
7	Missouri	Jefferson City	5,117,073	69,709	73.4		
8	New York	Albany	17,990,455	53,989	333.2		
9	North Carolina	Raleigh	6,628,637	52,672	125.8		
10	North Dakota	Bismarck	638,800	70,704	9.0		
11	Pennsylvania	Harrisburg	11,881,643	45,759	259.7		
12	Rhode Island	Providence	1,003,464	1,231	815.2		
13	Texas	Austin	16,986,510	267,277	63.6		
14	Virginia	Richmond	6,187,358	42,326	146.2		
15	Washington	Olympia	4,866,692	70,637	68.9		
16							
17							
Ready							

Figure 5.6 – The list of states after activating the Filter/AutoFilter feature

We then click on the downward-pointing arrow in cell E2 and select Top 10 from the dropdown menu, as shown in Fig. 5.7. This results in another dialog box, from which we can select various groupings, such as top 5, bottom 10, bottom 3, and so on. In our case, we simply accept the Top 10 default selection.

The results, showing the states with the top 10 population densities, are shown in Fig. 5.8. Notice that Fig. 5.8 contains only 10 records, corresponding to the 10 states having the highest population densities. The records are *not* automatically sorted by population density. We could, of course, sort them using the techniques discussed in Sec. 5.2 if we wished.

When finished, we click on the arrow in cell E2 again and select (All), thereby restoring the original list in preparation for the next data-retrieval operation.

	E3	▼	=	=C3/D3			

	A	B	C	D	E	F	G
1				Area	Population		
2	State ▼	Capital ▼	Populatio ▼	(sq. mile ▼	Density ▼		
3	Alaska	Juneau	550,043	615,230	(All)		
4	California	Sacramento	29,760,021	158,869	(Top 10...)		
5	Colorado	Denver	3,294,394	104,100	(Custom...) 0.9		
6	Florida	Tallahassee	12,937,926	59,988	9.0		
7	Missouri	Jefferson City	5,117,073	69,709	31.6 63.6		
8	New York	Albany	17,990,455	53,989	68.9		
9	North Carolina	Raleigh	6,628,637	52,672	73.4 125.8		
10	North Dakota	Bismarck	638,800	70,704	146.2		
11	Pennsylvania	Harrisburg	11,881,643	45,759	187.3		
12	Rhode Island	Providence	1,003,464	1,231	215.7 259.7		
13	Texas	Austin	16,986,510	267,277	333.2		
14	Virginia	Richmond	6,187,358	42,326	815.2		
15	Washington	Olympia	4,866,692	70,637	68.9		
16							
17							

Ready

Figure 5.7 – Preparing to select the 10 states with the highest population density

	E3	▼	=	=C3/D3			

	A	B	C	D	E	F	G
1				Area	Population		
2	State ▼	Capital ▼	Populatio ▼	(sq. mile ▼	Density ▼		
4	California	Sacramento	29,760,021	158,869	187.3		
6	Florida	Tallahassee	12,937,926	59,988	215.7		
7	Missouri	Jefferson City	5,117,073	69,709	73.4		
8	New York	Albany	17,990,455	53,989	333.2		
9	North Carolina	Raleigh	6,628,637	52,672	125.8		
11	Pennsylvania	Harrisburg	11,881,643	45,759	259.7		
12	Rhode Island	Providence	1,003,464	1,231	815.2		
13	Texas	Austin	16,986,510	267,277	63.6		
14	Virginia	Richmond	6,187,358	42,326	146.2		
15	Washington	Olympia	4,866,692	70,637	68.9		
16							
17							
18							
19							
20							

10 of 13 records found

Figure 5.8 – The 10 states with the highest population density

To determine which state has its capital in Richmond, we click on the downward-pointing arrow in cell B2 and select Richmond from the resulting dropdown menu (see Fig. 5.9). The results are shown in Fig. 5.10. Thus, Virginia is the state whose capital is Richmond.

		E3	▼	=	=C3/D3		
	A	B	C	D	E	F	G
1				Area	Population		
2	State ▼	Capital ▼	Populatio ▼	(sq. mile ▼	Density ▼		
3	Alaska	(All)	550,043	615,230	0.9		
4	California	(Top 10...)	29,760,021	158,869	187.3		
5	Colorado	(Custom...) / Albany	3,294,394	104,100	31.6		
6	Florida	Austin	12,937,926	59,988	215.7		
7	Missouri	Bismarck / Denver	5,117,073	69,709	73.4		
8	New York	Harrisburg	17,990,455	53,989	333.2		
9	North Carolina	Jefferson City / Juneau	6,628,637	52,672	125.8		
10	North Dakota	Olympia	638,800	70,704	9.0		
11	Pennsylvania	Providence	11,881,643	45,759	259.7		
12	Rhode Island	Raleigh / Richmond	1,003,464	1,231	815.2		
13	Texas	Sacramento	16,986,510	267,277	63.6		
14	Virginia	Tallahassee	6,187,358	42,326	146.2		
15	Washington	Olympia	4,866,692	70,637	68.9		
16							
17							
Ready							

Figure 5.9 – Preparing to find the state whose capital is Richmond

		E3	▼	=	=C3/D3		
	A	B	C	D	E	F	G
1				Area	Population		
2	State ▼	Capital ▼	Populatio ▼	(sq. mile ▼	Density ▼		
14	Virginia	Richmond	6,187,358	42,326	146.2		
16							
17							
18							
1 of 13 records found							

Figure 5.10 – The state whose capital is Richmond

Now let us consider which states have areas exceeding 100,000 square miles. To do so, we restore the original list, click on the downward-pointing arrow in cell D2, and select Custom, resulting in the dialog box shown in Fig. 5.11.

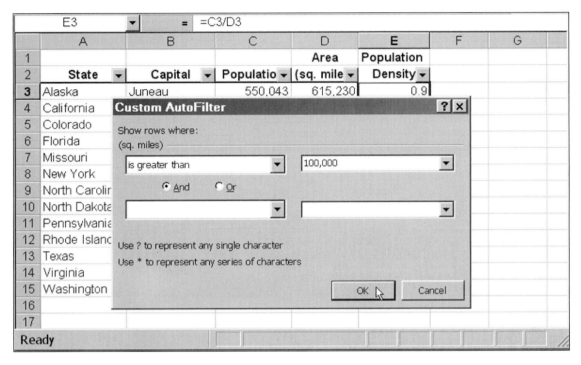

Figure 5.11 – The Custom AutoFilter dialog box

Within this dialog box, we select is greater than from the selections available in the upper left data entry area. (Click on the downward-pointing arrow to see the selections.) We then *type* the value 100,000 in the upper right data entry area and click on OK.

Figure 5.12 shows the results. Thus, Alaska, California, Colorado, and Texas have areas that exceed 100,000 square miles. Note that the states are not ranked by size; rather, they retain their original alphabetical sort order.

	E17	▼	=				
	A	B	C	D	E	F	G
1				Area	Population		
2	State ▼	Capital ▼	Populatio ▼	(sq. mile ▼	Density ▼		
3	Alaska	Juneau	550,043	615,230	0.9		
4	California	Sacramento	29,760,021	158,869	187.3		
5	Colorado	Denver	3,294,394	104,100	31.6		
13	Texas	Austin	16,986,510	267,277	63.6		
16							
17							

4 of 13 records found

Figure 5.12 – The states whose areas exceed 100,000 square miles

Finally, we consider which states have populations between 10 and 20 million people. We again restore the original list, click on the downward-pointing arrow in cell C2, and select Custom from the resulting dropdown menu. When the Custom AutoFilter dialog box appears, we select is greater than or equal to in the upper left area and enter 10,000,000 in the upper right area. Then we select And beneath the upper left area. Finally, we select is less than or equal to in the lower left area and enter 20,000,000 in the lower right area. The completed dialog box is shown in Fig. 5.13.

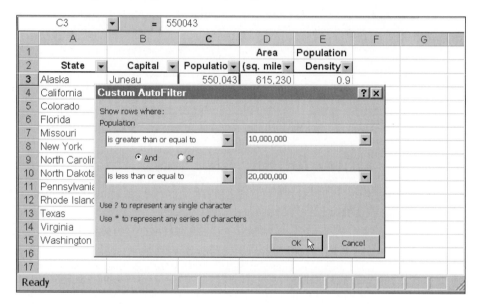

Figure 5.13 – The Custom AutoFilter dialog box

Clicking OK provides the desired results, as shown in Fig. 5.14. The figure shows us that Florida, New York, Pennsylvania, and Texas have populations between 10 and 20 million people. The records are shown in their original alphabetical order; they are not ranked by population.

	C3		=	550043			
	A	B	C	D	E	F	G
1				Area	Population		
2	State	Capital	Populatio	(sq. mile	Density		
6	Florida	Tallahassee	12,937,926	59,988	215.7		
8	New York	Albany	17,990,455	53,989	333.2		
11	Pennsylvania	Harrisburg	11,881,643	45,759	259.7		
13	Texas	Austin	16,986,510	267,277	63.6		
16							
17							
4 of 13 records found							

Figure 5.14 – The states whose populations are between 10 and 20 million people

Problems

5.4   Carry out the following filtering operations on the list containing the data given in Example 5.1.

(*a*)   Determine the five states having the smallest area.

(*b*)   Determine which state has its capital in Jefferson City.

(*c*)   Determine the total population of the three states having the largest area. (*Hint*: use AutoSum after the filtering operation.)

(*d*)   Determine the total area of the four states having the largest population.

(*e*)   Determine the average population density of the five states having the smallest population.

(*f*)   Determine which states have populations exceeding eight million people.

(*g*)   Determine which states have areas less than 60,000 square miles.

(*h*)   Determine which states have populations exceeding 16 million people or areas less than 70,000 square miles.

(*i*)   Determine which states have population densities between 65 and 200 persons per square mile.

(*j*)   Determine which states have state capitals whose names begin with either A or B.

(*k*)   Determine which states have populations between three and nine million people and areas between 60,000 and 80,000 square miles.

5.5   Enter Professor Boehring's class roster, given in Prob. 5.2, into an Excel worksheet. Then carry out the following filtering operations:

(*a*)   Determine which students are industrial engineering (IE) majors.

(*b*)   Determine which students are majoring in either chemical engineering (ChE) or mechanical engineering (ME).

(*c*)   Determine the average scores of the civil engineering (CE) students.

(*d*)   Determine which students have final grades of C or better.

(*e*)   Determine which students have final scores less than 70.

(*f*)   Determine which electrical engineering (EE) students have final scores less than 70.

(*g*)   Determine which students have final scores between 70 and 90.

(*h*)   Determine which students are either industrial engineering majors or have final scores of at least 90.

5.6   Carry out the following filtering operations on the list of droplet vaporization times given in Prob. 5.3.

(*a*)   List the vaporization times with their corresponding drop sizes for ethyl alcohol.

(*b*)   List the fluid and vaporization times for those droplets whose initial size is 0.0177 mL.

(*c*)   Determine which droplets vaporize in 10 seconds.

(*d*)   Determine which droplets vaporize between 10 and 18 seconds.

(*e*)   Determine which droplets vaporize in 12 seconds or less.

(*f*)   Determine which droplets vaporize in more than 75 seconds.

5.4 PIVOT TABLES

Ordinary lists of data are often referred to as "flat" lists or "linear" lists, because all of the information is entered in adjacent columns (or, less often, in adjacent rows). Some lists, however, can be recast into a two-dimensional structure that may provide better insight into the interrelationships between the data. This is particularly true of lists that contain repeated data fields (i.e., multiple cells containing the same information). These restructured two-dimensional lists are known as *pivot tables*.

Consider, for example, the linear list of states shown in Table 5.1. This list includes the names of several states, together with the geographic region that each state falls in, and four population values for each state, determined in different years. Notice that the name of each state, its geographical region, and the population years are duplicated repeatedly throughout the list. This causes the list to be very lengthy and awkward to work with. It is especially difficult to see trends associated with each geographical region and each census year. It would be nice if there were a more concise way to organize and present the data.

Table 5.1 – A linear list of states and related information

State	*Region*	*Year*	*Population*
Alaska	Northwest	1970	302,583
Alaska	Northwest	1980	401,851
Alaska	Northwest	1990	550,043
Alaska	Northwest	2000	626,932
California	Southwest	1970	19,971,069
California	Southwest	1980	23,667,764
California	Southwest	1990	29,760,021
California	Southwest	2000	33,871,648
(list continues on next page)			

Table 5.1 – Linear list of states (continued)

State	Region	Year	Population
Colorado	Southwest	1970	2,209,596
Colorado	Southwest	1980	2,889,735
Colorado	Southwest	1990	3,294,394
Colorado	Southwest	2000	4,301,261
Florida	South	1970	6,791,418
Florida	South	1980	9,746,961
Florida	South	1990	12,937,926
Florida	South	2000	15,982,378
Missouri	Midwest	1970	4,677,623
Missouri	Midwest	1980	4,916,762
Missouri	Midwest	1990	5,117,073
Missouri	Midwest	2000	5,595,211
New York	Northeast	1970	18,241,391
New York	Northeast	1980	17,558,165
New York	Northeast	1990	17,990,455
New York	Northeast	2000	18,976,457
North Carolina	South	1970	5,084,411
North Carolina	South	1980	5,880,415
North Carolina	South	1990	6,628,637
North Carolina	South	2000	8,049,313
North Dakota	Midwest	1970	617,792
North Dakota	Midwest	1980	652,717
North Dakota	Midwest	1990	638,800
North Dakota	Midwest	2000	642,200
Pennsylvania	Northeast	1970	11,800,766
Pennsylvania	Northeast	1980	11,864,720
Pennsylvania	Northeast	1990	11,881,643
Pennsylvania	Northeast	2000	12,281,054
Rhode Island	Northeast	1970	949,723
Rhode Island	Northeast	1980	947,154
Rhode Island	Northeast	1990	1,003,464
Rhode Island	Northeast	2000	1,048,319
Texas	Southwest	1970	11,198,655
Texas	Southwest	1980	14,225,513
Texas	Southwest	1990	16,986,510
Texas	Southwest	2000	20,851,820
Virginia	South	1970	4,651,448
Virginia	South	1980	5,346,797
Virginia	South	1990	6,187,358
Virginia	South	2000	7,078,515
Washington	Northwest	1970	3,413,244
Washington	Northwest	1980	4,132,353
Washington	Northwest	1990	4,866,692
Washington	Northwest	2000	5,894,121

A pivot table provides a much better way to represent data with repeated fields. Thus, Table 5.2 shows a pivot table containing the same data as Table 5.1. Now, however, the two-dimensional format eliminates the repeated data fields and makes the data much easier to interpret. In particular, notice that we have eliminated the repeated listing of geographical regions and the repeated listing of the state names by placing these items in separate columns. Similarly, we have eliminated the repeated listing of the census years by placing them in a separate row, as column headings for the census values.

Table 5.2 – The list of states, organized more concisely with a pivot table

Region	State		Population		
		1970	*1980*	*1990*	*2000*
Midwest	Missouri	4,677,623	4,916,762	5,117,073	5,595,211
	North Dakota	617,792	652,717	638,800	642,200
Northeast	New York	18,241,391	17,558,165	17,990,455	18,976,457
	Pennsylvania	11,800,766	11,864,720	11,881,643	12,281,054
	Rhode Island	949,723	947,154	1,003,464	1,048,319
Northwest	Alaska	302,583	401,851	550,043	626,932
	Washington	3,413,244	4,132,353	4,866,692	5,894,121
South	Florida	6,791,418	9,746,961	12,937,926	15,982,378
	North Carolina	5,084,411	5,880,415	6,628,637	8,049,313
	Virginia	4,651,448	5,346,797	6,187,358	7,078,515
Southwest	California	19,971,069	23,667,764	29,760,021	33,871,648
	Colorado	2,209,596	2,889,735	3,294,394	4,301,261
	Texas	11,198,655	14,225,513	16,986,510	20,851,820

To create a pivot table within Excel, follow the steps given below.

1. Be sure that the list containing the data is entered into a block of contiguous cells, with a single-cell heading over each column.

2. Activate any cell within the block. Then select PivotTable and PivotChart Report... from the Data menu.

3. Complete the three PivotTable and PivotChart Wizard dialog boxes, indicating the location of the source data and the location of the resulting pivot table. This will result in an empty worksheet such as that shown in Fig. 5.15. The worksheet will contain the following three objects:

   (*a*) A PivotTable Field List, containing the column headings included in the original data set.

   (*b*) An empty template, which is used to lay out the pivot table.

(*c*)   The *PivotTable Bar,* which includes a drop-down menu of special pivot table commands as well as icons for certain frequently used features. (Most of these same commands and features can be accessed by right-clicking on the actual pivot table.)

4.   The actual pivot table is constructed by dragging each of the headings from the PivotTable Field List (shown in the right portion of Fig. 5.15) to the appropriate place in the empty template (shown in the left portion of Fig. 5.15). You can also create a *pivot chart* (i.e., a bar chart associated with the pivot table entries) while the pivot table is being constructed, or after the pivot table has been built. The details associated with pivot table and pivot chart construction are illustrated in the following example.

Once the pivot table has been completed, it can be edited in the same manner as any other Excel worksheet.

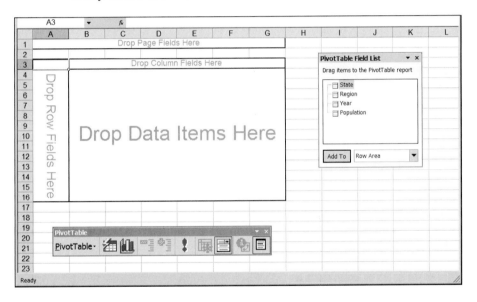

Figure 5.15 – Preparing to create a pivot table

Example 5.4   Creating a Pivot Table in Excel

Enter the state population data presented earlier in this section into an Excel worksheet and create the accompanying pivot table. Add a pivot chart after the table has been completed.

We begin by entering the data into an Excel worksheet, as shown in Fig. 5.16. Note that the data are entered into a four-column list, and that each column includes a single-cell heading.

	A	B	C	D	E	F
1						
2	**State**	**Region**	**Year**	**Population**		
3	Alaska	Northwest	1970	302,583		
4	Alaska	Northwest	1980	401,851		
5	Alaska	Northwest	1990	550,043		
6	Alaska	Northwest	2000	626,932		
7	California	Southwest	1970	19,971,069		
8	California	Southwest	1980	23,667,764		
9	California	Southwest	1990	29,760,021		
10	California	Southwest	2000	33,871,648		
11	Colorado	Southwest	1970	2,209,596		
12	Colorado	Southwest	1980	2,889,735		
13	Colorado	Southwest	1990	3,294,394		
14	Colorado	Southwest	2000	4,301,261		
15	Florida	South	1970	6,791,418		
16	Florida	South	1980	9,746,961		
17	Florida	South	1990	12,937,926		
18	Florida	South	2000	15,982,378		
19	Missouri	Midwest	1970	4,677,623		
20	Missouri	Midwest	1980	4,916,762		
21	Missouri	Midwest	1990	5,117,073		
22	Missouri	Midwest	2000	5,595,211		
23	New York	Northeast	1970	18,241,391		
24	New York	Northeast	1980	17,558,165		
25	New York	Northeast	1990	17,990,455		
26	New York	Northeast	2000	18,976,457		
27	North Carolina	South	1970	5,084,411		
28	North Carolina	South	1980	5,880,415		
29	North Carolina	South	1990	6,628,637		
30	North Carolina	South	2000	8,049,313		
31	North Dakota	Midwest	1970	617,792		
32	North Dakota	Midwest	1980	652,717		
33	North Dakota	Midwest	1990	638,800		
34	North Dakota	Midwest	2000	642,200		
35	Pennsylvania	Northeast	1970	11,800,766		
36	Pennsylvania	Northeast	1980	11,864,720		
37	Pennsylvania	Northeast	1990	11,881,643		
38	Pennsylvania	Northeast	2000	12,281,054		
39	Rhode Island	Northeast	1970	949,723		
40	Rhode Island	Northeast	1980	947,154		
41	Rhode Island	Northeast	1990	1,003,464		
42	Rhode Island	Northeast	2000	1,048,319		
43	Texas	Southwest	1970	11,198,655		
44	Texas	Southwest	1980	14,225,513		
45	Texas	Southwest	1990	16,986,510		
46	Texas	Southwest	2000	20,851,820		
47	Virginia	South	1970	4,651,448		
48	Virginia	South	1980	5,346,797		
49	Virginia	South	1990	6,187,358		
50	Virginia	South	2000	7,078,515		
51	Washington	Northwest	1970	3,413,244		
52	Washington	Northwest	1980	4,132,353		
53	Washington	Northwest	1990	4,866,692		
54	Washington	Northwest	2000	5,894,121		

Figure 5.16 – A list of states and associated population data

We then click on any cell within the list (we could also have highlighted the entire list), and then select PivotTable and PivotChart Report... from the Data menu. This results in the first PivotTable and PivotChart Wizard dialog box shown in Fig. 5.17(*a*). Note that an Excel worksheet is indicated as the data source. Also, note that a pivot table is indicated as the desired objective. (We could have requested both a pivot table and a corresponding pivot chart, if we had wished.)

Clicking on the Next button produces the second dialog box, shown in Fig. 5.17(*b*). This dialog box shows the range of the data, including the column headings. The range is determined automatically, though we may override the automatic list selection if we wish.

Clicking on the Next button results in the third dialog box, shown in Fig. 5.17(*c*). This dialog box allows us to select the location of the pivot table. We are choosing a new worksheet in this example, though we could have placed the pivot table in the original worksheet.

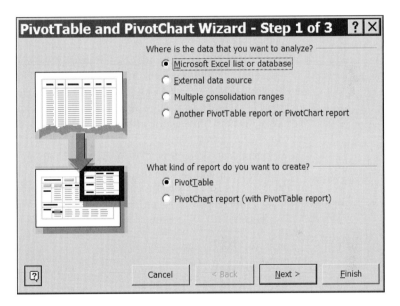

Figure 5.17(*a*) – PivotTable Wizard, first dialog box

Figure 5.17(*b*) – PivotTable Wizard, second dialog box

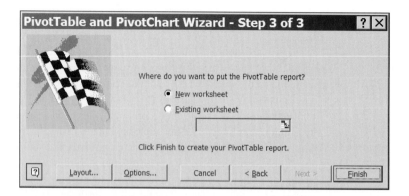

Figure 5.17(*c*) – PivotTable Wizard, third dialog box

We then click on the Finish button, resulting in the empty pivot-table worksheet shown previously in Fig. 5.15. This worksheet includes a PivotTable Field List, an empty pivot table template, and the *PivotTable Bar*. The PivotTable Field List includes the column headings from the original data list: State, Region, Year, and Population.

We now drag Region from the PivotTable Field List to the portion of the empty template labeled Drop Row Fields Here, resulting in the partially completed template shown in Fig. 5.18(*a*). (Instead of dragging, we could have clicked on Region within the PivotTable Field List, then selected Row Area in the lower right corner, and then clicked on Add To in the lower left corner.)

We then drag State from the PivotTable Field List to the same area within the template, resulting in Fig. 5.18(*b*). Dragging Year from the PivotTable Field List to the portion of the template labeled Drop Column Fields Here results in Fig. 5.18(*c*).

Finally, dragging Population from the PivotTable Field List to the portion of the template labeled Drop Data Items Here results in the completed pivot table, shown in Fig. 5.18(*d*).

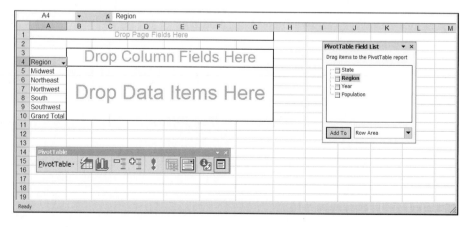

Figure 5.18(*a*) – Partially completed pivot table template, step one

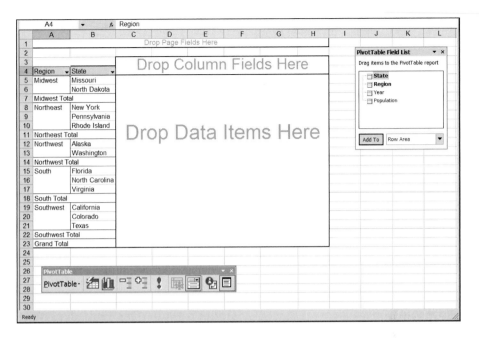

Figure 5.18(*b*) – Partially completed pivot table template, step two

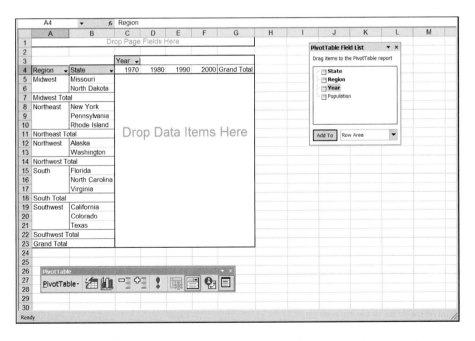

Figure 5.18(*c*) – Partially completed pivot table template, step three

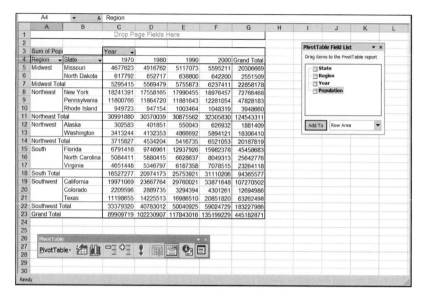

Figure 5.18(*d*) – The final pivot table

We could have generated a somewhat different pivot table, if we had wished, by dragging Year to the portion of the template labeled Drop Page Fields Here rather than Drop Column Fields Here. Figure 5.18(*e*) shows the resulting pivot table. Note that the state populations are totaled for all years in this pivot table. We could also display population totals for a single year from the drop-down menu in cell B1.

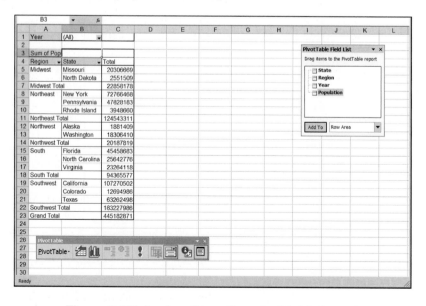

Figure 5.18(*e*) – An alternative pivot table layout

Figure 5.18(*f*) shows an edited version of the pivot table originally presented in Fig. 5.18(*d*). The various row and column headings have been placed in boldface type to enhance their readability. The grand totals shown in column H (which are meaningless in this example), the regional totals in rows 6, 10, 13, 17, and 21, and the grand totals shown in row 22 have been added automatically. These totals can be removed if desired, using various Excel editing techniques. The grand totals in Column H, for example, can be removed simply by right-clicking on the column heading in cell H3 and then selecting Hide from the resulting menu. The row totals can be removed in the same manner.

	A	B	C	D	E	F	G	H	I	J
1										
2		**Sum of Population**		Year						
3		**Region**	**State**	**1970**	**1980**	**1990**	**2000**	**Grand Total**		
4		**Midwest**	Missouri	4677623	4916762	5117073	5595211	20306669		
5			North Dakota	617792	652717	638800	642200	2551509		
6		**Midwest Total**		5295415	5569479	5755873	6237411	22858178		
7		**Northeast**	New York	18241391	17558165	17990455	18976457	72766468		
8			Pennsylvania	11800766	11864720	11881643	12281054	47828183		
9			Rhode Island	949723	947154	1003464	1048319	3948660		
10		**Northeast Total**		30991880	30370039	30875562	32305830	124543311		
11		**Northwest**	Alaska	302583	401851	550043	626932	1881409		
12			Washington	3413244	4132353	4866692	5894121	18306410		
13		**Northwest Total**		3715827	4534204	5416735	6521053	20187819		
14		**South**	Florida	6791418	9746961	12937926	15982378	45458683		
15			North Carolina	5084411	5880415	6628637	8049313	25642776		
16			Virginia	4651448	5346797	6187358	7078515	23264118		
17		**South Total**		16527277	20974173	25753921	31110206	94365577		
18		**Southwest**	California	19971069	23667764	29760021	33871648	107270502		
19			Colorado	2209596	2889735	3294394	4301261	12694986		
20			Texas	11198655	14225513	16986510	20851820	63262498		
21		**Southwest Total**		33379320	40783012	50040925	59024729	183227986		
22		**Grand Total**		89909719	102230907	117843016	135199229	445182871		
23										
24										

Figure 5.18(*f*) – An edited version of the final pivot table

Once a pivot table has been created, it can be rearranged by dragging the headings from one location to another. This feature enables you to reorganize the pivot table, in various ways, for more meaningful interpretations.

You can also create a *pivot chart* (i.e., a bar chart associated with the pivot table entries) while the pivot table is being constructed, or after the pivot table has been built. To create a pivot chart at the same time as the original pivot table, select the appropriate button in Step 1 of the PivotTable and PivotChart Wizard, as shown in Fig. 5.17(*a*). Or, you may construct a pivot chart after the pivot table has been created by selecting any cell within the pivot table and then clicking on the Chart Wizard button within the *PivotTable Bar* (see Fig. 5.19).

Figure 5.20 shows a pivot chart derived from the pivot table shown in Fig. 5.18(*f*), based upon the year 2000 census figures.

Chart Wizard

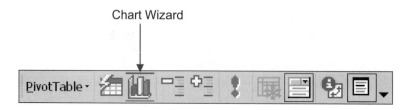

Figure 5.19 – The Pivot Table Bar

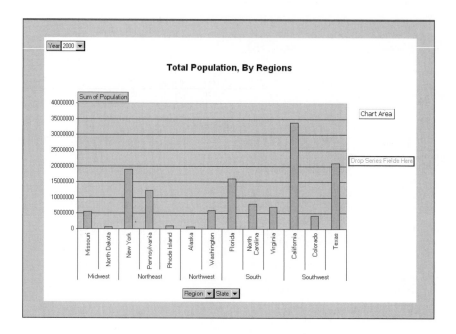

Figure 5.20 – A pivot chart

Problems

5.7 Enter the data shown in Fig. 5.16 into an Excel worksheet. Use the data to reconstruct the pivot table shown in Fig. 5.18(*f*). Create a pivot chart at the same time as the pivot table by making the appropriate entries within the PivotTable and PivotChart Wizard dialog boxes. Remove the column labeled Grand Total after the pivot table has been completed.

5.8 Rearrange the pivot table created in the last problem so that each census year corresponds to a separate table. To do so, drag the year heading to the portion of the template labeled Drop Page Fields Here. The final result should resemble the pivot table shown in Fig. 5.18(*f*).

5.9   Rearrange the pivot table created in Prob. 5.7 so that the regions and states are displayed as columns and the census years are displayed as rows. To do so, drag the headings in the original pivot table to their new locations.

5.10   Enter the data shown in Fig. 5.16 into an Excel worksheet. Use the data to construct a new pivot table with the regions and states displayed as columns and the census years displayed as years. Create a pivot chart at the same time as the pivot table by making the appropriate entries within the PivotTable and PivotChart Wizard dialog boxes.

5.11   Here, once again, are the final scores for Professor Boehring's *Introduction to Engineering* course, originally shown in Prob. 5.2.

Student	Departmental Major	Final Score	Final Grade
Barnes	EE	87.2	B
Davidson	ChE	93.5	A
Edwards	ME	74.6	C
Graham	ME	86.2	B
Harris	ChE	63.9	D
Jones	IE	79.8	B
Martin	EE	99.2	A
O'Donnell	CE	80.0	B
Prince	ChE	69.2	C
Roberts	EE	48.3	F
Thomas	ME	77.5	C
Williams	IE	94.5	A
Young	CE	73.2	C

Create a pivot table with the departmental majors listed by rows, the grades listed by columns, and the final scores shown within the data area.

5.12   Using the data given in Prob. 5.11, create a pivot table showing the student names and the final grades listed by rows, the departmental majors listed by columns, and the final scores shown within the data area. (You may either create this pivot table directly, as a new pivot table, or rearrange the pivot table created for the last problem by dragging the column headings to the appropriate places.) Compare with the pivot table created for Prob. 5.11. How well does this application lend itself to the use of pivot tables?

5.13   Using the data given in Prob. 5.3, create a pivot table showing the droplet vaporization times in the data area. Arrange the pivot table three different ways:

(*a*) With the fluid and initial drop size listed by rows and the temperature difference listed by columns.

(*b*) With the temperature difference listed by rows and the fluid and initial drop size listed by columns.

(*c*) With the fluid and temperature difference listed by rows and the initial drop size listed by columns.

Hide the sums within each pivot table.

CHAPTER 6

TRANSFERRING DATA

Some applications require that data from a *text file* be entered (*imported*) into an Excel worksheet. The text file may have been created with a text editor or a word processor (without any special formatting), or it may have originated from a computer program, an e-mail message, or a specialized shop-floor device. Data of this type can easily be entered into the traditional row/column spreadsheet format, where it may then be processed as an Excel worksheet.

Similarly, some applications require us to save (*export*) spreadsheet data in the form of a text file. Such data may then be entered into a report or an e-mail message, or it may be read by a customized computer program. Some care must be taken in doing this, however, so that the row/column layout of the data is preserved within the text file.

In this chapter, we will see how these operations are carried out within Excel. We will also see how Excel worksheets can be saved as an *HTML (HyperText Markup Language) file* that can be viewed by an Internet web browser, and how HTML files can be entered into Excel. In addition, we will see how Excel worksheets can be transferred between Excel and other Microsoft Office applications, such as Microsoft *Word* and Microsoft *PowerPoint*.

6.1 IMPORTING DATA FROM A TEXT FILE

To import data from a text file (i.e., to enter data from a text file into an Excel worksheet), follow the steps outlined below:

1. First, be sure that the file containing the data really is a *text* file; i.e., that it contains no formatting from a word processor, etc. This can usually be verified by viewing the file within the computer's operating system. (In Microsoft Windows, for example, open the file in a simple text editor, such as Notepad. If the file is readable, then it is a text file. Text files typically have the extension *.txt*, *.csv*, or *.prn*.)

2. If the file contains multiple data items within each line, be certain that the data items are separated by blank spaces, commas, semicolons, or tabs (or some combination thereof).

3. Enter Excel. Select Open from the File menu. When the Open dialog box appears, be sure to select Text Files as the file type (at the bottom of the dialog box, as shown in Fig. 6.1). Locate the correct file folder, using the Look in: drop-down menu and/or the "up" button 🔼 in the toolbar at the top of the dialog box. Then select the desired text file from the available list.

Figure 6.1 – The Open dialog box, configured to open a text file

4. The Text Import Wizard, consisting of three consecutive steps, will then appear. Provide the information requested in each dialog box. In particular, be sure to specify the correct delimiters (separators) in step 2.

Example 6.1 Importing Data from a Text File

Example 4.1 presented an Excel worksheet showing the following set of data describing the voltage of a capacitor as a function of time:

Time	*Voltage*
0	10.000
1	6.065
2	3.679
3	2.231
4	1.353
5	0.821
6	0.498
7	0.302
8	0.183
9	0.111
10	0.067

Let us assume that the data had originally been entered into a text file, with each pair of data items (i.e., time and the accompanying voltage) entered on a separate line and separated by a comma and a blank space, as shown in Fig. 6.2:

```
Time, Voltage
0, 10.
1, 6.065
2, 3.679
3, 2.231
4, 1.353
5, 0.821
6, 0.498
7, 0.302
8, 0.183
9, 0.111
10, 0.067
```

Figure 6.2 – A text file containing a set of data

The steps required to transfer the data from the text file into an Excel worksheet are shown below.

Suppose the data have been entered into a text file called Capacitor Data.txt. We begin by entering Excel, selecting Open from the File menu, and then selecting the proper file type (text files), file folder (Worksheets), and file name (Capacitor Data.txt), as illustrated in Fig. 6.3.

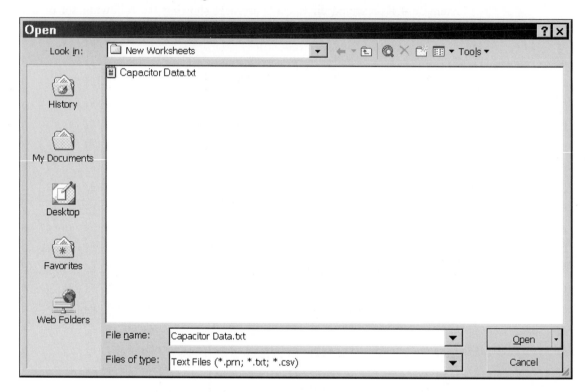

Figure 6.3 – Opening the text file Capacitor Data.txt

Once the file is opened, the first of three Text Import Wizard dialog boxes appears, as shown in Fig. 6.4. Notice that Delimited is selected as the original data type. Also, note that the first few lines of the text file are displayed in the lower portion of the dialog box.

The second Text Import Wizard dialog box is shown in Fig. 6.5. The appropriate delimiters (in this case, Comma and Space) must be specified in this dialog box. In addition, note that the box labeled Treat consecutive delimiters as one has been selected. The appearance of the data within the worksheet is then shown in the lower portion of the dialog box. Notice the arrangement into distinct columns. *This will occur only if the delimiters have been specified correctly.*

In the third dialog box, shown in Fig. 6.6, we specify the manner in which the data will be displayed within each worksheet column. The default selection is General, which automatically recognizes numerical values, dates, and labels (strings) and enters them correctly within the worksheet. In most situations, this will be the most convenient selection.

The resulting Excel worksheet is shown in Fig. 6.7. This worksheet may now be reformatted to take on the enhanced appearance shown in Fig. 4.8.

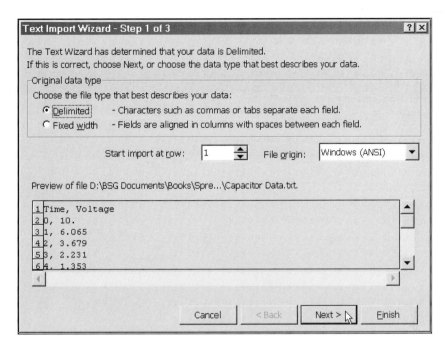

Figure 6.4 – Text Import Wizard, first dialog box

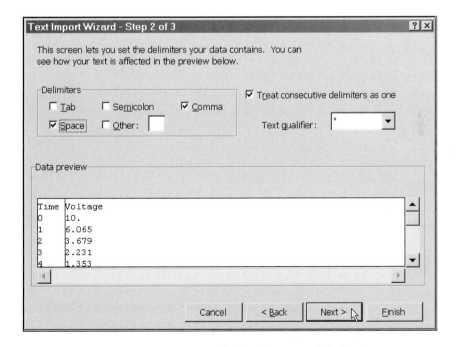

Figure 6.5 – Text Import Wizard, second dialog box

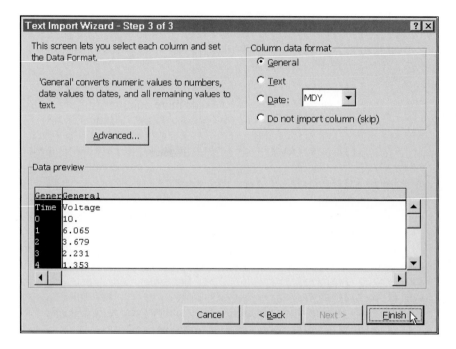

Figure 6.6 – Text Import Wizard, third dialog box

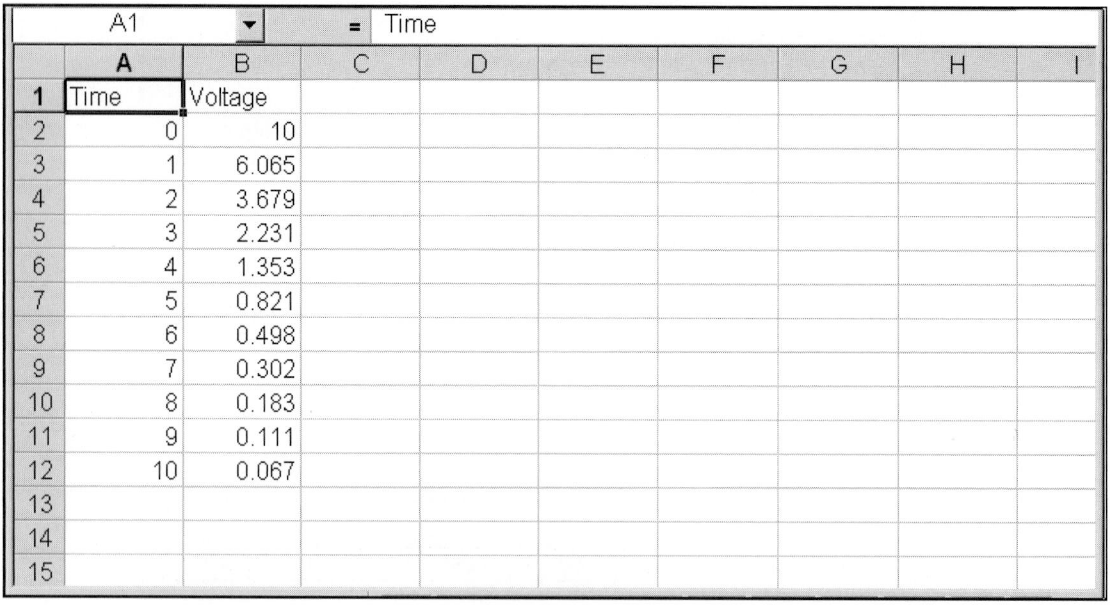

Figure 6.7 – An Excel worksheet after importing data from a text file

6.2 EXPORTING DATA TO A TEXT FILE

To export data from an Excel worksheet to a text file, follow the steps outlined below:

1. If the worksheet contains two or more columns, the resulting text file will have multiple data items within each line. Within the text file, these data items can be separated by either tabs or commas. Decide which type of separator you wish to use. This will determine the text file type.

2. Within Excel, select Save As from the File menu.

   (*a*) If you want the data items within each line to be separated by tabs, select Text (Tab delimited) as the file type (i.e., as the Save as type:, at the bottom of the dialog box), as shown in Fig. 6.8. Locate the correct folder, using the Save in: drop-down menu and/or the "up" button in the toolbar at the top of the dialog box. Then specify a file name in the space provided near the bottom. (Do not include an extension with the file name; the extension .txt will automatically be added at the end, even if you have included your own extension.)

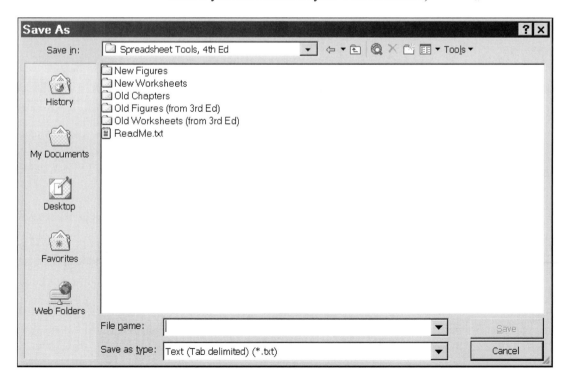

Figure 6.8 – The Save As dialog box, configured to save a tab-delimited text file

(*b*) If you want the data items within each line to be separated by commas rather than tabs, select CSV (Comma delimited) rather than Text (Tab delimited) as the file type, as shown in Fig. 6.9. Then locate a folder and name the file, as explained above.

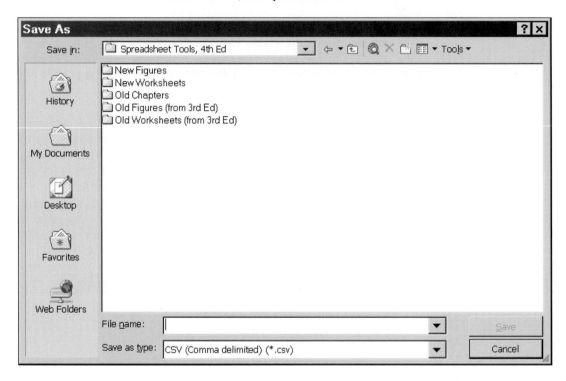

Figure 6.9 – The Save As dialog box, configured to save a comma-delimited text file

Example 6.2 Exporting Data to a Text File

Example 2.4 presents an Excel worksheet containing several student exam scores, with individual and overall averages (see Fig. 2.23). Note that the worksheet includes numerical constants, numerical data generated from formulas, and labels (strings). Save the worksheet data in two different text files, using tabs as separators in the first text file and commas in the second.

Figure 6.10 shows the worksheet with the Save As dialog box superimposed over the top. Note that the file will be saved as a Text (Tab delimited) file, it will be named Ex2-3.txt, and it will be stored in the New Worksheets folder.

Figure 6.11 shows the resulting text file. The individual data items are separated by tabs within each line. Note that the lines containing the Richardson and Williams data extend out to the right because the names extend beyond the first tab space.

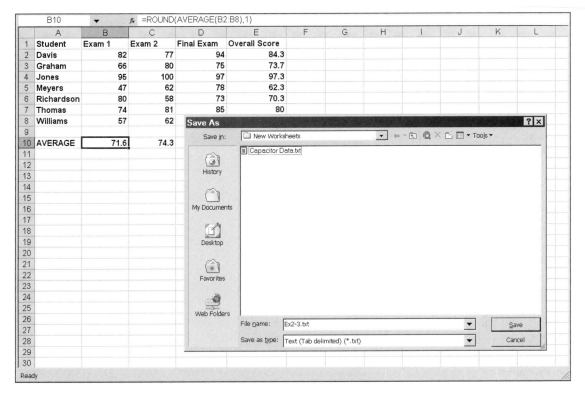

Figure 6.10 – Preparing to save a tab-delimited text file

Student	Exam 1	Exam 2	Final Exam	Overall Score
Davis	82	77	94	84.3
Graham	66	80	75	73.7
Jones	95	100	97	97.3
Meyers	47	62	78	62.3
Richardson	80	58	73	70.3
Thomas	74	81	85	80
Williams	57	62	67	62
AVERAGE	71.6	74.3	81.3	75.7

Figure 6.11 – The tab-delimited text file

If the worksheet data were stored as a .csv (comma delimited) file rather than a .txt (tab delimited) file, the Save As dialog box would appear as shown in Fig. 6.12. The file will now be named Ex2-3.csv.

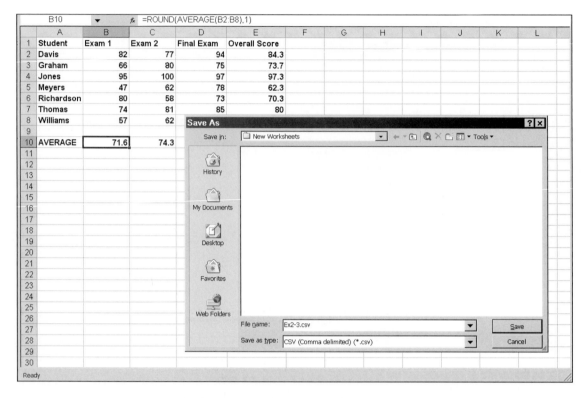

Figure 6.12 – Preparing to save a comma-delimited text file

Figure 6.13 shows the resulting .csv file. Notice that the individual data items within each line are now separated by commas.

```
Student,Exam 1,Exam 2,Final Exam,Overall Score
Davis,82,77,94,84.3
Graham,66,80,75,73.7
Jones,95,100,97,97.3
Meyers,47,62,78,62.3
Richardson,80,58,73,70.3
Thomas,74,81,85,80
Williams,57,62,67,62
,,,,
AVERAGE,71.6,74.3,81.3,75.7
```

Figure 6.13 – The comma-delimited text file

Problems

6.1 The data given below are taken from Prob. 4.14. Enter the data, including headings, into a text file using a text editor. Within each line, separate time from bacteria concentration with a blank space. Then import the text file into an Excel worksheet.

Time (min)	Bacteria Concentration (ppm)
0	6
1	9
2	15
3	19
4	32
5	42
6	63
7	102
8	153
9	220
10	328

6.2 Repeat Prob. 6.1, using a comma rather than a blank space to separate the data items within each line of the text file.

6.3 The final scores for Professor Boehring's *Introduction to Engineering* course are shown below (all names are fictitious):

Student	Departmental Major	Final Score	Final Grade
Barnes	EE	87.2	B
Davidson	ChE	93.5	A
Edwards	ME	74.6	C
Graham	ME	86.2	B
Harris	ChE	63.9	D
Jones	IE	79.8	B
Martin	EE	99.2	A
O'Donnell	CE	80.0	B
Prince	ChE	69.2	C
Roberts	EE	48.3	F
Thomas	ME	77.5	C
Williams	IE	94.5	A
Young	CE	73.2	C

Enter the data, including the column headings, into an Excel worksheet. Then export the worksheet to a text file. Configure the text file two different ways (i.e., create two different text files):

(*a*) With tabs separating multiple data items within each line.

(*b*) With commas separating multiple data items within each line.

6.4 An engineer is responsible for monitoring the quality of a batch of 1000-ohm resistors. To do so, the engineer must accurately measure the resistance of a number of randomly selected resistors within the batch. Use a text editor to enter the data, including the headings, into a text file. Within each line, separate the data items with commas. Then import the text file into an Excel worksheet.

Sample No.	Resistance (ohms)	Sample No.	Resistance (ohms)
1	1006	16	960
2	1006	17	976
3	978	18	954
4	965	19	1004
5	988	20	975
6	973	21	1014
7	1011	22	955
8	1007	23	973
9	935	24	993
10	1045	25	1023
11	1001	26	992
12	974	27	981
13	987	28	991
14	966	29	1013
15	1013	30	998

6.5 Modify the worksheet prepared for Prob. 6.4 in the following ways:

(*a*) Place all of the data in two columns, if you have not already done so. The first column should contain the sample number, and the second column should contain the resistance.

(*b*) Determine the average resistance for all 30 resistors. Place this value at the bottom of the second column.

(*c*)   Add a third column to the worksheet, labeled *Deviation*. Within this column, show the difference between the resistance for each sample and the average resistance, as determined in part (*b*).

(*d*)   Export the worksheet to a text file, using blank spaces to separate the data items within each line.

6.3 TRANSFERRING HTML DATA

An Excel worksheet can easily be saved as an *HTML* (*HyperText Markup Language*) file, thus enabling it to be posted to the Internet. The worksheet may include graphs. The procedure is similar to that used when saving a worksheet as a text file. In particular, select Save As Web Page… from the File menu (or select Save As from the file menu, and specify Web Page as the file type). Then locate the correct folder and specify a file name, as shown in Fig. 6.14. (Do not include an extension with the file name; the extension .htm will automatically be added at the end.) The resulting file can then be viewed with an Internet browser and posted to the Internet as a page on the World Wide Web. The familiar row-and-column spreadsheet format will be preserved.

Figure 6.14 – The Save As dialog box, configured to save a worksheet as an HTML file

Example 6.3 Exporting Data to an HTML File

Save the worksheet containing student exam scores, originally developed in Example 2.3 and shown in Example 6.2 (see Fig. 6.10), as an HTML file.

Figure 6.15 shows the worksheet with the Save As dialog box (resulting from the Save As Web Page... selection) superimposed over the top. Note that the file will be saved as a Web Page file named Ex2-3.htm, and it will be stored in the New Worksheets folder.

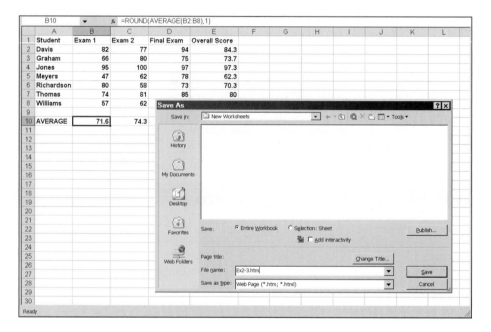

Figure 6.15 – Preparing to save a worksheet as an HTML file

Figure 6.16 shows the resulting HTML file, as viewed from a web browser. Notice that the original spreadsheet layout shown in Fig. 6.15 has been preserved.

The resulting HTML file could then be transferred across the Internet (via the Internet protocol FTP) and incorporated into a web site. It could then be viewed by others and altered (with the proper software). Hence, multiple individuals, located at several different sites, could contribute to the content and appearance of the final worksheet. The details of how this would be done are, however, beyond the scope of our present discussion.

A worksheet that has been saved as an HTML file can easily be read into Excel, following a procedure that is very similar to that used to import a text file. Simply select Open from the File menu, and specify Web Pages as the file type. Then select the desired .htm file from the appropriate source folder, as shown in Fig. 6.17.

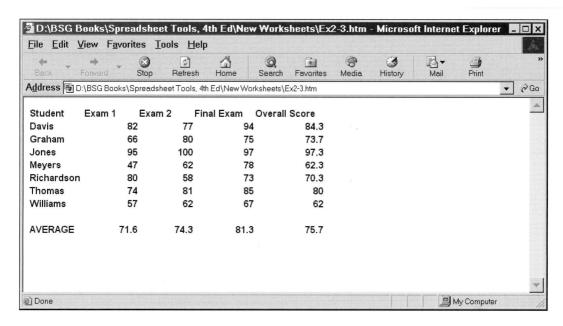

Figure 6.16 – An Excel worksheet, saved as an HTML file and viewed within a browser

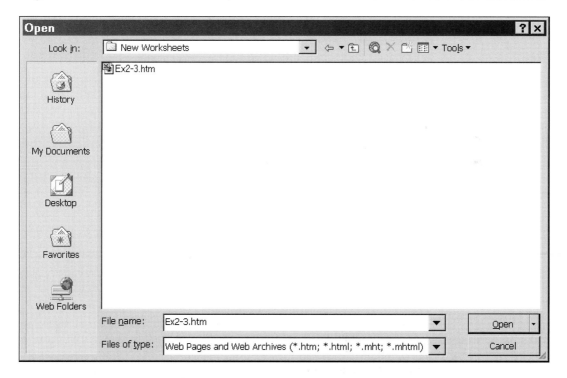

Figure 6.17 – Preparing to read an HTML file into Excel

6.4 TRANSFERRING DATA TO MICROSOFT WORD

An Excel worksheet or a portion of a worksheet can easily be transferred into Microsoft Word, thus allowing Excel data to be included in a report, newsletter, or other type of Word document. There are several ways to carry out the transfer, depending on whether the data will be *copied* directly to Word, *embedded* within Word, or *linked* back to Excel.

Copying Directly to Word

The most direct way to transfer an Excel worksheet into Word is to insert it directly using the Cut and Paste commands. Numerical and string data will then appear within a table. (If a numerical value is generated by a formula, the calculated *value* will appear, not the formula.) Graphs will appear as pictures.

To copy a block of cell values directly to Word, proceed as follows:

1. Within Excel, highlight the cells that you wish to transfer. Then select Copy (or Cut, if you wish to *move* the cells) from the Edit menu, or click on the Copy (or the Cut) icon within the Standard Toolbar (see Sec. 3.1, particularly Fig. 3.5).

2. Now switch to Word. Position the cursor where you want the data to appear. Then select Paste from the Edit menu, or click on the Paste icon within the Word Standard Toolbar.

Once the worksheet data have been copied to Word in the form of a table, you can convert the table to ordinary text if you wish. To do so, highlight the table and choose Convert/Table to Text... from Word's Table menu. Then, in the resulting dialog box, specify how you want the data items to be separated (commas, tabs, etc.).

Note that tabular data within Word can be copied to Excel by reversing the above procedure.

Example 6.4 Copying Data to Microsoft Word

Copy the worksheet containing student exam scores, originally developed in Example 2.4 and shown in Fig. 2.23, to a Microsoft Word document.

We begin by opening the Excel worksheet and highlighting cells A1 through E10, as shown in Fig. 6.18. We then copy the highlighted cells to the *clipboard* by selecting Copy from the Edit menu (or by clicking on the Copy icon in the Excel Standard Toolbar).

We then enter Microsoft Word. Figure 6.19 shows the Word document that is to contain the worksheet data. Note that a space (a single line, shown as a centered paragraph marker) has been created to accept the worksheet data, and that the cursor has been placed at this location. Selecting Paste from Word's Edit menu then produces the table shown in Fig. 6.20. Or, you may click on the Paste icon in the Excel Standard Toolbar.

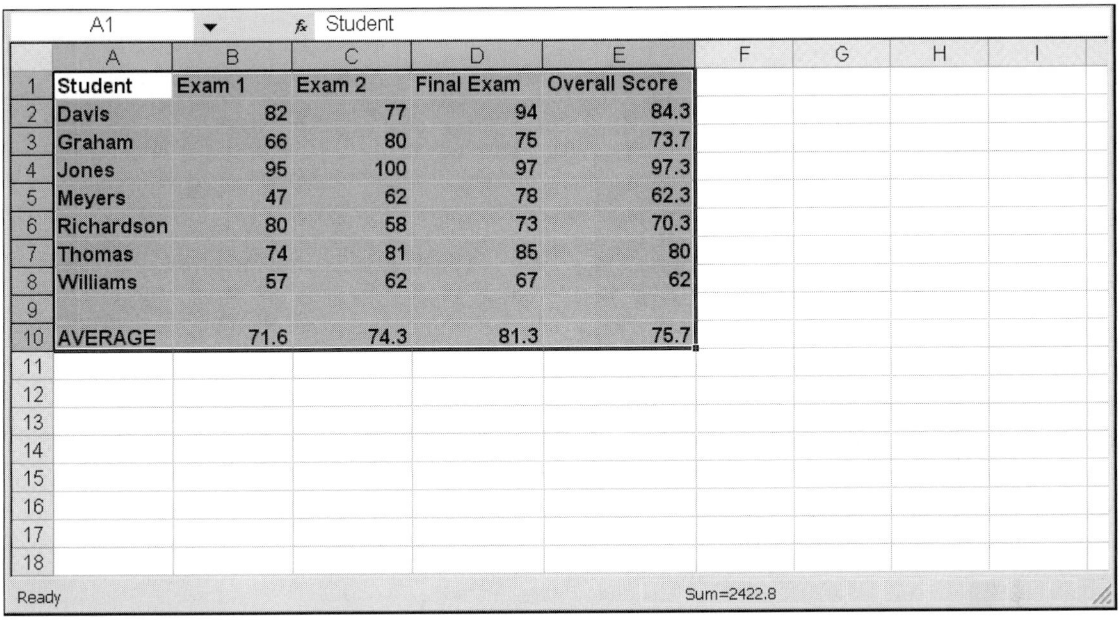

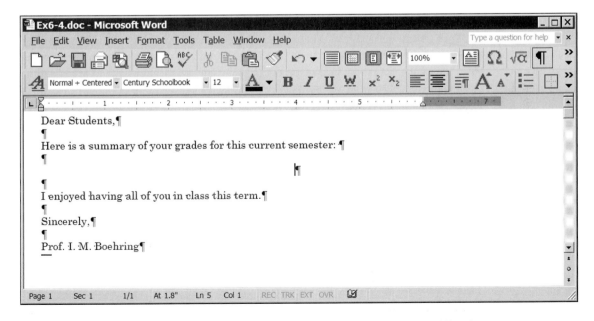

Figure 6.18 – The original Excel worksheet

Figure 6.19 – Preparing to copy the cell values into Word

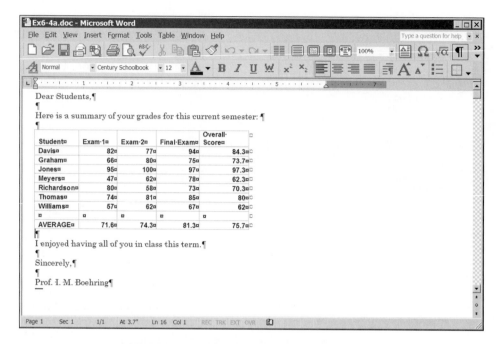

Figure 6.20 – The Word document showing the copied cell values as a table

Finally, let us convert the table to ordinary tab-delimited text within Word. This is accomplished by highlighting the table, then selecting Convert/Table to Text… from Word's Table menu. We then select Tabs from the resulting dialog box, as shown in Fig. 6.21. The final Word document is shown in Fig. 6.22. Notice that all formatting marks have been hidden in Fig. 6.22. Also, note that the font associated with the Excel data may be changed to match the original Word font if desired.

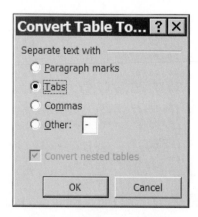

Figure 6.21 – The Convert/Table to Text dialog box within Word (from the Table menu)

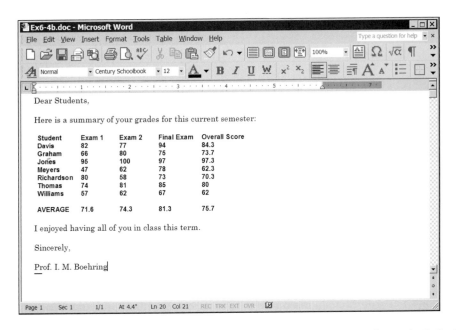

Dear Students,

Here is a summary of your grades for this current semester:

Student	Exam 1	Exam 2	Final Exam	Overall Score
Davis	82	77	94	84.3
Graham	66	80	75	73.7
Jones	95	100	97	97.3
Meyers	47	62	78	62.3
Richardson	80	58	73	70.3
Thomas	74	81	85	80
Williams	57	62	67	62
AVERAGE	71.6	74.3	81.3	75.7

I enjoyed having all of you in class this term.

Sincerely,

Prof. I. M. Boehring

Figure 6.22 – The Word document with the worksheet data converted to tab-delimited text

Copying an Excel object, such as a graph, into Word can be carried out in the same way (Copy from Excel, Paste at the desired location within Word). When an Excel object is copied directly into Word in this manner, it appears as a *picture* within the Word document. The picture can then be resized, moved to another location, etc.

Embedding within Word

An *embedded* worksheet is copied to Word as a graphical object. If the embedding is carried out properly, it is possible to edit the data within Word using menus and toolbars copied from Excel. The original Excel data will remain unchanged.

To embed a block of cells directly to Word, proceed as follows:

1. Within Excel, highlight the cells that you wish to transfer. Then select Copy (Cut) from the Edit menu, or click on the Copy (Cut) icon within the Standard Toolbar.

2. Now switch to Word. Position the cursor where you want the data to appear. Then select Paste Special (not Paste) from the Edit menu. This will result in the dialog box shown in Fig. 6.23, allowing you to specify how the data will be transferred to Word. Notice the two buttons located near the left edge of the dialog box. To embed the data, be sure that Paste: (not Paste link:) is selected, as shown.

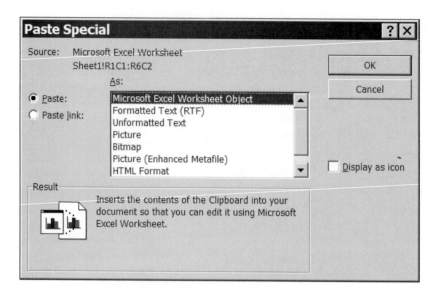

Figure 6.23 – The Paste Special dialog box, configured to embed an Excel Worksheet Object

3. Within the center window of the Paste Special dialog box, select Microsoft Excel Worksheet Object. Once you click on OK, worksheet data will appear within Word as a graphical object, as shown in Fig. 6.24. (Remember that formulas will be replaced by their calculated values.) If you double click on this object, it will become active and the normal Word menus and toolbars will be replaced by Excel menus and toolbars. You may then proceed to edit the data as though you were in Excel. Note, however, that these changes affect only the object embedded within Word. The original Excel worksheet will remain unchanged.

As a matter of interest, if you select Formatted Text (RTF) instead of Microsoft Excel Worksheet Object in the Paste Special dialog box, the cell values will appear as a table within Word, as shown in Fig. 6.20. You may edit this table within Word, but you will not have access to the Excel menus and toolbars, and you will not be able to edit the table as though you were in Excel.

Similarly, if you select Unformatted Text instead of Microsoft Excel Worksheet Object, the cell values will appear as ordinary tab-delimited text, as shown in Fig. 6.22. You will not have access to the Excel menus and toolbars, and you will not be able to edit the table as though you were in Excel.

Other Excel objects, such as graphs, may also be embedded within Word as described above. Double clicking on an embedded object will allow you to edit the object as you would in Excel, using Excel menus and toolbars, even though you are still in Word.

Excel need not be running in order to make use of this feature. However, Excel must be installed on the same computer, or, if you are using a local area network, it must be accessible on the network.

Example 6.5 Embedding Data within Microsoft Word

In this example, we will again transfer the worksheet containing student exam scores to a Microsoft Word document, as we did in Example 6.4. (See Fig. 2.23 in Example 2.4 for the original worksheet.) Now, however, we will transfer the worksheet as an embedded object rather than simply copying it to Word.

We begin by opening the Excel worksheet and highlighting cells A1 through E10, as we did in Example 6.4 (see Fig. 6.18). We then copy the highlighted cells to the clipboard, enter Microsoft Word, and position the cursor at the proper location to accept the worksheet data within the Word file (see Fig. 6.19).

So far, the procedure is the same as in Example 6.4. Now, however, we select Paste Special from Word's Edit menu, and we choose Paste: and As: Microsoft Excel Worksheet Object, as shown in Fig. 6.23. This causes the highlighted cells to be embedded in the Word document in the form of an Excel object, as shown in Fig. 6.24. This document appears similar to the Word document shown in Fig. 6.20. But this document contains an *embedded object*, not a *copied table*. If we double click on the object, it becomes active and can be edited directly in Word, as shown in Fig. 6.25. Notice that the menus and toolbars have changed from Word menus and toolbars (shown in Figs. 6.24 and 6.26) to Excel menus and toolbars (shown in Fig. 6.25), even though we are still in Word.

Now let us edit the embedded worksheet cells by changing Kim Davis's first exam score from 82 to 100. The editing is carried out as though we were in Excel, so changing Kim's exam score also changes her overall score as well as the class averages in the bottom row. The results of the changes are shown in Fig. 6.26. Remember that all of this was carried out in Word. The original Excel worksheet will remain unchanged.

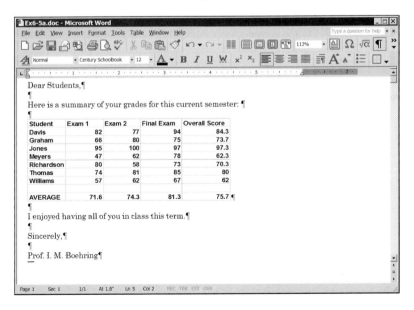

Student	Exam 1	Exam 2	Final Exam	Overall Score
Davis	82	77	94	84.3
Graham	66	80	75	73.7
Jones	95	100	97	97.3
Meyers	47	62	78	62.3
Richardson	80	58	73	70.3
Thomas	74	81	85	80
Williams	57	62	67	62
AVERAGE	71.6	74.3	81.3	75.7

Figure 6.24 – The Word document showing the embedded cell values as a graphical object

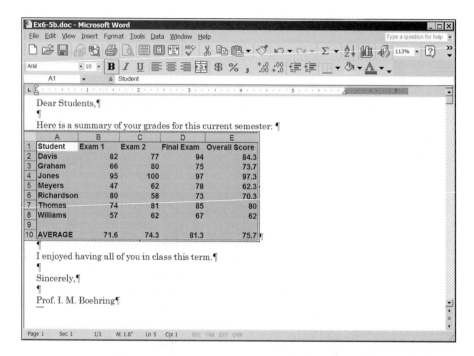

Figure 6.25 – Activating the embedded worksheet for editing within Word

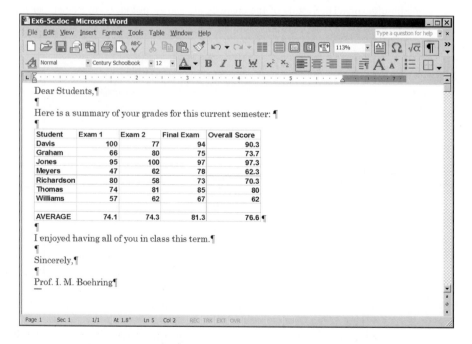

Figure 6.26 – The edited Excel object within Word

Linking between Word and Excel

We can also copy a worksheet to Word as an object that is *linked* to Excel. The worksheet will again appear as a graphical object within Word, but any changes that are made to the worksheet within Excel can easily be transferred to the graphical object within Word. Moreover, activating the graphical object in Word will cause Excel to open, showing the linked worksheet. This is a very convenient way to edit a worksheet that is included in a Word document and subject to periodic changes, like a newsletter or a monthly report.

To link a block of cells between Word and Excel, proceed as follows:

1. Within Excel, highlight the cells that you wish to transfer. Then select Copy (Cut) from the Edit menu, or click on the Copy (Cut) icon within the Standard Toolbar.

2. Now switch to Word. Position the cursor where you want the data to appear. Then select Paste Special (not Paste) from the Edit menu. This will again result in the Paste Special dialog box, which we originally saw in Fig. 6.23. Now, however, we must select Paste link: rather than Paste:. We must again choose Microsoft Excel Worksheet Object as the type of object.

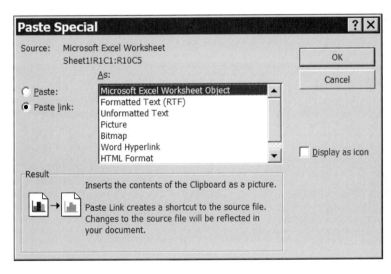

Figure 6.27 – The Paste Special dialog box, linking a worksheet between Word and Excel

3. Once you close the Paste Special dialog box, the block of data will appear within Word as a graphical object. Double clicking this object will open Excel, containing the original worksheet. You may then edit the worksheet however you wish. When you are finished, save the worksheet. You may then leave Excel if you wish. The changes made in the original Excel worksheet can be activated in Word by selecting Update Link from Word's Edit menu, or by right-clicking on the linked data and selecting Update Link.

Example 6.6 Linking Data between Excel and Microsoft Word

Now let's transfer the worksheet containing student exam scores to a Word document, as we did in the previous two examples. This time we will transfer the data to Word as a linked graphical object. We will see how a change in the original Excel worksheet is recognized in the linked graphical object within Word.

We begin as we did in the previous two examples, by opening the Excel worksheet and highlighting cells A1 through E10 (see Fig. 6.18). We then copy the highlighted cells to the clipboard, enter Microsoft Word, and position the cursor at the proper location to accept the worksheet data within the Word file (see Fig. 6.19).

So far, the procedure is the same as in the previous examples. Now, however, we select Paste Special from Word's Edit menu, and we choose Paste link: and As: Microsoft Excel Worksheet Object, as shown in Fig. 6.27. This causes the highlighted cells to be placed in the Word document as a linked graphical object, as shown in Fig. 6.28. This document appears to be identical to the Word document shown in Fig. 6.24. But this document contains a *linked object* rather than an embedded object or a copied table. If we double click on the worksheet object, Excel opens the worksheet, as shown in Fig. 6.29. We may now edit the worksheet if we wish. Any changes made to this Excel worksheet will show up in the linked object within Word once we select Update Link from the Edit menu, or right-click on the object and select Update Link.

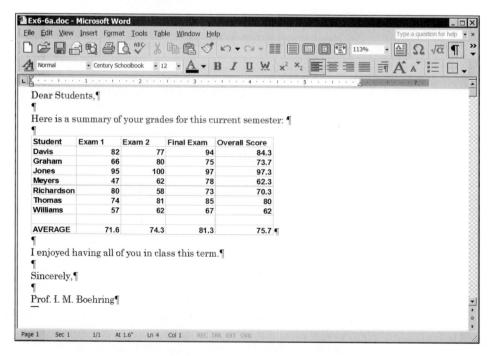

Figure 6.28 – The Word document showing the linked worksheet data as a graphical object

To see this more clearly, let us again change Kim Davis's first exam score from 82 to 100. We begin by double clicking on the worksheet object in Word. This opens Excel, as shown in Fig. 6.29. Thus, the actual editing will take place in Excel, not in Word. Figure 6.30 shows the result of changing Kim's exam score. Note that this change also changes her overall score and some of the class averages shown in the bottom row. We then save the altered Excel worksheet.

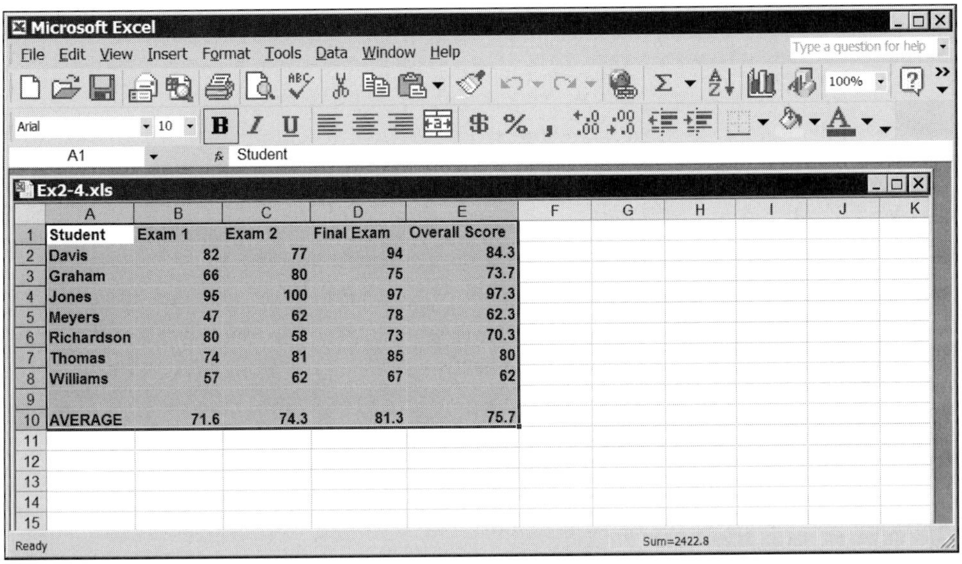

Figure 6.29 – Double clicking on the Word object causes Excel to open

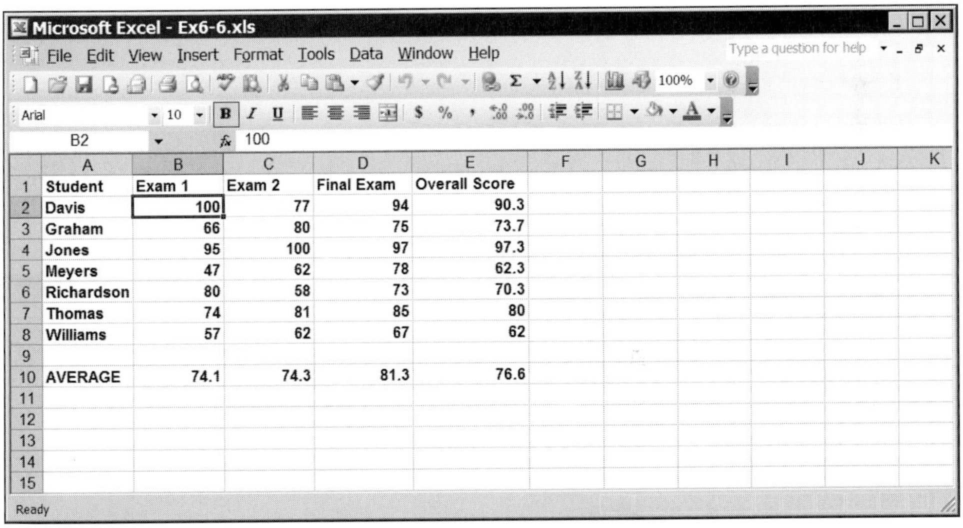

Figure 6.30 – Changing an exam score affects the entire worksheet

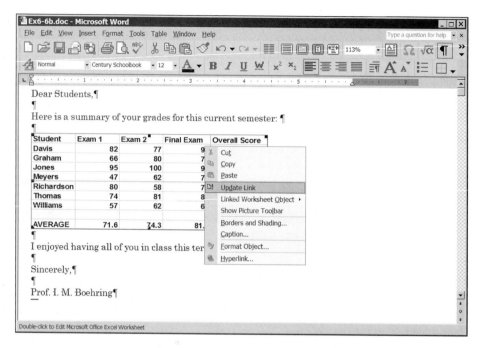

Figure 6.31 – Preparing to update the worksheet object within Word

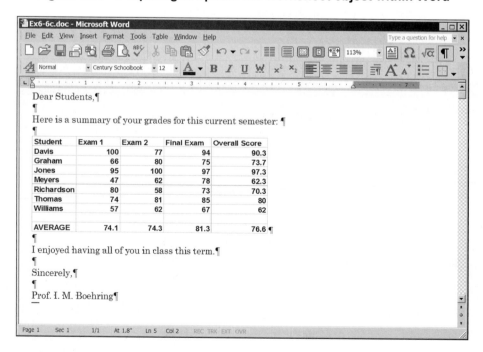

Figure 6.32 – The Word document with the updated worksheet object

We now close Excel and return to Word (Excel may remain open if we wish). To update the linked worksheet object, we right-click on the object to display the drop-down menu shown in Fig. 6.31. (A similar menu is obtained by choosing Edit from the menu bar.) Selecting Update Link from this menu will cause the worksheet object to be updated, as shown in Fig. 6.32.

Transferring a Graph to Word

Graphs (charts) are independent objects, even if they appear to be a part of a worksheet within Excel. Hence, they must be transferred separately. Graphs, like cell values, can be copied to Word three different ways. The simplest is to use copy and paste (Copy from Excel, Paste at the desired location within Word.) The graph will then be copied to Word as a *picture*, which can be resized or moved. Often, this is sufficient for the purpose at hand.

A graph can also be embedded in or linked to a Word document, as with cellular values. The procedures are the same as described in the preceding sections. The Paste Special dialog boxes are slightly different, however, as illustrated in Figs. 6.33(*a*) and 6.33(*b*) below. In particular, notice that the appropriate selections are Paste: and As: Microsoft Excel Chart Object when embedding a graph in Word, and Paste Link: and As: Microsoft Excel Chart Object when linking a graph between Excel and Word.

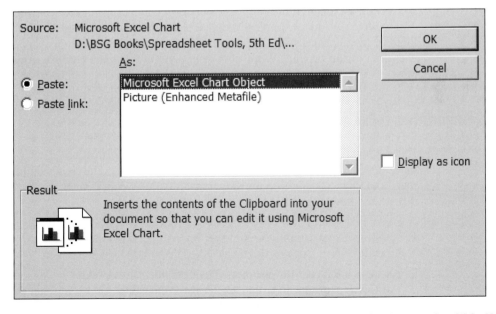

Figure 6.33(*a*) – The Paste Special dialog box, configured to embed a graph within Word

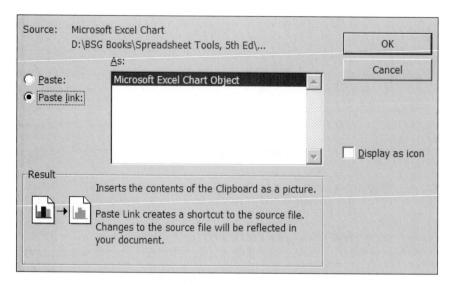

Figure 6.33(*b*) – The Paste Special dialog box, linking a graph between Word and Excel

Transferring a Table from Word to Excel

A block of data can also be transferred from a Word document to Excel, where it will be displayed as an Excel worksheet. Essentially, the procedure involves converting the data within the Word document to a table (select Convert/Text to Table... from Word's Edit menu), copying the table, and then pasting the copied table to the appropriate location within Excel. The individual data items within each line of the original Word document should be separated by commas or by tabs.

6.5 TRANSFERRING DATA TO MICROSOFT POWERPOINT

When preparing a presentation, it is often convenient to transfer a graph or a portion of an Excel worksheet to a PowerPoint Slide. This is accomplished in much the same manner as a transfer from Excel to Word. Graphs transfer slightly differently, however, than cell values.

To transfer a block of cell values to PowerPoint, proceed as follows:

1. Within Excel, highlight the portion of the worksheet that you wish to transfer. Then Copy or Cut the selected cells in the usual manner.

2. Now switch to PowerPoint. Position the cursor where you want the worksheet to appear within the PowerPoint slide. You may then Paste the cell values directly into the slide if you wish. This may be adequate if your presentation is not subject to periodic revisions. Or, if you wish, you may

choose Paste Special to embed the cell values within the slide, or link them between PowerPoint and Excel. If you decide to use Paste Special, the resulting dialog box will appear similar to those shown in Figs. 6.23 and 6.27, depending whether you choose the Paste: or Paste link: options.

3. Edit the PowerPoint slide as required. The manner in which the editing is carried out will vary, depending on how the cell values were transferred. (Remember that cells containing formulas will show the resulting *values*, not the formulas, when they are transferred to PowerPoint.) Data that were pasted directly into a slide can be edited as ordinary text. Embedded data can be edited using Excel's menus and toolbars, even though you are still in PowerPoint. Linked data will actually be edited within Excel; the corresponding PowerPoint slide can then be updated to display the changes made to the original worksheet.

Transferring a graph to PowerPoint is similar, except that Paste causes the graph to be embedded within the slide. Thus, the use of Paste Special is unnecessary in this situation.

Example 6.7 Transferring Data to Microsoft PowerPoint

Figure 6.34 shows a variation of the worksheet containing Professor Boehring's student exam scores (see Fig. 6.18). We now see a bar graph showing the students' overall scores, in addition to the original tabular data. Let us transfer the block of cells containing the numerical exam scores to a PowerPoint slide, and the bar graph to a separate PowerPoint Slide.

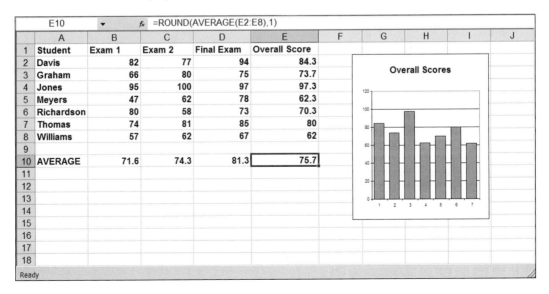

Figure 6.34 – The original Excel worksheet plus a bar graph showing the overall scores

We begin by highlighting cells A1 through E10, as in the previous examples, and we select Copy from Excel's Edit menu. We then switch to PowerPoint and select a slide layout (Format/Slide Layout...), as shown in Fig. 6.35. Next we add the title Semester Grades, click on the blank area below the title, and select Paste from PowerPoint's Edit menu. The result, after some minor editing, is shown in Fig. 6.36.

Note that we used Paste to copy the cell values directly into the PowerPoint slide. We could have used Paste Special to embed the cell values in the slide or to link them with the original Excel worksheet.

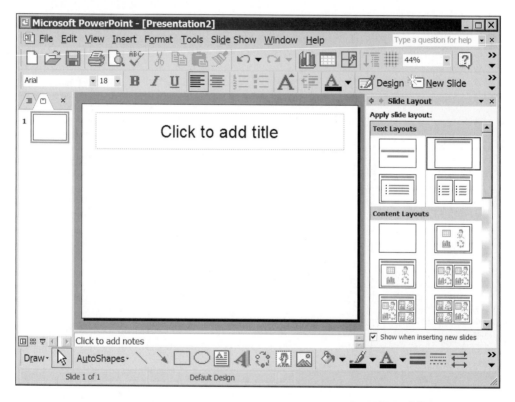

Figure 6.35 – Preparing to copy the cell values into PowerPoint

We now turn our attention to the bar graph. We begin by clicking on the bar graph within Excel to make it an active object and then select Copy from Excel's Edit menu, as shown in Fig. 6.37. We then click on the New Slide button within PowerPoint, add the title Distribution of Overall Scores, and paste the bar graph into the area beneath the title. Figure 6.38 shows the final result, after enlarging the bar graph (by dragging a corner) to fill the slide.

Note the similarity between this process and that used with Microsoft Word, in Example 6.4.

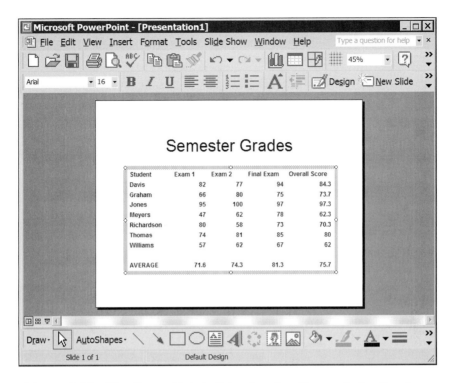

Figure 6.36 – The PowerPoint slide showing the copied cell values

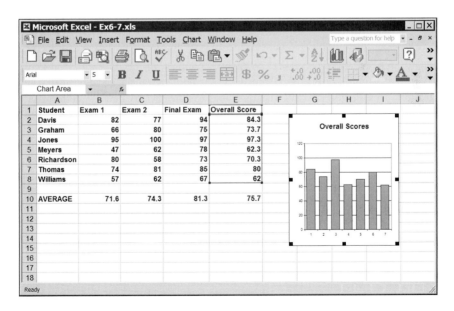

Figure 6.37 – Preparing to copy the bar graph from the original Excel worksheet

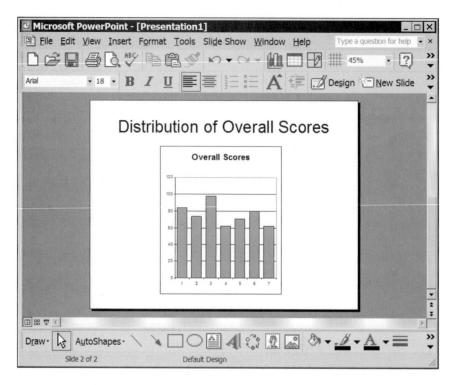

Figure 6.38 – The PowerPoint slide containing the copied bar graph

Problems

6.6 Prepare an Excel worksheet containing the data given in Prob. 6.1. Include an *x-y* graph of the data, showing bacteria concentration versus time. Be sure to format the data and the graph, so that everything is attractive and readable. Then:

(*a*) Save the worksheet as an HTML file.

(*b*) View the HTML file, using your favorite web browser, and verify that the information shown is basically the same as that seen initially in the Excel worksheet.

(*c*) Read the HTML file back into Excel, and verify that the information shown is the same as the original worksheet.

6.7 Prepare an Excel worksheet showing the information given in Prob. 6.3 for Professor Boehring's *Introduction to Engineering* course. Include a pie chart, showing the distribution of letter grades. Then:

(*a*) Save the worksheet as an HTML file.

(*b*)   View the HTML file, using your favorite web browser, and verify that the information shown is basically the same as that seen initially in the Excel worksheet.

(*c*)   Read the HTML file back into Excel, and verify that the information shown is the same as the original worksheet.

6.8   Prepare an Excel worksheet containing the data given in Prob. 6.1 (bacteria concentration vs. time). Format the worksheet so that it is attractive and readable. Save the worksheet. Then:

(*a*)   Copy the cell values to a Word document using Copy and Paste.

(*b*)   Copy the cell values to the same Word document using Copy and Paste Special, with the Paste: option active. Copy the cell values five different times, using each of the following formats:

(*i*)   Microsoft Excel Worksheet Object

(*ii*)   Formatted Text (RTF)

(*iii*) UnformattedText

(*iv*)   Picture

(*v*)   Bitmap

(*c*)   If you have carried out steps (*a*) and (*b*) correctly, you will have six different copies of the cell values in the same Word document. Notice the characteristics of each copy. Label each copy, identifying how it was created.

(*d*)   Double click on each copy of the cell values. Observe what happens, and notice how each set of cell values can be edited.

6.9   Repeat Prob. 6.8 in its entirety, using the Paste Link: option within the Paste Special dialog box. Observe how the use of Paste Link: differs from Paste: when copying to Word.

6.10 Add an *x-y* graph to the Excel worksheet that you prepared for Prob. 6.8 showing bacteria concentration versus time. Format the graph so that it is clear and attractive. Save the entire worksheet. Then activate the graph and:

(*a*)   Copy it to a Word document using Copy and Paste.

(*b*)   Copy the graph to the same Word document using Copy and Paste Special, with the Paste: option active. Copy the graph as a Microsoft Excel Chart Object.

(*c*)   Copy the graph to the same Word document, again using Copy and Paste Special, with the Paste: option active. But this time copy the graph as an Enhanced Metafile Picture.

(*d*)   If you have carried out steps (*a*) and (*c*) correctly, you will have three different copies of the graph in the same Word document. Notice the characteristics of each copy. Label each copy, identifying how it was created.

(*e*)   Double click on each copy of the graph. Observe what happens, and notice how each graph can be edited.

6.11   Repeat Prob. 6.8, but copy the worksheet to PowerPoint rather than Word.

6.12   Repeat Prob. 6.10, but copy the graph to PowerPoint rather than Word.

6.13   Prepare an Excel worksheet containing the final exam scores given in Prob. 6.3. Copy the data to a Word document and format it as a table.

6.14   Enter the list of resistances given in Prob. 6.4 into a Word document. Add a title, and edit the resulting Word document as required for maximum legibility. Then convert the data into a table, copy the table, and paste it into an empty Excel worksheet. Edit the worksheet as required.

6.15   Prepare an Excel worksheet containing the list of resistances given in Prob. 6.4. Format the worksheet so that it is clear and attractive. Then modify the worksheet by adding a pie chart showing the distribution of departmental majors, and a second pie chart showing the distribution of final grades. Transfer each pie chart to a separate PowerPoint slide. Add an appropriate title to each slide.

6.16   Open the worksheet created for Prob. 6.15 and link the cell values to a Word document. Then change the following values in the original worksheet:

Sample No.	Resistance (ohms)
2	1022
5	1013
8	951

Save the worksheet and close Excel after you have made the changes. Then update the Word document to include the changes.

6.17   Add a bar graph to the worksheet created for Prob. 6.15. Link the bar graph to a Word document. Then modify the cell values, as described in Prob. 6.16. Notice what happens to the updated bar graph in the Word document after the data are changed.

6.18   Link the initial bar graph created for Prob. 6.17 to a PowerPoint slide. Then modify the cell values, as described in Prob. 6.16. Notice what happens to the updated bar graph in the PowerPoint slide after the data are changed.

PART 2

ENGINEERING APPLICATIONS

CHAPTER 7

CONVERTING UNITS

Engineers are frequently required to convert from one system of units to another. Most beginning engineering students learn to do this using *unit equivalences* that are obtained from tabulated conversion factors. However, unit conversions can also be carried out within a spreadsheet program, eliminating the need to carry out hand calculations based upon tabulated values.

In this chapter, we will see how unit conversions are carried out in Excel. In particular, we will see how the CONVERT library function can be used to carry out simple and complex unit conversions.

7.1 SIMPLE CONVERSIONS

Let us refer to conversions that involve only single units as *simple conversions*. Thus, a conversion from feet to meters is a simple conversion, since it involves only units of length. Similarly, a conversion from horsepower to watts is a simple conversion, since it involves only units of power (even though power is derived from other units; such as mass $\times$ length^2/time^3, or force $\times$ length/time). Simple conversions can be carried out directly in Excel using the CONVERT library function. Before discussing this, however, let us review the traditional approach to such problems.

Simple conversions are carried out by multiplying the original quantity by an appropriate *unit equivalence factor*. The unit equivalence is set up so that the original units cancel one another, leaving the new unit in the final result.

Example 7.1 A Simple Conversion (Feet to Meters)

Convert 2.5 feet into an equivalent number of meters, using the conversion factor 1 ft = 0.3048 m.

To carry out the conversion, we write

$$L = 2.5\,\text{ft} \times \frac{0.3048\,\text{m}}{1\,\text{ft}} = 0.762\,\text{m}$$

where the results are rounded to three significant figures.

The term (0.3048 m/1 ft) is a *unit equivalence factor,* based upon the given conversion factor. We could also have written the unit equivalence as (1 m/3.28084 ft), which states that 1 meter is equivalent to 3.28084 feet. The resulting quotient (0.3048) is the same in either case.

Note that *the unit equivalence is always written so that the original units* (in this case, ft) *cancel one another, leaving the new unit* (m) *in the final result.*

7.2 SIMPLE CONVERSIONS IN EXCEL

In Excel, we use the CONVERT function in place of the unit equivalence factor. This function is included in Excel's Analysis ToolPak, which must be activated when Excel is first installed. To activate the Analysis ToolPak, select Add-Ins... from Excel's Tools menu, as shown in Fig. 7.1. Then select Analysis ToolPak from the Add-Ins dialog box, as shown in Fig. 7.2.

The CONVERT function is written with three arguments, which represent the original quantity, the original unit (written as an abbreviation), and the final unit (also written as an abbreviation), respectively. Thus, to convert 2.5 feet to meters, we would write the CONVERT function as

=CONVERT(2.5,"ft","m")

The original quantity (i.e., the first argument) can also be expressed as a cell address or an expression; e.g.,

=CONVERT(A1,"ft","m")

The unit abbreviations (i.e., the second and third arguments) must be one of the permissible abbreviations associated with the CONVERT function, as listed in Table 7.1. (A few nontechnical abbreviations are omitted from this table.) These abbreviations are also listed if you access Help, as shown in Fig. 7.3. Note that the unit abbreviations are *case sensitive*; hence, you must be careful to use the proper upper- and lowercase letters. Also, each abbreviation must be enclosed in double quotes, as shown in the table.

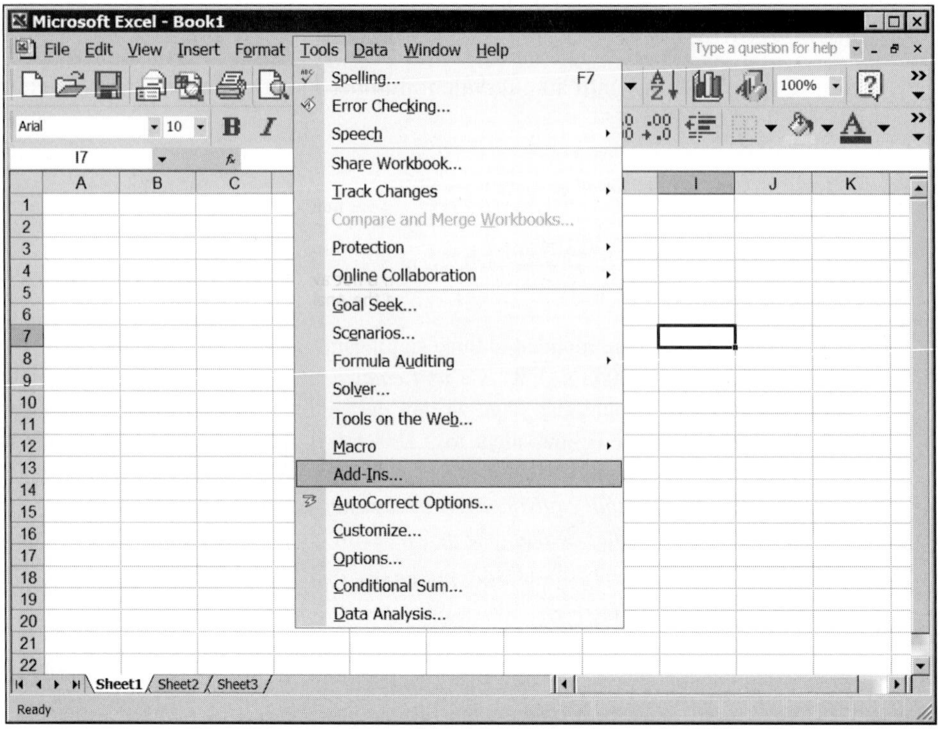

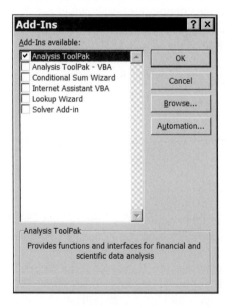

Figure 7.1 – Selecting Add-Ins… from Excel's Tools menu

Figure 7.2 – Selecting the Analysis ToolPak from the Add-Ins dialog box

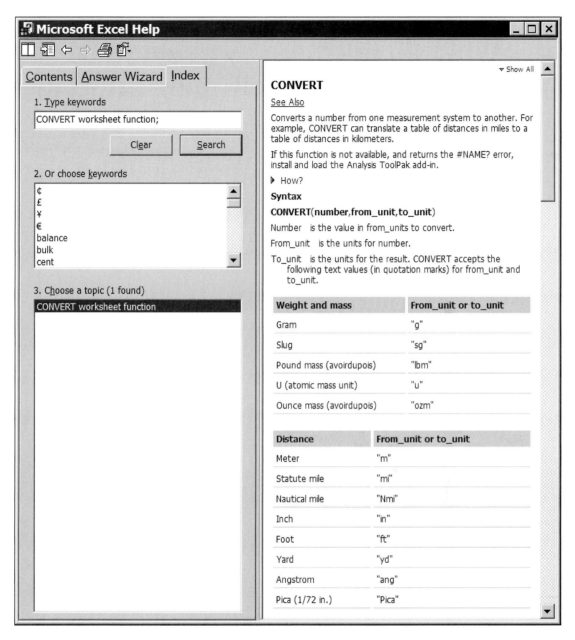

Figure 7.3 – Accessing the CONVERT function in Excel's Help

Note that the CONVERT function will accept only abbreviations with consistent dimensions. Thus, it will allow you to convert feet to meters (since both are units of length), but not BTU (energy units) to newtons (force).

Table 7.1   Excel Unit Abbreviations

Dimension	Unit	Abbreviation
Energy	BTU	"BTU"
	Calorie (IT)	"cal"
	Calorie (thermodynamic)	"c"
	Electron volt	"ev"
	Erg	"e"
	Foot-poundal	"flb"
	Horsepower-hour	"HPh"
	Joule	"J"
	Watt-hour	"Wh"
Force	Dyne	"dyn"
	Newton	"N"
	Pound force	"lbf"
Length	Angstrom	"ang"
	Foot	"ft"
	Inch	"in"
	Meter	"m"
	Mile, statute (5,280 feet)	"mi"
	Mile, nautical	"Nmi"
	Pica (1/72 inch)	"Pica"
	Yard	"yd"
Magnetism	Gauss	"ga"
	Tesla	"T"
Mass	Gram	"g"
	Ounce mass	"ozm"
	Pound mass	"lbm"
	Slug	"sg"
	U (atomic mass unit)	"u"
Power	Horsepower	"HP"
	Watt	"W"
Pressure	Atmosphere	"atm"
	mm of Mercury	"mmHg"
	Pascal	"Pa"
Temperature	Degree Celsius	"C"
	Degree Fahrenheit	"F"
	Degree Kelvin	"K"

(Table continues on next page)

Table 7.1   Excel Unit Abbreviations (*continued*)

Dimension	*Unit*	*Abbreviation*
Time	Day	"day"
	Hour	"hr"
	Minute	"mn"
	Second	"sec"
	Year	"yr"
Volume	Fluid ounce	"oz"
	Gallon	"gal"
	Liter	"l"
	Quart	"qt"
	U.K. pint	"uk_pt"
	U.S. pint	"pt"

Notes: Abbreviations are case sensitive.
Quotation marks must be included.

Example 7.2 A Simple Conversion (Feet to Meters) in Excel

Create an Excel worksheet that will convert feet to meters. Use the worksheet to convert 2.5 feet into an equivalent number of meters.

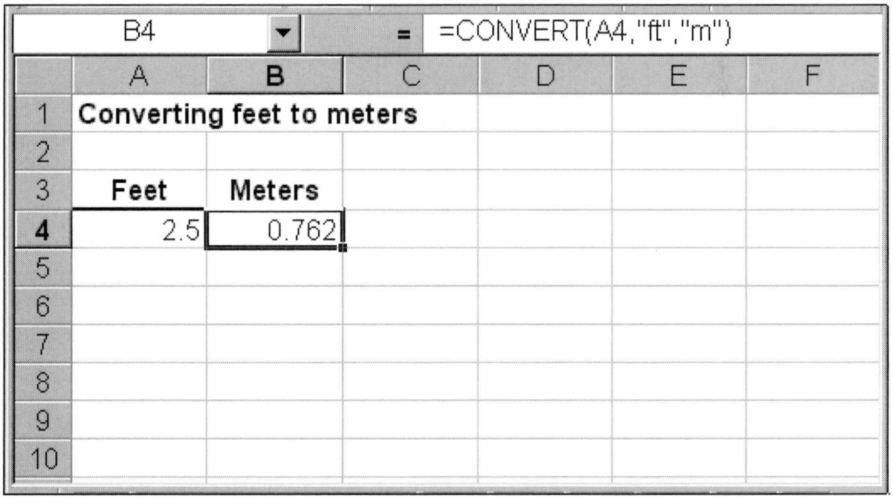

B4	▼	=	=CONVERT(A4,"ft","m")			
	A	B	C	D	E	F
1	Converting feet to meters					
2						
3	Feet	Meters				
4	2.5	0.762				
5						
6						
7						
8						
9						
10						

Figure 7.4 – Converting feet to meters with the CONVERT function

Figure 7.4 shows the desired worksheet. Notice the value 2.5, which appears in cell A4. This is the given value, in feet. The equivalent value, 0.762 meters, is shown in cell B4. Note the use of the CONVERT function to obtain the desired result, as shown in the formula bar.

This worksheet is set up so that it can convert *any* value in feet to an equivalent number of meters. Simply enter the given value in cell A4, and the resulting equivalent value will be displayed in cell B4. Thus, the approach to the problem is much more general than if we had simply written =CONVERT(2.5,"ft","m") in a single cell.

A metric abbreviation may be preceded by a *unit prefix*, which multiplies the given value by some power of 10. Thus, "cm" refers to centimeters, "kg" refers to kilograms, and so on. A list of permissible unit prefixes is shown in Table 7.2.

Table 7.2   Unit Prefixes

Prefix	*Value*	*Abbreviation*	*Prefix*	*Value*	*Abbreviation*
exa	10^{18}	"E"	deci	10^{-1}	"d"
peta	10^{15}	"P"	centi	10^{-2}	"c"
tera	10^{12}	"T"	milli	10^{-3}	"m"
giga	10^{9}	"G"	micro	10^{-6}	"u"
mega	10^{6}	"M"	nano	10^{-9}	"n"
kilo	10^{3}	"k"	pico	10^{-12}	"p"
hecto	10^{2}	"h"	femto	10^{-15}	"f"
deka	10	"e"	atto	10^{-18}	"a"

Example 7.3 Converting Feet to Millimeters

Create an Excel worksheet that will convert feet to millimeters. Use the worksheet to convert 2.5 feet into an equivalent number of millimeters.

This example is identical to Example 7.2, except that we are now converting feet to millimeters rather than feet to meters. To carry out the conversion, we write the CONVERT function as CONVERT(A4,"ft","mm") rather than CONVERT(A4,"ft","m"), as in Example 7.2.

The result, shown in Fig. 7.5, reveals that 2.5 feet is equivalent to 762 millimeters. Notice the formula =CONVERT(A4,"ft","mm") which has been entered into cell B4. We see that that the prefix "m" has been included in the CONVERT function's last argument.

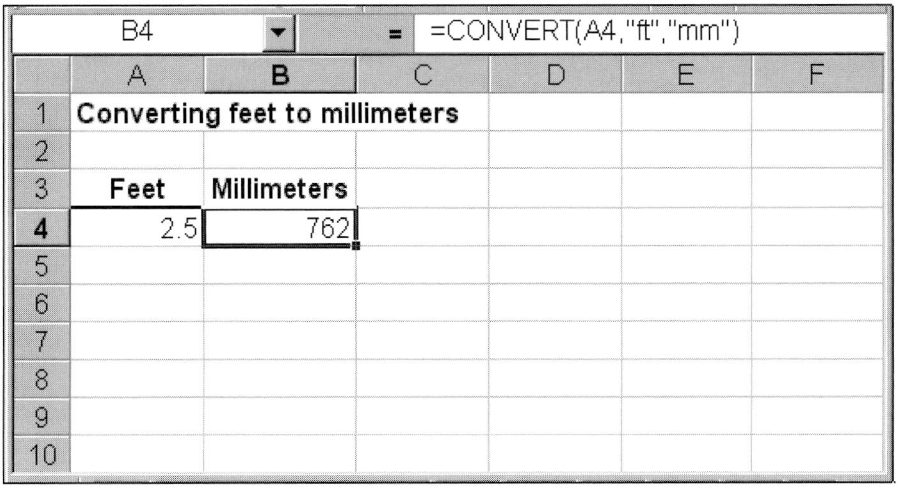

Figure 7.5 – Converting feet to millimeters with the CONVERT function

7.3 CONVERTING TEMPERATURES

Temperature conversions require special care, because conversions involving the *temperature differences* are not the same as unit conversions for single temperatures. This distinction is important for certain complex unit conversions such as heat capacities, discussed later in this chapter. The CONVERT function can convert a single temperature from one temperature unit to another. However, the temperature differences must be converted manually, using one of the following temperature difference size equivalences:

1 Celsius degree (C°) = 1.8 Fahrenheit degrees (F°)

1 Celsius degree (C°) = 1 Kelvin degree (K°)

1 Fahrenheit degree (F°) = 1 Rankine degree (R°)

1 Kelvin degree (K°) = 1.8 Rankine degrees (R°)

Note that we refer to a Celsius temperature difference as *Celsius degrees, not degrees Celsius*. Degrees Celsius would be used to refer to a single temperature. The same is true for the other temperature units.

Also, note that physicists refer to temperature measurements as Kelvins (K), not degrees Kelvin (°K). We use the common engineering designation, °K, in this chapter to maintain consistency with the other temperature designations.

To understand the distinction between single temperatures and temperature differences more clearly, recall that, at a pressure of 1 atmosphere, water freezes at 32°F and boils at 212°F. Thus, water will remain a liquid through a temperature range of 180 Fahrenheit degrees; i.e., (212 – 32) = 180. If we switch to the Celsius temperature scale, we note that water freezes at 0°C and boils at 100°C. Thus, water will remain a liquid through a temperature range of 100 Celsius degrees. Clearly, then, a temperature difference of 100 Celsius degrees is equivalent to a temperature difference of 180 Fahrenheit degrees. Or, dividing by 100, we see that *1 Celsius degree is equivalent to 1.8 Fahrenheit degrees*. This establishes the equivalent sizes of the two units.

Example 7.4 Converting Celsius to Fahrenheit

Create an Excel worksheet that will convert a temperature reading from degrees Celsius to degrees Fahrenheit. Use the worksheet to convert 22°C to an equivalent Fahrenheit temperature.

Figure 7.6 shows the desired worksheet. Notice the value 22, which appears in cell A4. This is the given value, in degrees Celsius. The equivalent value, 71.6 degrees Fahrenheit, is shown in cell B4. Notice the use of the CONVERT function to obtain the desired result, as shown in the formula bar.

As in the previous examples, the worksheet is set up to convert *any* Celsius temperature to an equivalent Fahrenheit temperature. We simply enter the Celsius temperature in cell A4, and the resulting Fahrenheit temperature will be displayed in cell B4. Thus, the approach to the problem is much more general than if we had simply written =CONVERT(22,"C","F") in a single cell.

B4	▼		fx	=CONVERT(A4,"C","F")	
A	B	C	D	E	F
1	**Converting temperatures from Celsius to Fahrenheit**				
2					
3	**°C**	**°F**			
4	22	71.6			
5					
6					
7					
8					
9					
10					

Figure 7.6 – Converting temperatures with the CONVERT function

Example 7.5 Converting a Temperature Difference

Create an Excel worksheet that will convert a temperature difference of 22 Celsius degrees to Fahrenheit degrees.

If we were to carry out this conversion by hand using a unit equivalence factor, as described in Ex. 7.1, we would write

$$F° = 22\,C° \times \frac{1.8\,F°}{1\,C°} = 39.6\,F°$$

Thus, we multiply the given temperature difference by 1.8 to obtain the desired result, which is 39.6 F°.

The corresponding Excel worksheet is shown in Fig. 7.7. Notice the value 22, which appears in cell A4. This is the given temperature difference in Celsius degrees. The equivalent temperature difference in Fahrenheit degrees is 39.6, as shown in cell B4. This result was obtained by multiplying the value in cell A4 by 1.8, as seen in the formula bar. Note that the formula shown in cell B4 does not make use of the CONVERT function. (Compare with the method and the result obtained in Example 7.4.)

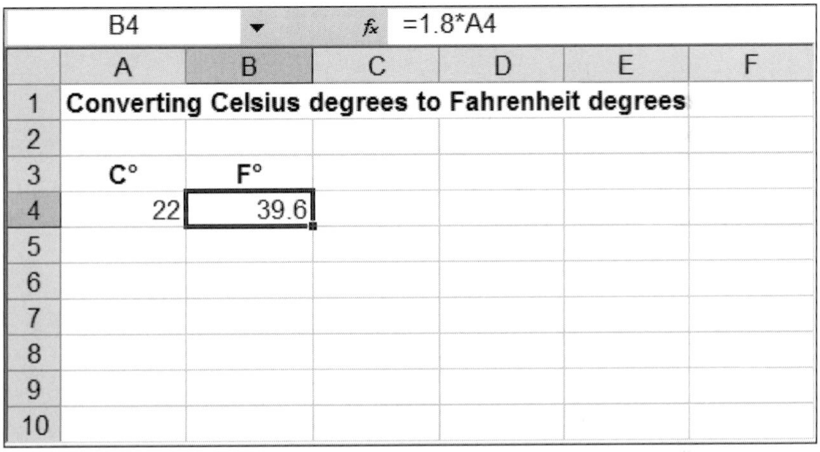

	B4	▼		ƒx	=1.8*A4	
	A	B	C	D	E	F
1	Converting Celsius degrees to Fahrenheit degrees					
2						
3	C°	F°				
4	22	39.6				
5						
6						
7						
8						
9						
10						

Figure 7.7 – Converting a temperature difference from Celsius to Fahrenheit

7.4 COMPLEX CONVERSIONS

We refer to conversions involving multiple units as *complex conversions*. Thus, a conversion from pounds per square inch (psi) to Newtons per square meter (Pascals) can be considered a complex conversion, since it involves units of force and units of length squared. (This particular conversion can also be treated as a simple conversion if a unit equivalence factor is available for pressure units, converting psi directly to Pascals.)

Complex conversions can be carried out with either a single complex unit equivalence factor (psi to Pascals, for example), or a series of simple unit equivalence factors. Both approaches are illustrated in the following example.

Example 7.6 A Complex Conversion (psi to Pascals)

Convert a pressure of 6.3 pounds per square inch (psi) to an equivalent number of newtons per square meter (Pa).

One way to solve this problem is to use the single unit equivalence factor 1 lb_f/in^2 = 6894.8 N/m^2. Thus, we can write

$$P = 6.3 \,(lb_f /in^2) \times \frac{6894.8 \,(N/m^2)}{1\,(lb_f /in^2)} = 43{,}437 \; N/m^2$$

or, simply

$$P = 6.3 \; psi \times \frac{6894.8 \; Pa}{1 \; psi} = 43{,}437 \; Pa$$

Hence, 6.3 lb_f/in^2 (psi) is equivalent to 43,437 N/m^2 (Pascals).

The problem can also be solved using a series of simple unit equivalence factors. Thus,

$$P = \frac{6.3 \; lb_f}{in^2} \times \frac{4.44822 \; N}{1 \; lb_f} \times \left(\frac{1 \; in}{0.0254 \; m}\right)^2 = 43{,}437 N/m^2$$

or

$$P = \frac{6.3 \; lb_f}{in^2} \times \frac{4.44822 \; N}{1 \; lb_f} \times \left(\frac{39.37 \; in}{1 \; m}\right)^2 = 43{,}437 N/m^2$$

This latter method parallels the approach used in Excel.

In both cases, the unit equivalences have been written in such a manner that the original units cancel out, leaving only the new units.

7.5 COMPLEX CONVERSIONS IN EXCEL

Recall that Excel can only carry out simple unit conversions using the CONVERT function. Hence, a complex conversion must be carried out as a series of simple conversions, using separate unit conversion factors.

A unit conversion factor can be obtained in Excel by writing 1 as the first argument in the CONVERT function; e.g., CONVERT(1,"ft","m"). Particular care must be used in writing the second and third arguments so that the correct units appear in the numerator and denominator of each unit equivalence. For example, when writing the conversion between feet and meters, the function

CONVERT(1,"ft","m")

is equivalent to the unit conversion

$$\frac{0.3048 \text{ m}}{1 \text{ ft}}$$

whereas the function

CONVERT(1,"m","ft")

is equivalent to

$$\frac{3.28084 \text{ ft}}{1 \text{ m}}$$

Thus, we see that *the second argument in the* CONVERT *function appears in the denominator of the corresponding unit equivalence factor, whereas the third argument appears in the numerator.* Moreover, we see that *reversing the order of the last two arguments in the* CONVERT *function is equivalent to computing a reciprocal.* For example,

$$\text{CONVERT(1,"ft","m")} = 1 / \text{CONVERT(1,"m","ft")} \tag{7.1}$$

These properties of the CONVERT function must be utilized when carrying out complex conversions in an Excel worksheet.

Example 7.7 A Complex Excel Conversion (psi to Pascals)

Create an Excel worksheet that will convert pounds per square inch (psi) to newtons per square meter (Pa). Use the worksheet to convert 6.3 psi into an equivalent number of pascals.

The desired worksheet is shown in Fig. 7.8. Notice that the original quantity (6.3 psi) is entered in cell A4. Cell B4 contains the formula

=CONVERT(A4,"lbf","N")*CONVERT(1,"m","in")^2

Within this formula, the first CONVERT function converts 6.4 pounds force to newtons. The second CONVERT function generates the unit equivalence factor between meters and inches. Note that the second argument ("m") corresponds to the denominator of the unit equivalence factor, and the third argument ("in") corresponds to the numerator.

Also, note that the second CONVERT function is squared. Hence the original units (square inches) will cancel out, leaving the desired new units (square meters) in the denominator.

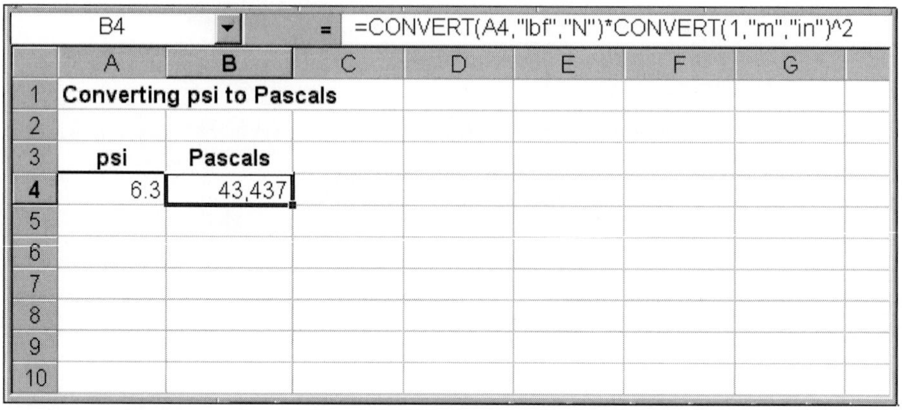

B4		▼	=	=CONVERT(A4,"lbf","N")*CONVERT(1,"m","in")^2		
	A	B	C	D	E	F	G
1	Converting psi to Pascals						
2							
3	psi	Pascals					
4	6.3	43,437					
5							
6							
7							
8							
9							
10							

Figure 7.8 – Carrying out a complex unit conversion in Excel

Another way to carry out this conversion is to write the formula in cell B4 as

$$=CONVERT(A4,"lbf","N") / CONVERT(1,"in","m")^2$$

as shown in Fig. 7.9. Now the second CONVERT function appears in the denominator, but the order of the last two arguments is interchanged; i.e., the second argument is "in" and the third is "m". When written in this form, the formula makes use of the reciprocal property expressed by Equation (7.1) above.

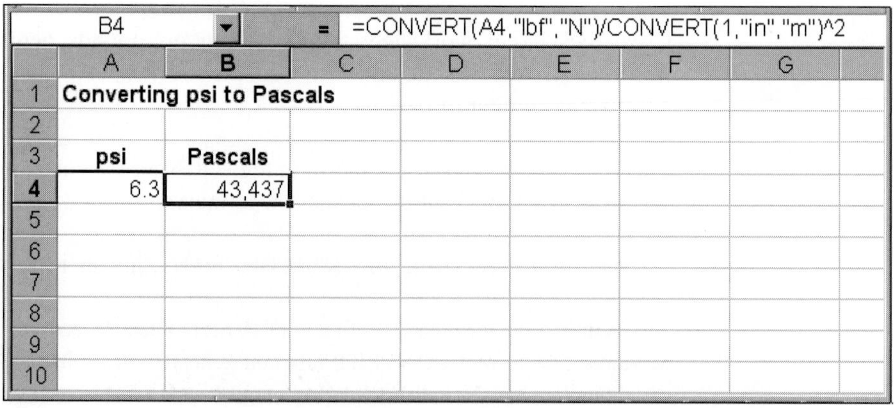

B4		▼	=	=CONVERT(A4,"lbf","N")/CONVERT(1,"in","m")^2		
	A	B	C	D	E	F	G
1	Converting psi to Pascals						
2							
3	psi	Pascals					
4	6.3	43,437					
5							
6							
7							
8							
9							
10							

Figure 7.9 – Another approach to complex unit conversions in Excel

Problems

7.1   Use Excel to carry out each of the following simple unit conversions:

(*a*)   Convert 118.7 feet to meters.

(*b*)   Convert 118.7 feet to centimeters.

(*c*)   Convert 32 meters to inches.

(*d*)   Convert 15.8 millimeters to angstroms.

(*e*)   Convert 8 (statute) miles to kilometers.

(*f*)   Convert 400 grams to slugs.

(*g*)   Convert 1.71 ounces to pounds mass.

(*h*)   Convert 4.9 slugs to kilograms.

(*i*)   Convert 200 days to minutes.

(*j*)   Convert 70 years to seconds.

(*k*)   Convert 0.02 seconds to hours.

(*l*)   Convert 64.8 Newtons to pounds force.

(*m*)   Convert 38.3 pounds force to newtons.

(*n*)   Convert 12.8 atmospheres to newtons per square meter (pascals).

(*o*)   Convert 0.7 atmospheres to millimeters of mercury.

(*p*)   Convert 91.2 pascals to atmospheres.

(*q*)   Convert 12 fluid ounces to liters.

(*r*)   Convert 8.5 liters to gallons.

(*s*)   Convert 1000 BTU to thermodynamic calories.

(*t*)   Convert 44.7 foot-poundals to BTU.

(*u*)   Convert 10.2 kilocalories (IT) to joules.

(*v*)   Convert 800 watt-hours to foot-poundals.

(*w*)   Convert 200 horsepower to watts.

(*x*)   Convert 1000 megawatts to horsepower.

(*y*)   Convert 87°F to degrees Kelvin.

(*z*)   Convert 54°C to degrees Fahrenheit.

7.2   Use Excel to evaluate each of the following unit equivalences:

(*a*)   Pounds force and newtons.

(*b*)   Slugs and kilograms.

(c) Feet and kilometers.

(d) Hours and years.

(e) Atmospheres and millimeters of mercury.

(f) Electron volts and joules.

(g) Watts and horsepower.

(h) Kilowatt-hours and BTU.

7.3 Use Excel to carry out each of the following complex unit conversions:

(a) Convert an area of 5 km^2 to ft^2.

(b) Convert a velocity of 20.8 ft/sec to mi/hr.

(c) Convert an acceleration of 14 ft/sec^2 to m/hr^2.

(d) Convert a pressure of 8.8 N/cm^2 to lb$_f$/ft^2.

(e) Convert a heat capacity of 0.285 BTU/(lb$_m$)(F°) to J/(kg)(K°). (See the note below.)

(f) Convert a heat transfer coefficient of 500 BTU/(hr)(ft^2)(F°) to kcal/(min)(m^2)(K°). (See the note below.)

(g) Convert a viscosity of 78.8 g/(cm)(sec) to lb$_m$/(ft)(hr).

(h) Convert a volumetric flowrate of 0.44 cm^3/sec to m^3/hr.

(i) Convert a volumetric flowrate of 12,500 gal/hr to liters/min.

(j) Convert a density of 137 lb$_m$/ft^3 to kg/m^3.

(k) Convert a density of 0.281 g/cm^3 to slugs/ft^3.

Note: Heat capacities and heat transfer coefficients [Probs. 7.3(e) and (f)] involve *temperature differences* (i.e., the number of degrees), not actual temperatures. Remember that 1.8 Fahrenheit degrees (F°), or 1.8 Rankine degrees (R°), is equivalent to 1 Celsius degree (C°), or 1 Kelvin degree (K°), as discussed in Sec. 7.3.

CHAPTER **8**

ANALYZING DATA STATISTICALLY

Engineering analysis generally begins with the analysis of data. Engineers often gather data to measure *variability* or *consistency*. For example, a car manufacturer may want to determine the precise dimensions of several manufactured parts that are all of the same type, such as the diameters of several ball bearings coming off an assembly line. This information will provide an indication of the consistency with which the ball bearings are being manufactured. Similarly, a clothing manufacturer may want to study the variation in sizes among potential customers in order to determine how many items of each size should be manufactured.

Statistical data analysis can tell us a great deal about a given data set, and it provides a basis for comparing one data set with another. For example, we can determine the average value for a data set, the minimum and maximum values, the values that occur most frequently, the "half-way" point (half the data lie above, half below), and the degree of spread in the data. However, a great deal of tedious arithmetic must be carried out to obtain these results, particularly if the data set is large. But Excel can carry out the arithmetic for us, allowing us to concentrate on the interpretation of the results. We will see how this is done in this chapter.

8.1 DATA CHARACTERISTICS

There are several commonly used parameters that allow us to draw conclusions about the characteristics of a data set. They are the *mean, median, mode, min, max, variance,* and *standard deviation.* Let us discuss each of them individually.

Mean

The *mean* is the most commonly used characteristic of a data set. It is also referred to as the *average* or the *arithmetic average*. It is an indication of the *expected behavior* of a data set.

The mean is determined by the well-known formula

$$\bar{x} = \frac{(x_1 + x_2 + \cdots + x_n)}{n} = \frac{1}{n}\sum_{i=1}^{n} x_i \tag{8.1}$$

where $\bar{x}$ represents the mean, x_i represents an individual data value, and n represents the total number of data values. (Note that the subscript i ranges from 1 to n.) Statisticians refer to the mean as the *first moment about the origin*.

In Excel, the AVERAGE function is used to determine the mean. The argument contained in parentheses indicates a block of cells containing the values to be averaged. Thus, the expression =AVERAGE(B1:B12) will determine the mean of the values stored in cells B1 through B12. Within the indicated block of cells, zeros are included in the calculation but blank cells are ignored.

Median

The *median* is a value such that half the data values lie above and half lie below. If the number of data values is odd, the median coincides with one of the data values. For example, 3 is the median for the data set (2, 0, 8, 3, 5). If the number of data values is even, however, the median is usually taken as the average of the two centermost values. Thus, 4 is the median for the data set (2, 8, 3, 5).

In Excel, the MEDIAN function is used to determine the median. It is used in the same manner as the AVERAGE function described above. Thus, the expression =MEDIAN(B1:B12) will determine the median of the values stored in cells B1 through B12. The numerical values within the cells need *not* be sorted.

Mode

The *mode* is the value that occurs with the greatest frequency within a data set. Not all data sets have a mode. On the other hand, some data sets have multiple modes. The data set (1, 2, 3, 4, 5), for example, does not have a mode because no value occurs more frequently than any other. However, in the data set (1, 2, 2, 4, 5), the mode is 2. The data set (1, 2, 2, 3, 4, 4, 5) has two modes, 2 and 4.

In Excel, the mode can be determined with the MODE function. Again, the arguments indicate the block of cells containing the data. Thus, the expression =MODE(B1:B12) will determine the mode of the values stored in cells B1 through B12. The MODE function returns an error message (#N/A) if the data set does not have a mode.

If multiple modes are present, the MODE function will return the value that occurs with the greatest frequency, or the first value encountered if all of the modes occur with the same frequency. Thus, the MODE function will return 4 for the data set (1, 2, 2, 3, 4, 4, 4, 5), but it will return 2 for the data set (1, 2, 2, 3, 4, 4, 5).

Min and Max

The *min* and the *max* (i.e., the minimum and the maximum) simply indicate the extremities of the data set. In Excel, the MIN and MAX functions return these values. The arguments again indicate the block of cells containing the data. Thus, the expression =MIN(B1:B12) returns the smallest value within the cells B1 through B12, whereas =MAX(B1:B12) returns the largest value. Blank cells are ignored.

Note that the MIN and MAX functions return the values that are the smallest and the largest *algebraically*. They do *not* return the values that are the smallest and the largest in magnitude. Thus, for the data set (−5, −2, 1), the MIN function would return −5 (which is algebraically the smallest value), and the MAX function would return 1 (which is algebraically the largest).

Variance

The *variance* provides an indication of the degree of *spread* in the data. The greater the variance, the greater the spread.

The variance (more precisely, the *sample* variance) is defined as

$$s^2 = \frac{1}{n-1}\sum_{i=1}^{n}(x_i - \overline{x})^2 \qquad (8.2)$$

where s^2 represents the variance, x_i represents an individual data value, $\overline{x}$ represents the mean, and n represents the number of data values. Note that the equation involves summing the square of the difference between each data value and the calculated mean. Statisticians refer to the variance as the *second moment about the mean*.

To see why the variance provides a measure of the spread in the data, let us consider the summation term more carefully. Each of the terms being added will always be greater than or equal to zero, since it is the square of another number. If there is little spread in the data, each of the individual data values will be close to the calculated mean. Hence, each of the terms being added will be relatively small, resulting in a relatively small variance.

On the other hand, if there is considerable spread in the data, some of the data values will be relatively far from the calculated mean. Therefore, some of the terms in the summation will be relatively large positive numbers, resulting in a large variance. Thus, the greater the spread in the data, the larger the variance.

Excel includes the VAR function, which is used to determine the variance. As before, the argument indicates the block of cells containing the data. Thus, the expression =VAR(B1:B12) returns the variance of the numerical values in cells B1 through B12.

Standard Deviation

The *standard deviation* also provides a measure of spread in the data set. It is simply the square root of the variance. Thus, the *sample* standard deviation is

$$s = \sqrt{s^2} = \sqrt{\frac{1}{n-1}\sum_{i=1}^{n}(x_i - \bar{x})^2} \tag{8.3}$$

where s represents the sample standard deviation and all other symbols are as defined for the variance.

Clearly, the standard deviation will be large if the variance is large, and the standard deviation will be small if the variance is small. Why, then, are *both* of these parameters used to indicate the degree of spread? The answer is that the variance is the more fundamental quantity, though its dimensions are expressed in units *squared* rather than the same units used for the mean, median, mode, min, and max. For example, if the individual data values have the units of inches, the resulting mean, median, mode, min, and max will also be expressed in inches, but the variance will be expressed in square inches. The standard deviation, on the other hand, will be expressed in inches, which is consistent with the other parameters. By selecting the standard deviation to indicate spread, we therefore obtain consistent units among all of the parameters. To determine a value for the standard deviation, however, you must first calculate the variance. Hence, you should have some understanding of both parameters and their interrelationship.

In Excel, the function STDEV returns the standard deviation of its arguments. Again, the arguments indicate the block of cells containing the data. Hence the expression =STDEV(B1:B12) will return the standard deviation of the data in cells B1 through B12. We could, of course, first determine the variance and then calculate its square root, but the direct calculation is easier.

Recent versions of Excel (e.g., Excel 2002) extend the AutoSum feature (see Example 2.1) to permit automatic use of the AVERAGE, COUNT, MIN, MAX and STDEV functions, as well as the SUM function. To use any of these functions automatically, simply highlight a cell adjacent to a row or column of numbers and click on the AutoSum button Σ ▼ in the Standard Toolbar. Then click on the down arrow to select the desired function from the resulting drop-down list.

For those readers with some statistical background, Excel allows you to generate a summary of all applicable descriptive statistics for a given data set (i.e., one or more columns or rows of numerical data). This feature requires the Descriptive Statistics feature found in the Analysis ToolPak. To install the Analysis ToolPak, choose Add-Ins from the Tools menu. Then select Analysis ToolPak from the resulting Add-Ins dialog box (see Figs. 7.1 and 7.2). Once the Analysis ToolPak has been installed, it will remain installed unless it is removed, by reversing the above procedure.

To generate a table of descriptive statistics, proceed as follows:

1. Enter the basic data into one or more adjacent columns or rows within the worksheet.

2. Choose Data Analysis/Descriptive Statistics from the Tools menu. Then provide the required information in the resulting dialog box. In particular, you must specify the cell blocks that contain the data (called the Input Range) and the location of the output data (either the entire range or the address of the upper left corner of the block that will contain the table). You can specify a range either by typing it directly into the dialog box or by dragging the mouse over the block of cells containing the data (hold down the mouse button as you move the mouse).

3. Be sure that the box labeled Summary statistics is checked.

If the location of the Descriptive Statistics dialog box interferes with your dragging the mouse over the desired block of cells, you can move the dialog box by placing the mouse pointer in the dialog box title bar (i.e., Descriptive Statistics) and then dragging the dialog box to a more convenient location.

Example 8.1 Analyzing a Data Set

A car manufacturer wishes to determine how accurately the cylinders are being machined in several engine blocks. The design specifications call for a cylinder diameter of 3.500 inches, with a tolerance of ±0.005 inch.

To determine the accuracy of the cylinders, several engine blocks were taken from the assembly line during manufacture and one cylinder was measured in each block. For consistency, the measurement was always perpendicular to the axis of the engine block (i.e., perpendicular to the straight line connecting the centers of the cylinders).

Analyze the data by placing them in an Excel worksheet and then calculating the mean, median, mode, min, max, and standard deviation. Also, generate a table showing all summary statistics for the given data.

The data obtained for 20 engine blocks, selected at random, are shown in the following table. Note that four of the individual values exceed the allowable tolerance of ±0.005 inch (three above, one below).

Sample	Diameter (in)	Sample	Diameter (in)
1	3.502	11	3.497
2	3.497	12	3.504
3	3.495	13	3.498
4	3.500	14	3.499
5	3.496	15	3.501
6	3.504	16	3.500
7	3.509	17	3.503
8	3.497	18	3.494
9	3.502	19	3.499
10	3.507	20	3.508

Figure 8.1 shows the Excel worksheet containing the data and the calculated parameters. The calculated values are located to the right of the actual data. Note that cell E5, which contains the numerical value for the mean, is highlighted. The expression used to obtain this value, =AVERAGE(B4:B23), is shown in the formula bar at the top of the figure. Similar formulas were used to obtain the other five values.

The numerical values, other than the standard deviation, were *formatted* to display three decimal places by selecting Cells from the Format menu, selecting the Number tab, then the Number category, and, finally, the value 3 in the data area labeled Decimal places within the resulting dialog box.

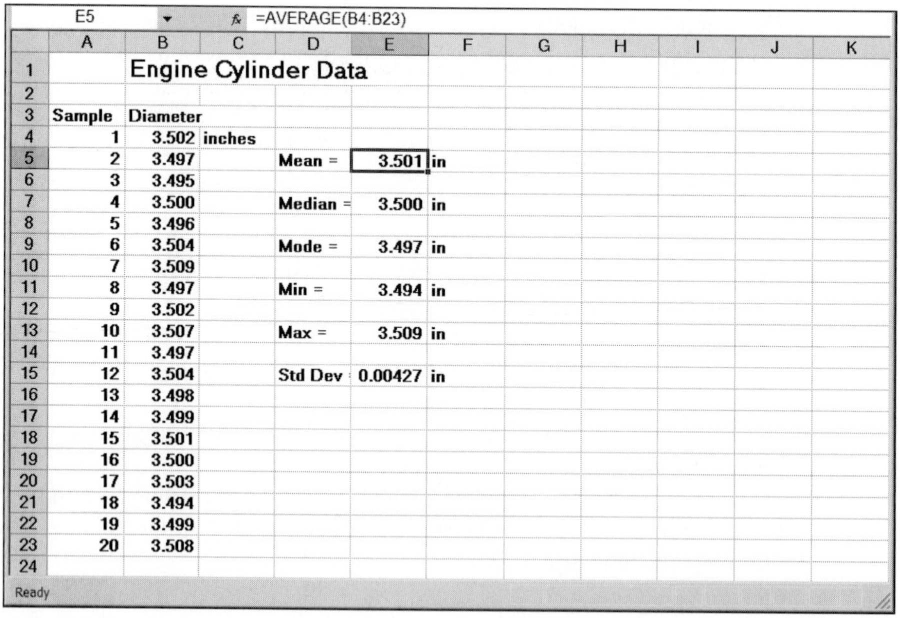

Figure 8.1 – A typical data set and a simple analysis of the data

From the worksheet, we obtain the following values (all units are in inches):

mean:	3.501	median:	3.500	mode:	3.497
min:	3.494	max:	3.509	std dev:	0.00427

Notice that the mean (3.501) is slightly higher than the median (3.500), whereas the mode (3.497) is somewhat lower. Also note that the standard deviation (0.00427) is only 0.122 percent of the mean.

In order to generate a table of summary statistics, we select Data Analysis/Descriptive Statistics from the Tools menu. This results in the dialog box shown in Fig. 8.2. Note that cells B3:B23 are designated as the input range. The box labeled Labels in first row is checked, to indicate that the content of cell B3 is a label rather than a numerical value.

Note also that Output Range is selected as the output option. This will cause the summary table to be displayed within the original worksheet, rather than a separate worksheet page or a separate workbook. Cell H3 is indicated as the output range. This cell address represents the upper left corner of the resulting summary table. (We could also have entered the complete range if we wished, though in most situations we would not know the exact size of the summary table in advance.)

The resulting worksheet, including the desired summary table, is shown in Fig. 8.3. Note that the numerical values shown in column I can be formatted if desired.

Within the list of parameters included in the summary statistics, the *standard error* is a measure of the amount of error in a predicted *y*-value corresponding to a given *x*-value, based upon a curve fit of the data (see Chap. 9). The *skewness* is a measure of the asymmetry of the data with respect to the mean, and the *kurtosis* is a measure of "flatness" or "pointedness" of the data distribution. The meaning of the remaining parameters should be straightforward.

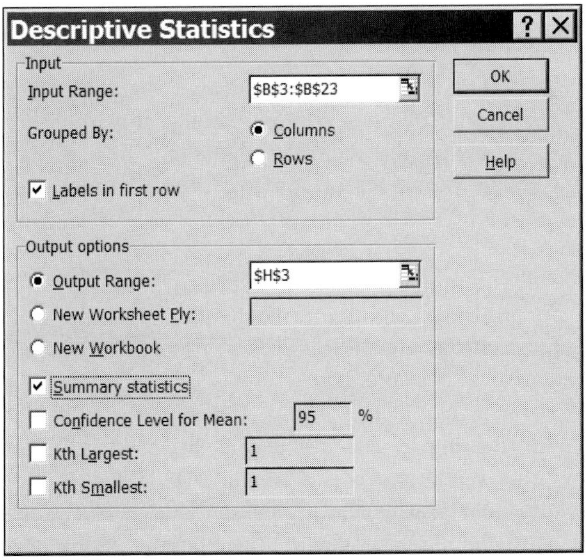

Figure 8.2 – The Descriptive Statistics dialog box

E5			▼	f_x	=AVERAGE(B4:B23)					
	A	B	C	D	E	F	G	H	I	J
1		**Engine Cylinder Data**								
2										
3	**Sample Diameter**							*Diameter*		
4	1	3.502	inches							
5	2	3.497		**Mean =**	3.501	in		Mean	3.5006	
6	3	3.495						Standard Error	0.000955318	
7	4	3.500		**Median**	3.500	in		Median	3.5	
8	5	3.496						Mode	3.497	
9	6	3.504		**Mode =**	3.497	in		Standard Deviation	0.00427231	
10	7	3.509						Sample Variance	1.82526E-05	
11	8	3.497		**Min =**	3.494	in		Kurtosis	-0.535038427	
12	9	3.502						Skewness	0.473530272	
13	10	3.507		**Max =**	3.509	in		Range	0.015	
14	11	3.497						Minimum	3.494	
15	12	3.504		**Std Dev**	0.004	in		Maximum	3.509	
16	13	3.498						Sum	70.012	
17	14	3.499						Count	20	
18	15	3.501								
19	16	3.500								
20	17	3.503								
21	18	3.494								
22	19	3.499								
23	20	3.508								
24										

Ready

Figure 8.3 – The collection of descriptive statistics for the given data set

Problems

8.1   Using your own version of Excel, reconstruct the worksheet shown in Fig. 8.1 showing the mean, median, etc. for the data given in Example 8.1. Change the values of some of the diameters and observe what effect this has on the mean, the median, etc.

8.2   Add a summary table of descriptive statistics to the worksheet developed for Prob. 8.1. Examine the results carefully, and determine which of the table entries are meaningful to you (this will depend on the extent of your statistical background).

8.3   The heights of 20 engineering students are given in the following table. Using Equations (8.1), (8.2), and (8.3), determine the mean, median, mode, min, max, variance, and standard deviation. Carry out the calculations by hand, using only a calculator, a pencil, and a piece of paper. (Do *not* use a spreadsheet to obtain a solution. Also, do *not* use any of the built-in statistical functions that may be present in your calculator.)

Student	Height (in)	Student	Height (in)
1	70.6	11	70.2
2	71.1	12	72.0
3	73.3	13	70.0
4	72.6	14	69.8
5	70.0	15	69.0
6	71.6	16	69.4
7	66.5	17	68.3
8	71.1	18	73.8
9	67.0	19	66.9
10	68.8	20	71.6

8.4   Enter the data given in Problem 8.3 into an Excel worksheet and determine the mean, median, mode, min, max, variance, and standard deviation. Be sure the worksheet is legible and clearly labeled.

8.2 HISTOGRAMS

Though the mean, median, mode, min, max, and standard deviation provide useful information about a data set, it is often desirable to plot the data in a manner that illustrates how the values are distributed within their range. This type of plot is called a *histogram*, or a *relative frequency* plot. From the relative frequency plot, we can obtain a plot of the *cumulative distribution*, which allows us to estimate the likelihood that a data value associated with an item drawn at random is less than or greater than some specified value. The details associated with each plot are discussed below.

Histogram Fundamentals

To create a histogram, you must first subdivide the range of the data into a series of adjacent, equally spaced intervals. The first interval must begin at or below the smallest data value (the min), and the last interval must extend to or beyond the largest data value (the max). Each interval will have some lower bound, x_i, and an upper bound, x_{i+1}, where $x_{i+1} = x_i + \Delta x$ and Δx represents the fixed interval width. These intervals are sometimes referred to as *class intervals*.

Once the intervals have been defined, you must determine how many data values fall within each interval. The *relative frequencies* are then obtained as

$$f_i = \frac{n_i}{n} \tag{8.4}$$

where f_i represents the relative frequency for the ith interval, n_i represents the number of data values in the ith interval, and n represents the total number of data values. Note: if k is the number of intervals, then

$$\sum_{i=1}^{k} n_i = n \tag{8.5}$$

and

$$\sum_{i=1}^{k} f_i = 1 \tag{8.6}$$

The histogram is usually expressed as a bar graph showing the values of the interval counts (n_i) or the relative frequencies (f_i). From this bar graph, it is easy to see how the data are distributed.

Example 8.2 Constructing a Histogram

Construct a histogram of the data given in Example 8.1. Choose 10 equally spaced intervals ranging from $x = 3.490$ to $x = 3.510$ inches. If a data value falls on an interval boundary, assign the data value to the lower interval. (*Note*: This boundary rule is arbitrary. Many authors suggest that a data value falling on an interval boundary be assigned to the *upper* interval. However, our choice of the lower interval is consistent with the rule employed in Excel.)

The first interval will extend from 3.490 to 3.492 inches. The second interval will extend from 3.492 to 3.494 inches, though it will actually include only those values that *exceed* 3.492 inches because a value of exactly 3.492 inches will be assigned to the first interval, and so on. For simplicity, however, we will write the interval bounds to four significant figures. This practice is customarily followed when constructing histograms.

The class intervals and their corresponding values are shown in the following table:

Interval Number	Interval Range	Number of Values	Relative Frequency
1	3.490 – 3.492	0	0
2	3.492 – 3.494	1	0.05
3	3.494 – 3.496	2	0.10
4	3.496 – 3.498	4	0.20
5	3.498 – 3.500	4	0.20
6	3.500 – 3.502	3	0.15
7	3.502 – 3.504	3	0.15
8	3.504 – 3.506	0	0
9	3.506 – 3.508	2	0.10
10	3.508 – 3.510	1	0.05

Thus, we see that 5 percent of the data values fall within the interval 3.492 – 3.494, 10 percent fall within the interval 3.494 – 3.496, and so on. (More precisely, 5 percent of the data exceed 3.492 but are less than or equal to 3.494, 10 percent exceed 3.494 but are less than or equal to 3.496, etc.)

Histogram Generation in Excel

When working with real data, the number of values within a data set is often quite large. Thus, it may be very tedious to count the number of values within each interval, as we have done in the previous example. Fortunately, however, most spreadsheets will do this counting for us once we have entered the data.

In Excel, histograms can be constructed very easily using the Histogram feature found in the Analysis ToolPak (see Sec. 8.1 for instructions on installing the Analysis ToolPak if it is not already available).

To construct a histogram in Excel, proceed as follows:

1. Enter the basic data and the interval bounds into the worksheet. (The intervals are called *bins* in Excel.) Typically, the data are entered in a single column (or a single row), and the interval bounds are entered in another column (or another row). Only the *right* interval bounds are entered. Hence, the last interval (rightmost interval) will be open-ended; that is, it will contain all data values that exceed the largest bound.

2. Choose Data Analysis/Histogram from the Tools menu. Then provide the required information in the resulting dialog box. In particular, you must specify the cell blocks that contain the data (called the Input Range) and the interval bounds (the Bin Range). You can specify each range either by typing it directly into the dialog box or by dragging the mouse over the block of cells containing the data (hold down the mouse button as you move the mouse). If the location of the Histogram dialog box interferes with your dragging the mouse over the desired block of cells, you can move the dialog box by placing the mouse pointer in the dialog box title bar (i.e., Histogram) and then dragging the dialog box to a more convenient location.

3. If the histogram is to be embedded within the current worksheet, you must also select the Output Range option and enter the cell address of the upper corner of the block that will contain the output data.

4. When building a conventional histogram, the Pareto box at the bottom of the dialog box should *not* be selected. You *may* select either of the remaining two boxes (Cumulative Percentage and Chart Output), though they should be left blank for now. We will say more about these two boxes later in this chapter.

Example 8.3 Generating a Histogram in Excel

Modify the worksheet developed in Example 8.1 (see Fig. 8.1) by adding a histogram of the data. Use the same class intervals as in Example 8.2.

We begin by creating some additional space in the worksheet, so that there is ample room for the histogram. To do so, we will shorten the label in column C of Fig. 8.1 and then decrease the width of columns C and F. We then add the upper interval bounds in column G, as shown in Fig. 8.4. The next step is to select Data Analysis from the Tools menu and then select Histogram from the Data Analysis dialog box.

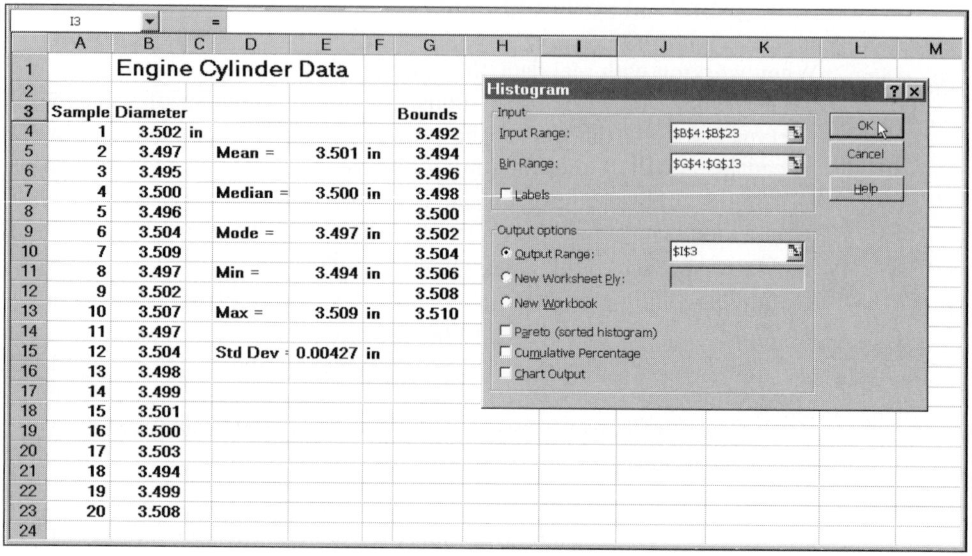

Figure 8.4 – Defining a histogram for the given data set using the Histogram dialog box

Once the Histogram selection is made, the Histogram dialog box will appear, as shown in Fig. 8.4. (In Fig. 8.4, the Histogram dialog box has been moved to one side of the worksheet so that the requested cell ranges can be entered by dragging the mouse pointer over the appropriate cell blocks.) We then proceed to fill in the requested information. The cells containing the data are specified by first clicking on the Input Range data area and then dragging the mouse pointer across cells B4 through B23. Similarly, the interval boundaries are specified by clicking on the Bin Range data area and then dragging the mouse pointer across cells G4 through G13. Finally, the upper left corner of the output area is specified by selecting Output Range as an output option and then selecting cell I3. Figure 8.4 shows that these selections have been made. Note that none of the three boxes at the bottom of the dialog box (Pareto, Cumulative Percentage, and Chart Output) have been selected.

When the required ranges have been specified, we click on the OK button, resulting in the table of frequencies shown in Fig. 8.5. Note that the intervals in column I are labeled *Bin*, and the interval values in column J are labeled *Frequency*. The values tabulated in the *Frequency* column are integer sums, not the relative frequencies (i.e., fractional values) described earlier. We can, of course, obtain a set of relative frequencies in the next column (column K) by dividing each of the cell values in column J by the total number of data points, as in Equation (8.4) above.

	K17	▼		=								
	A	B	C	D	E	F	G	H	I	J	K	L	M
1		Engine Cylinder Data											
2													
3	Sample	Diameter					Bounds		Bin	Frequency			
4	1	3.502 in					3.492		3.492	0			
5	2	3.497		Mean =	3.501 in		3.494		3.494	1			
6	3	3.495					3.496		3.496	2			
7	4	3.500		Median =	3.500 in		3.498		3.498	4			
8	5	3.496					3.500		3.500	4			
9	6	3.504		Mode =	3.497 in		3.502		3.502	3			
10	7	3.509					3.504		3.504	3			
11	8	3.497		Min =	3.494 in		3.506		3.506	0			
12	9	3.502					3.508		3.508	2			
13	10	3.507		Max =	3.509 in		3.510		3.510	1			
14	11	3.497							More	0			
15	12	3.504		Std Dev =	0.00427 in								
16	13	3.498											
17	14	3.499											
18	15	3.501											
19	16	3.500											
20	17	3.503											
21	18	3.494											
22	19	3.499											
23	20	3.508											
24													

Figure 8.5 – A histogram showing the frequencies for the given data

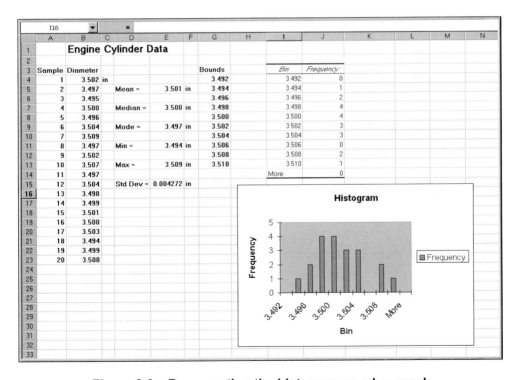

Figure 8.6 – Representing the histogram as a bar graph

A histogram is customarily plotted as a bar graph (i.e., an Excel *Column Chart*). This can be carried out manually, using the techniques described in Chapter 4, or it can be carried out automatically when the histogram is created. To create a bar graph automatically, simply select Chart Output at the bottom of the Histogram dialog box (see Fig. 8.4). When this option is selected, the bar graph will appear as an embedded bar graph, as shown in Fig. 8.6.

Figure 8.6 differs from a traditional histogram in two ways. First, the numerical labels shown beneath every other interval (the *bin* values) actually correspond to the interval *right boundaries* (which are shown as tick marks) rather than the intervals themselves. Thus, the first interval extends from a value of 3.490 to 3.492, and so on. When you are interpreting the histogram, each of these values should therefore be associated with the tick mark to its right.

The graph can be clarified somewhat by rotating the labels so that they are vertical, centered beneath each interval. To do so, click on the *x*-axis and choose Selected Axis/Alignment from the Format menu. Then either drag the alignment indicator to the 12 o' clock position or enter the value 90 in the area labeled Degrees. It will then be easier to associate each numerical label with the right boundary beneath each interval, as shown in Fig. 8.7.

The other difference concerns the manner in which the vertical bars are drawn within the bar graph. Since the vertical bars represent contiguous intervals, they should be drawn adjacent to each other, without any intervening space. This is easy to accomplish, as follows: Click on any of the vertical bars and choose Selected Data Series/Options from the Format menu. Then choose a value of 0 for the Gap width. The resulting histogram will then appear as shown in Fig. 8.7. (Note that the legend has been removed from this figure, and the title and axis labels have been altered for clarity.)

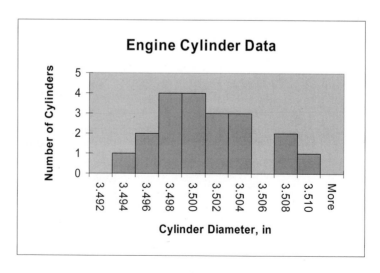

Figure 8.7 – A refined histogram bar graph

When constructing a histogram, the number of intervals is another important consideration. Too few intervals results in a histogram that lacks detail and consequently provides little information about how the data are distributed, as shown in Fig. 8.8. On the other hand, too many intervals results in gaps within the histogram. This distorts the overall shape of the histogram, as shown in Fig. 8.9. As a rule, a choice of 10 to 15 intervals works well with many data sets. Fewer intervals might be better, however, if the number of data values (n) is relatively small. If n is large, the square root of n often provides a useful starting point.

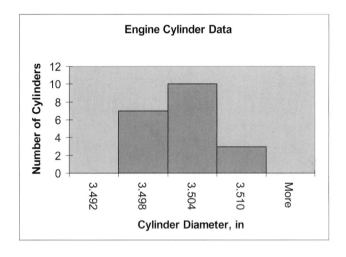

Figure 8.8 – A histogram with too few intervals

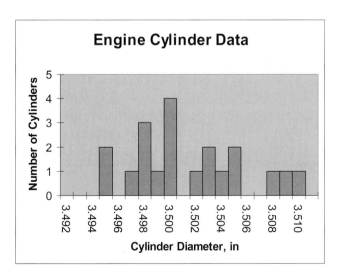

Figure 8.9 – A histogram with too many intervals

Problems

8.5   Using your own copy of Excel, generate the worksheets and figures shown in Figs. 8.5, 8.6, and 8.7 (see Example 8.3). Edit the graphs in Figs. 8.6 and 8.7 so that they fill the screen and are clearly labeled.

8.6   A class of 30 students obtained the following exam scores in their Introduction to Engineering class:

Student No.	Exam Score	Student No.	Exam Score
1	87	16	71
2	64	17	41
3	74	18	77
4	56	19	74
5	95	20	56
6	74	21	79
7	76	22	90
8	67	23	47
9	82	24	44
10	67	25	79
11	91	26	96
12	64	27	69
13	71	28	66
14	41	29	50
15	78	30	77

Enter the data into an Excel worksheet and do the following:

(*a*)   Determine the mean, median, min, and max. Explain the difference between the mean and the median.

(*b*)   Construct a 10-interval histogram spanning the range from 0 to 100.

(*c*)   Based upon this histogram, how many students have exam scores ranging from 71 to 80? How many have exam scores above 90? How many have exam scores of 50 or less?

8.7   Use the data given in Prob. 8.6 to construct the following histograms:

(*a*)   A five-interval histogram that spans the range of the data. (Note that this histogram need not necessarily begin at 0 and extend to 100. Choose any convenient values for the lower and upper bounds.)

(*b*)   A 15-interval histogram that spans the same range.

(*c*)   A 30-interval histogram that spans the same range.

(*d*)   Based upon these results, what would you suggest is the best number of intervals for this data set, and why?

8.3 CUMULATIVE DISTRIBUTIONS

We have seen that a histogram provides a graphical illustration of how a data set is distributed. Of equal importance is the *cumulative distribution*, which provides another way to view the manner in which the data are distributed. The cumulative distribution allows us to determine the likelihood that a particular value drawn at random is less than or greater than some specified value. For example, we might wish to determine the likelihood that a cylinder diameter within a randomly selected engine block is greater than some specified value, say 3.500 inches.

To construct a cumulative distribution, we must first construct a histogram and determine the relative frequency associated with each of the intervals, using Equation (8.4). We then determine the following cumulative values:

$$F_1 = f_1$$

$$F_2 = f_1 + f_2$$

$$F_3 = f_1 + f_2 + f_3$$

. . . . .

$$F_j = f_1 + f_2 + \cdots + f_j = \sum_{i=1}^{j} f_i$$

. . . . .

$$F_k = f_1 + f_2 + \cdots + f_k = \sum_{i=1}^{k} f_i \tag{8.7}$$

Note that the last term, F_k, will always equal 1 because of Equation (8.6). This provides a convenient means to check the accuracy of the individual f_i values. (Add up the f_i values and see if they sum to 1.)

Example 8.4 Constructing a Cumulative Distribution

Construct a set of numerical values for the cumulative distribution corresponding to the histogram developed in Example 8.2.

We have already determined the class intervals and their corresponding relative frequencies. In order to determine the cumulative distribution, we must determine the partial sums of the relative frequencies. Thus, we see that

$$F_1 = 0$$

$$F_2 = 0.05$$

$$F_3 = 0.05 + 0.10 = 0.15$$

$$F_4 = 0.05 + 0.10 + 0.20 = 0.35$$

$$F_5 = 0.05 + 0.10 + 0.20 + 0.20 = 0.55$$

and so on. The results are summarized below. (Note that $F_{10} = 1.00$, as required.)

Interval Number	Interval Range	Number of Values	Relative Frequency	Cumulative Distribution
1	3.490 – 3.492	0	0	0
2	3.492 – 3.494	1	0.05	0.05
3	3.494 – 3.496	2	0.10	0.15
4	3.496 – 3.498	4	0.20	0.35
5	3.498 – 3.500	4	0.20	0.55
6	3.500 – 3.502	3	0.15	0.70
7	3.502 – 3.504	3	0.15	0.85
8	3.504 – 3.506	0	0	0.85
9	3.506 – 3.508	2	0.10	0.95
10	3.508 – 3.510	1	0.05	1.00

Sometimes the cumulative distribution is expressed in terms of percentages rather than decimal quantities. The percentages are obtained by multiplying the decimal quantities by 100. Thus, in the previous example, we could have written $F_2 = 5$ percent, $F_3 = 15$ percent, and so on.

Cumulative Distributions in Excel

In Excel, the cumulative distribution can be obtained at the same time the histogram is generated. To do so, we simply select Cumulative Percentage in the Histogram dialog box. The distribution will be expessed in terms of percentages. Note that it is possible to generate the histogram, the cumulative percentages, and the associated graphs all at the same time. The graph of the cumulative percentages will be shown as a *line graph* (see Sec. 4.7) superimposed over the histogram bar graph, as illustrated in the next example.

Example 8.5 Generating a Cumulative Distribution in Excel

Modify the worksheet developed in Example 8.3 (see Figs. 8.5 and 8.6) to include the cumulative distribution, expressed as percentages, in addition to the histogram frequencies. Generate a combined graph of the frequencies and the cumulative percentages.

We begin as we did in Example 8.3, with the worksheet shown in Fig. 8.4. (Recall that the Histogram dialog box was obtained by selecting Data Analysis/Histogram from the Tools menu.) Now, however, we select both the Cumulative Percentage and the Chart

Output features at the bottom of the Histogram dialog box. Clicking on the OK button produces the histogram shown in Fig. 8.10. The resulting worksheet is identical to that shown in Fig. 8.5 except that column K now contains the cumulative percentages in addition to the information generated earlier. The graphical display now shows a bar graph of the histogram and a plot of the cumulative percentages superimposed as a line graph. (You may have to move around in the worksheet in order to see the graphical display.)

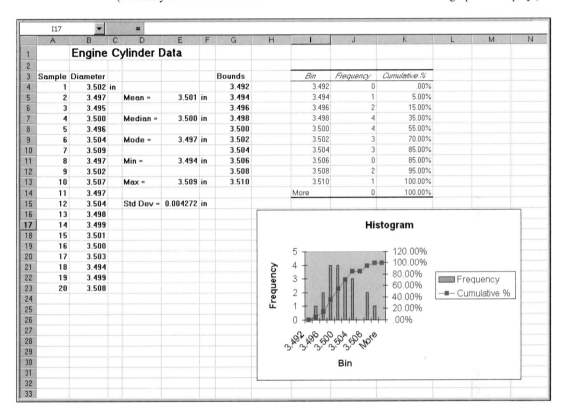

Figure 8.10 – Superimposing a graph of the cumulative distribution on the histogram

Note that the graph in Fig. 8.10 contains *two* y-axis scales, one to the left of the graph and the other to the right. The axis on the left, labeled Frequency, applies to the bar graph showing the frequencies, whereas the axis on the right, labeled with percent signs, applies to the superimposed line graph showing the cumulative percentages.

There are three problems associated with histograms that include a cumulative distribution, such as the graph shown in Fig. 8.10. First, as discussed earlier, the bar graph intervals are not contiguous. Second, the line graph showing the cumulative percentages is valid only for equally spaced intervals. And finally, the line graph connects points drawn at the *centers* of the intervals, whereas we would prefer to draw this line graph from the right interval boundaries.

We have already discussed a solution to the first problem; namely, click on a vertical bar, choose Selected Data Series/Options from the Format menu, and specify a value of 0 for the Gap width. The only way to correct the remaining problems, however, is to plot the cumulative distribution as a separate x-y graph, using the methods described in Secs. 4.2 and 4.3. This permits us much greater flexibility in the appearance of the final graph. In particular, it allows us to plot the cumulative distribution with the cumulative values associated with the right interval bounds rather than the centers of the intervals.

The general procedure is to assign a y-value of zero to the left boundary of the first nonempty interval and the first cumulative value as the y-value for the right boundary. Each successive cumulative value is then assigned as the y-value for the right boundary of its corresponding interval. The last cumulative value (either 1.0 or 100 percent) will then be assigned as the y-value corresponding to the right boundary of the last interval. The method is illustrated in the following example.

Example 8.6 Plotting the Cumulative Distribution

Construct a separate x-y graph of the cumulative distribution generated in Example 8.5. Express the cumulative values as fractional values ranging from 0 to 1 rather than as percentages ranging from 0 to 100 percent. Associate the cumulative values with the right interval boundaries, in accordance with traditional procedures for displaying a cumulative distribution.

We begin with the worksheet shown in Fig. 8.10. We will proceed by first deleting the original graph and then copying the interval bounds (the *Bin* values) into a new location. Let us choose column I, beneath the original interval bounds (cells I17 through I27). These values will form the labels for the x-axis of the new graph. We then place the corresponding cumulative values within adjoining cells in column J (cells J17 through J27). Note that the cumulative values are now shown as fractions rather than percentages (choose Cells/Number from the Format menu, then select the desired numerical format).

It is now quite easy to generate the desired x-y graph, using the methods described in Secs. 4.2 and 4.3. Figure 8.11 shows the resulting worksheet, containing the new x-y graph. Note that the title, labels, and background color have been changed, grid lines have been added, and the legend has been removed. In addition, the entire graph has been enlarged so that it is more legible.

We can obtain some very useful information from the cumulative distribution once it is plotted in the form shown in Fig. 8.11. In particular, if we choose any value along the x-axis, the corresponding y-value indicates the likelihood that a single sample selected at random from the given population will have a value less than or equal to the x-value. Information of this type is of great importance to engineers involved in manufacturing processes or other processes in which quality control is an important consideration.

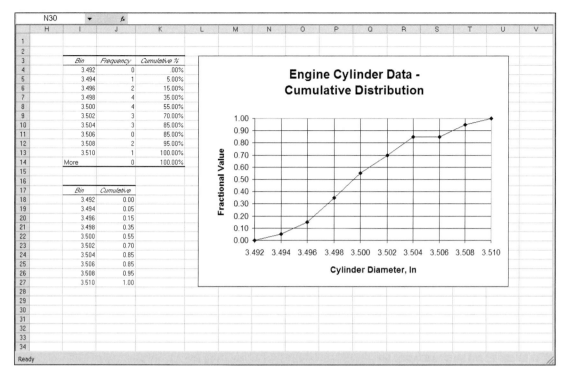

Figure 8.11 – A refined *x-y* graph of the cumulative distribution

Example 8.7 Drawing Inferences from the Cumulative Distribution

From the cumulative distribution plotted in Fig. 8.11, estimate the likelihood that an arbitrary cylinder diameter within a randomly selected engine block will not exceed 3.503 inches.

To solve this problem, locate the value 3.503 along the *x*-axis in Fig. 8.9. The corresponding *y*-value is approximately 0.77. (The value of the gridlines should now be readily apparent.) Hence, we conclude that there is a 77 percent likelihood that the diameter of an arbitrarily selected cylinder does not exceed 3.503 inches. Conversely, there is a 23 percent likelihood (1 − 0.77) that the cylinder diameter *does* exceed 3.503 inches. The manufacturing engineer must now exercise some judgment in deciding whether or not the 77 percent value is sufficiently high.

The information obtained from the graph would probably be more accurate if a smooth curve were drawn through the aggregate of the data points (assuming that a larger number of data points would result in a smoother and more accurate graph). This can, of course, always be done by hand. However, Excel can also pass a smooth curve (a *trendline*) through a set of data without going through the individual data points. We will see how this is accomplished in the next chapter.

The subject of data analysis can be studied much more extensively, though the material is beyond the scope of our present discussion. You might consider taking a course in probability and statistics if you wish to pursue this topic further. For now, however, let us recognize that Excel includes many other features that are statistical in nature. You might wish to explore some of them by examining the entries listed in the Data Analysis dialog box under the Tools menu, as well as the various statistical functions listed in the Insert Function dialog box under the Insert/Function... menu.

Problems

8.8   Using your own version of Excel, verify the solution to each of the following problems:

(*a*)   Re-create the cumulative distribution and the corresponding graph shown in Fig. 8.10 (see Example 8.5).

(*b*)   Modify the worksheet developed in Prob. 8.8(*a*) to obtain the worksheet and the graph shown in Fig. 8.11.

8.9   Using Equations (8.4) through (8.7), determine the relative frequencies and the cumulative distribution for the data given in Problem 8.2. Carry out the calculations by hand, using only a calculator, a pencil, and a piece of paper. (Do *not* use a spreadsheet to obtain a solution.) Do not plot the data.

8.10   Using the Analysis Toolpak/Histogram features included in Excel, carry out the following operations on the data given in Prob. 8.3.

(*a*)   Enter a series of interval (*bin*) bounds based upon one-inch interval spacing.

(*b*)   Determine the frequencies associated with the intervals.

(*c*)   Determine the cumulative percentages associated with the intervals.

(*d*)   Create a bar graph of the frequencies with adjacent intervals, as shown in Fig. 8.7.

8.11   Using the histogram data developed in Prob. 8.6, construct a graph of the cumulative distribution in each of the following ways:

(*a*)   Using the Analysis Toolpak/Histogram features included in Excel, develop a combined bar graph/line graph with separate *y*-axes similar to that shown in Fig. 8.10.

(*b*)   Develop a separate *x-y* graph of the cumulative distribution, as described in Example 8.6. Include a set of grid lines and appropriate labels, as shown in Fig. 8.11.

8.12 Plot the cumulative distribution data developed in Prob. 8.10(c). Then use the graph to answer the following questions:

(a) What is the likelihood that the height of an engineering student selected at random will not exceed 5 feet 10 inches?

(b) What is the likelihood that an engineering student selected at random will be at least 6 feet tall?

(c) Suppose the data used for Prob. 8.10(c) applies only to male students and that female students are assumed to be 10 percent shorter than male students. What is the likelihood that a female engineering student selected at random will be at least 5 feet 4 inches tall?

(d) Suppose you were selecting one male and one female engineering student to pose as a couple for a picture in the student newspaper. If the students are selected at random, what is the likehood that the height of the female student will not exceed 5 feet 5 inches and the height of the male student will be at least 5 feet 9 inches? (*Hint:* Obtain the product of the individual probabilities.)

8.13 The U.S. Environmental Protection Agency has tested the average fuel efficiency of 24 late-model cars equipped with V-6 engines and automatic transmissions. The results obtained are shown below.

Sample	*Mileage (mpg)*	*Sample*	*Mileage (mpg)*
1	22.9	13	25.5
2	23.9	14	22.2
3	21.4	15	21.7
4	25.4	16	23.5
5	23.9	17	27.1
6	24.4	18	23.0
7	23.1	19	23.9
8	22.0	20	23.6
9	25.4	21	19.2
10	20.7	22	22.7
11	21.4	23	26.0
12	22.8	24	21.3

Enter the data into an Excel spreadsheet and then analyze the data as follows:

(a) Determine the mean, median, mode, min, max, and standard deviation.

(b)   Construct a histogram, based upon a reasonable interval width.

(c)   Construct a cumulative distribution. Show the cumulative distribution in the form of an x-y graph.

8.14   Use the results of the previous problem to answer the following questions:

(a)   Explain the difference between the mean and the median.

(b)   From an examination of the histogram, what would you conclude about the data?

(c)   If a car of this type is chosen at random, what is the likelihood that the fuel efficiency of this car will not exceed 20 mpg? What is the likelihood that the fuel efficiency will not exceed 22 mpg? What is the likelihood that the fuel efficiency *will* exceed 25 mpg?

8.15   An engineer is responsible for monitoring the quality of a batch of 1000-ohm resistors. To do so, the engineer must accurately measure the resistance of a number of resistors within the batch, selected at random. The results obtained are shown below.

Sample No.	Resistance (ohms)	Sample No.	Resistance (ohms)
1	1006	16	960
2	1006	17	976
3	978	18	954
4	965	19	1004
5	988	20	975
6	973	21	1014
7	1011	22	955
8	1007	23	973
9	935	24	993
10	1045	25	1023
11	1001	26	992
12	974	27	981
13	987	28	991
14	966	29	1013
15	1013	30	998

Enter the data into an Excel spreadsheet and then analyze the data as follows:

(a) Determine the mean, median, mode, min, max, and standard deviation.

(b) Construct a histogram, based upon a reasonable interval width.

(c) Construct a cumulative distribution. Show the cumulative distribution in the form of an *x-y* graph.

(d) Based upon the cumulative distribution for this random sample, how likely is it that a resistor selected at random will deviate from the target value of 1000 ohms by more than 2 percent, either above or below?

CHAPTER 9

FITTING EQUATIONS
TO DATA

In Chapter 8, we considered the analysis of single-valued data (i.e., $x_1, x_2, x_3, \cdots$, and so on). We now turn our attention to the analysis of *paired* data values (i.e., $P_1 = (x_1, y_1)$, $P_2 = (x_2, y_2)$, $\cdots$, and so on). Engineers frequently collect paired data in order to understand the characteristics of an object or the behavior of a system. The data may indicate a *spatial profile* (for example, temperature versus distance) or a *time history* (voltage versus time). Or the data may indicate *cause-and-effect relationships* (for example, force as a function of displacement) or *system output as a function of a changing input parameter* (yield of a chemical reaction as a function of temperature).

Of particular interest, when working with paired data, are methods for estimating a value for y (the dependent variable) corresponding to some specified value of x (the independent variable) when x falls between two of the given data points. The simplest way to do this is to connect the two data points surrounding the point of interest with a straight line segment. This procedure is known as *linear interpolation*. This method may be satisfactory if the data points are relatively close together or if they do not exhibit much curvature. Another approach, which is more accurate but also more complicated, is to pass a curve (e.g., a polynomial) through several of the data points. Either method allows us to calculate a value for y corresponding to the given x with reasonable accuracy.

Measured data usually show some *scatter*, which is due to fluctuations or errors in the measurements. Therefore, when fitting a curve through measured data, we usually pass the curve through the *aggregate* of the data rather than the individual data points (though it is common practice to disregard "outliers," that is, occasional isolated data points that are far removed from the main cluster, possibly because of erroneous measurements). This procedure, based upon the *method of least squares*, allows us to capture the overall trend reflected by the entire data set.

In this chapter, we will see how we can determine the equation for a curve that passes through two or more data points, or the aggregate of the data when scatter is present. In Excel, we will do this by first plotting the data as an *x-y* graph and then passing a curve through the graph.

9.1 LINEAR INTERPOLATION

Linear interpolation allows us to connect two adjacent data points with a straight line segment. We can then use the equation for the line segment to estimate a value of *y* corresponding to a given value of *x,* provided *x* falls between the given data points.

Let us refer to the two given data points as $P_1 = (x_1, y_1)$ and $P_2 = (x_2, y_2)$. We wish to determine the value of *y* corresponding to some specified value of *x*, where *x* lies between the two given data points; i.e., $x_1 < x < x_2$. One way to approach this problem is to connect the given data points with a straight line. The desired value of *y* can then be determined from the straight line, as illustrated in Fig. 9.1.

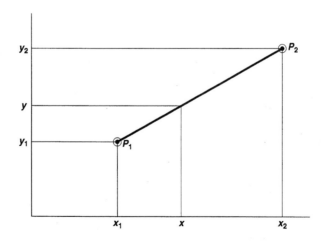

Figure 9.1 – Connecting two data points with a straight line segment

Mathematically, let us proceed as follows. From simple proportionality, we can write

$$\frac{y - y_1}{x - x_1} = \frac{y_2 - y_1}{x_2 - x_1} \tag{9.1}$$

Solving for y, we obtain

$$y = y_1 + \frac{y_2 - y_1}{x_2 - x_1}(x - x_1)$$ (9.2)

Equation (9.2) is the traditional form of the linear interpolation formula. It is sometimes written as

$$y = y_1 + \frac{\Delta y}{h}(x - x_1)$$ (9.3)

where Δy represents $y_2 - y_1$ and h represents the interval spacing, $x_2 - x_1$. This form of the interpolation equation may be easier to work with when carrying out a hand calculation, since there is less of a tendency to confuse the various y- and x-values.

This technique is useful when only two data points are known, or when there are multiple data points that clearly define a straight line with no significant scatter. If scatter is present, however, the use of a trendline (described in Secs. 9.3 and 9.4) is preferable.

Example 9.1 Linear Interpolation

Suppose we are given the two data points (2.0, 5.5) and (2.1, 7.2). Determine the value of y corresponding to $x = 2.03$ using linear interpolation.

Let us use the form of the linear interpolation formula given by Equation (9.3) to solve this problem. We begin by identifying the various terms in the equation:

$$x_2 = 2.1 \qquad\qquad y_2 = 7.2$$

$$x_1 = 2.0 \qquad\qquad y_1 = 5.5$$

$$h = (2.1 - 2.0) = 0.1 \qquad \Delta y = (7.2 - 5.5) = 1.7$$

$$x = 2.03$$

We can now calculate the desired value of y as

$$y = y_1 + \frac{\Delta y}{h}(x - x_1) = 5.5 + \frac{1.7}{0.1}(x - 2.0) = 17x - 28.5$$

When $x = 2.03$, we can write

$$y = 17(2.03) - 28.5 = 6.01$$

Hence, the value of y corresponding to $x = 2.03$ is 6.01. That is, the interpolated data point is (2.03, 6.01), based upon linear interpolation.

In Excel, the use of Equation (9.2) may be more straightforward than Equation (9.3), since it does not involve the calculation of any intermediate parameters. The recommended procedure is to enter the tabulated x-values in one column and the corresponding y-values in another, as in earlier chapters of this book. The given x-value, corresponding to the unknown y-value, can be entered into a separate cell. It is then a simple matter to determine the interpolated y-value using an Excel formula to represent Equation (9.2).

As a practical matter, it is easier to carry out linear interpolation using Excel's FORECAST function. This function is written as FORECAST(x, *range of known y-values*, *range of known x-values*), where x represents the value of the independent variable whose corresponding y-value is to be determined. The procedure is illustrated in the following example.

Example 9.2 Linear Interpolation in Excel

Solve the linear interpolation problem described in Example 9.1 using Excel.

Figure 9.2(a) shows an Excel worksheet in which the tabulated x-values (2.0 and 2.1) are entered into cells A4 and A5, and the corresponding y-values (5.5 and 7.2) are entered into cells B4 and B5. Cell A7 contains the x-value (2.03) whose y-value is to be determined. Cell B5 contains the desired interpolated y-value (6.01). Note that cell B7 is a calculated value, as indicated by the formula shown in the formula bar at the top of the figure. Examination of this formula reveals that it is equivalent to Equation (9.2).

Figure 9.2(b) shows a similar Excel worksheet in which the dependent variable in cell B7 is determined using the formula =FORECAST(A7,B4:B5,A4:A5), as shown in the formula bar at the top of the figure. The interpolated result is $y = 6.01$, as expected.

B7			▼	f_x =B4+(B5-B4)/(A5-A4)*(A7-A4)		
	A	B	C	D	E	F
1	**Linear Interpolation**					
2						
3	**x**	**y**				
4	2.0	5.5				
5	2.1	7.2				
6						
7	2.03	6.01				
8						
9						
10						
Ready						

Figure 9.2(a) – Linear interpolation within an Excel worksheet

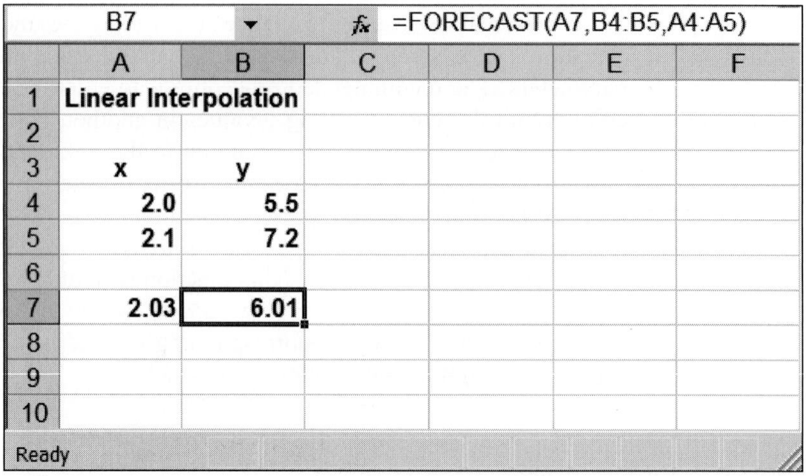

B7	▼		f_x =FORECAST(A7,B4:B5,A4:A5)			
	A	B	C	D	E	F
1	Linear Interpolation					
2						
3	x	y				
4	2.0	5.5				
5	2.1	7.2				
6						
7	2.03	6.01				
8						
9						
10						
Ready						

Figure 9.2(b) – Linear interpolation using the Forecast function

Problems

9.1 The heat capacity of iron is given below as a function of temperature.

Temperature (°C)	Heat Capacity (kcal/kg·C°)
0	0.1055
100	0.1168
200	0.1282
300	0.1396
400	0.1509
500	0.1623
600	0.1737
700	0.1805

Determine the heat capacity at each of the following temperatures, using linear interpolation. (*Note*: Heat capacities involve *temperature differences* (C°), not actual temperatures (°C), as explained in Chap. 7).

(*a*)   80 °C   (*c*)   410 °C

(*b*)   335 °C   (*d*)   675 °C

9.2 The temperature of air in a "standard atmosphere" is given below as a function of altitude.

Altitude (ft)	Temperature (°F)
0	59.0
5,000	41.2
10,000	23.3
15,000	5.5
20,000	−12.3
25,000	−30.2
30,000	−48.0
35,000	−65.8
40,000	−67.0
50,000	−67.0

Determine the temperature at each of the following altitudes using linear interpolation:

(a)   6,530 ft   (c)   18,800 ft

(b)   12,400 ft   (d)   25,300 ft

9.3   The gravitational acceleration varies with latitude (distance from the equator) and altitude above sea level. The following table gives the gravitational acceleration at various latitudes, at sea level.

Latitude (°)	Gravitational Acceleration (m/s^2)
0	0.978039
10	0.978195
20	0.978641
30	0.979329
40	0.980171
50	0.981071
60	0.981918
70	0.982608
80	0.983059
90	0.983217

The following table gives corrections as a function of altitude. These values are independent of latitude.

Altitude (m)	Correction (m/s^2)
200	-0.617×10^{-4}
300	-0.926×10^{-4}
400	-1.234×10^{-4}
500	-1.543×10^{-4}
600	-1.852×10^{-4}
700	-2.160×10^{-4}
800	-2.469×10^{-4}
900	-2.777×10^{-4}

Using these two tables and linear interpolation, answer each of the following questions:

(a) What is the gravitational acceleration at sea level and a latitude of 18.5°?

(b) What is the correction term for an altitude of 275 m above sea level?

(c) What is the gravitational acceleration at a latitude of 18.5° and an altitude of 275 m above sea level?

(d) What is the gravitational acceleration in Pittsburgh, which is located at a latitude of 40.5° and an altitude of 235 m above sea level?

9.4 Compound interest factors are used to determine the increase in a sum of money if the money is allowed to accumulate for several years with interest payable at a certain rate and compounded annually. If P represents the initial sum of money and F is the future sum after n years, we can write

$$F = fP$$

where f is the compound interest factor for n years at a specified interest rate. Several values for the compound interest factor are given below as a function of n, based upon an interest rate of 10 percent per year.

n	f	n	f
0	1.000	25	12.183
5	1.649	30	20.086
10	2.718	35	33.115
15	4.482	40	54.598
20	7.389	45	90.017

(a) Suppose $1000 is deposited in a bank account that pays 10 percent interest, compounded annually. Using the tabulated values and linear interpolation, determine how much money will accumulate after eight years.

(b) How much money will accumulate after 12 years?

(c) How long will it take your money to triple if it earns interest at 10 percent per year, compounded annually?

(d) Suppose you plan to retire in 42 years. If you continue to earn interest at 10 percent per year, compounded annually, how much will have accumulated by the time you are ready to retire?

9.5 Prob. 4.6 presented data for the current, in milliamps, passing through an electronic device as a function of time. The data are reproduced below, for your convenience.

Time (sec)	Current (mA)	Time (sec)	Current (mA)
0	0	10	0.64
1	1.06	12	0.44
2	1.51	14	0.30
3	1.63	16	0.20
4	1.57	18	0.14
5	1.43	20	0.091
6	1.26	25	0.034
8	0.92	30	0.012

Use linear interpolation to answer the following questions, based upon these tabulated values:

(a) What current will flow through the device after 0.7 second?

(b) What current will flow through the device after 12.75 seconds?

(c) When will the current reach a value of 1.00 milliamp?

(d) When will the current reach a value of 0.1 milliamp? When will it reach 0.01 milliamp?

9.2 THE METHOD OF LEAST SQUARES

The method of least squares is intended to fit a curve (including a straight line, as a special case) through an *aggregate* of data. It is widely used when working with measured data, where data scatter is an important consideration.

To understand the method of least squares, consider the graph shown in Fig. 9.3. This graph contains four data points and a curve passing through the aggregate of the data. The individual data points are represented as $P_1 = (x_1, y_1)$, $P_2 = (x_2, y_2)$, $P_3 = (x_3, y_3)$, and $P_4 = (x_4, y_4)$, and the equation for the curve is represented in general terms as $y = f(x)$.

For each data point, $P_i = (x_i, y_i)$, we can define an *error*, e_i, as the difference between y_i, the actual y-value, and $f(x_i)$, the corresponding calculated y-value. Thus, we can write

$$e_i = y_i - f(x_i) \tag{9.4}$$

for each of the data points; that is, for $i = 1, 2, \ldots, n$, where n is the total number of data points. In Fig. 9.3, for example, we can see that e_2 is defined as the difference between y_2, the actual data point, and $f(x_2)$, the corresponding point on the curve.

Remember that the curve that we refer to in general terms as $y = f(x)$ is actually represented by some specific equation. The exact equation depends upon the specific curve. For example, if we wish to fit the straight line $y = ax + b$ to the data, our goal is to determine the values of a and b that will result in a good curve fit. Similarly, if we wish to fit the polynomial $y = c_0 + c_1x + c_2x^2 + c_3x^3$ to the data, we must determine the values of c_0, c_1, c_2, and c_3 that will result in a good fit. Thus, *the overall strategy in fitting a curve to a set of data points is to determine an appropriate set of values for the coefficients in the equation of the chosen curve.*

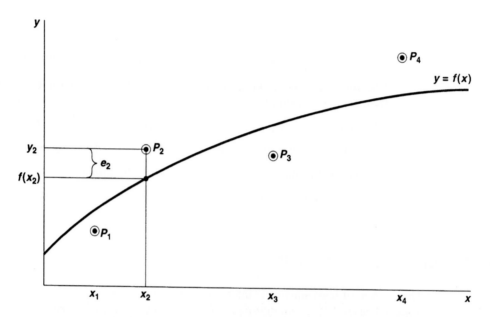

Figure 9.3 – Passing a curve through the aggregate of three data points

Intuitively, we would like to select the unknown coefficients in such a manner that the error terms, e_i, will be as small *in magnitude* as possible (see Fig. 9.3). One approach might be to select the coefficients so that the sum of the errors is minimized. The problem with this approach, however, is that the individual errors might be large in magnitude but opposite in sign (i.e., several *positive* errors that are large in magnitude and several *negative* errors that are large in magnitude), so that the errors would tend to cancel each other when added together. This could result in some very bad curve fits, even though the *sum* of the errors might be very small.

A better approach is to select the coefficients in such a manner that the sum of the *squares* of the errors is minimized. Keep in mind that the square of the error will always be nonnegative, regardless of whether the individual error is positive or negative. Hence, by summing the squares of the errors, we will always be summing positive numbers, thus eliminating the possibility that large errors might cancel one another because of differing signs. In other words, the only way the sum of the *squares* of the errors can be minimized is by making the individual errors as small in magnitude as possible, thus assuring a good fit. This is the idea behind the method of least squares.

The procedure, then, is to write an equation for the sum of the square errors in terms of the unknown coefficients. We can then use calculus to determine the values of the coefficients that minimize the sum of the errors.

9.3 FITTING A STRAIGHT LINE TO A SET OF DATA

We now make use of the least squares concept to fit a straight line to a set of data. To do so, we must determine the values of a and b in the equation $y = ax + b$ in such a manner that the sum of the squares of the errors will be minimized. In Equation (9.4), we wrote each error term as $e_i = y_i - f(x_i)$. Now let us substitute the equation for the straight line into the general expression for $f(x)$. Hence, we can write the error term as

$$e_i = y_i - (ax_i + b) \tag{9.5}$$

If we let z represent the sum of the squares of the errors (SSE), we can write

$$z = e_1^2 + e_2^2 + e_3^2 + \cdots$$

$$z = [y_1 - (ax_1 + b)]^2 + [y_2 - (ax_2 + b)]^2 + [y_3 - (ax_3 + b)]^2 + \cdots$$

$$z = \sum_{i=1}^{n} [y_i - (ax_i + b)]^2 \tag{9.6}$$

Our goal is to determine the values of a and b that will minimize z in Equation (9.6). To do so, we must set the derivatives of z with respect to a and b equal to zero. Thus, we first take the derivative of z with respect to a, holding b constant, and set the result equal to zero. We then take the derivative of z with respect to b, holding a constant, and set it equal to zero. (These are called *partial derivatives*.) We can write these derivatives as

$$\frac{\partial z}{\partial a} = -2 \sum_{i=1}^{n} x_i [y_i - (ax_i + b)] = 0$$

$$\frac{\partial z}{\partial b} = -2 \sum_{i=1}^{n} [y_i - (ax_i + b)] = 0$$

Now let us divide each equation by 2 and then write the second equation ahead of the first equation, resulting in

$$\sum_{i=1}^{n} [ax_i + b - y_i] = 0$$

$$\sum_{i=1}^{n} [ax_i^2 + bx_i - x_i y_i] = 0$$

Finally, we use the distributive rule for addition and then factor the constants a and b out of the resulting summation terms. This results in the final form of the least squares equations:

$$a\sum_{i=1}^{n} x_i + bn = \sum_{i=1}^{n} y_i \tag{9.7}$$

$$a\sum_{i=1}^{n} x_i^2 + b\sum_{i=1}^{n} x_i = \sum_{i=1}^{n} x_i y_i \tag{9.8}$$

Thus, we have two equations in two unknowns. The unknown quantities are a and b. Once we have determined a and b, we can substitute their values into the equation $y = ax + b$ to determine a specific equation for the desired straight line. We can then plot the line, and we can use it to estimate a value for y that corresponds to a specified value for x.

Remember that Equations (9.7) and (9.8) apply only when fitting a *straight line* to a set of data. The method of least squares can also be used to fit other types of equations to a data set, though the individual equations will differ from Equations (9.7) and (9.8). The method used to obtain these equations is the same, however, as that described above. We will say more about the use of other types of equations with the method of least squares later in this chapter.

Example 9.3 Fitting a Straight Line to a Set of Data

An engineer measured the force exerted by a spring as a function of its displacement from the equilibrium position. Here are the data:

Data Point No.	Distance (cm)	Force (N)
1	2	2.0
2	4	3.5
3	7	4.5
4	11	8.0
5	17	9.5

Determine a straight line that passes through the aggregate of the data using the method of least squares. Then plot the data points and the resulting straight line. Use the equation of the straight line to estimate the force corresponding to a displacement of 8.5 cm.

Note that distance is the independent variable (x) and force is the dependent variable (y). Thus, we are seeking an equation for force as a function of distance. Note that there are five data points; hence, $n = 5$ in the least squares equations.

In order to apply the method of least squares, we must expand the above table as follows:

i	x_i	y_i	x_i^2	$x_i y_i$
1	2.0	2.0	4.0	4.0
2	4.0	3.5	16.0	14.0
3	7.0	4.5	49.0	31.5
4	11.0	8.0	121.0	88.0
5	17.0	9.5	289.0	161.5
Sums:	41.0	27.5	479.0	299.0

If we substitute these values into Equations (9.7) and (9.8), we obtain the following two simultaneous equations for a and b:

$$41a + 5b = 27.5$$

$$479a + 41b = 299$$

These equations can be solved by a variety of techniques, such as direct substitution or the use of Cramer's rule. They can also be solved using Excel, as explained in Chapter 11. For now, however, we simply state the solution as $a = 0.514706$ and $b = 1.279412$. (You can verify that these solutions are valid by substituting them into the equations and recalculating the right-hand values.) Thus, the equation of the line that passes through the data points is

$$y = 0.514706x + 1.279412$$

Figure 9.4 shows a plot of the given data and the line. Notice the scatter in the data.

In order to estimate the force (y) corresponding to a displacement (x) of 8.5 cm, we can write

$$y = (0.514706)(8.5) + 1.279412 = 5.654413$$

Thus, the force is approximately 5.7 N.

Once a line passing through the data has been determined, it is generally helpful to evaluate each of the error terms and then compute a numerical value for the sum of the squares of the errors (SSE), using the formula

$$\text{SSE} = \sum_{i=1}^{n} [y_i - f(x_i)]^2 \tag{9.9}$$

SSE is an indication of the quality of the curve fit – the smaller the sum of the square errors, the better the fit. This procedure is particularly useful when fitting several different curves to the same set of data. The curve resulting in the smallest value for the sum of the square errors will provide the best fit.

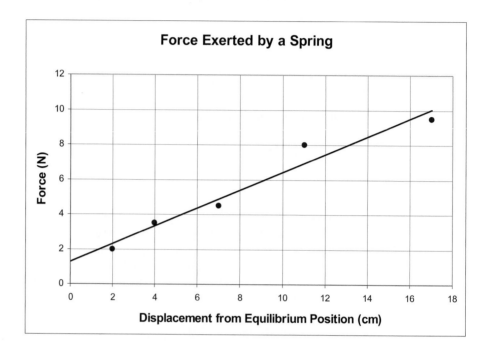

Figure 9.4 – Passing a straight line through a set of measured data

Another indication of the quality of the curve fit is the so-called *r-squared* value, which is defined as

$$r^2 = 1 - \frac{\text{SSE}}{\text{SST}} \tag{9.10}$$

where SST represents the sum of the squares of the deviations about the mean, given by

$$\text{SST} = \sum_{i=1}^{n} [y_i - \bar{y}]^2 \tag{9.11}$$

The *r-squared* value varies between 0 and 1. Note that r^2 will equal 1 when SSE equals zero. Hence, an r^2 value close to 1 (which means that the sum of the square errors is small) generally indicates a good fit.

Example 9.4 Assessing a Curve Fit

Assess the quality of the curve fit obtained in the last example by computing the individual error terms and the sum of the square errors.

From the last example, the equation of the straight line that best represents the aggregate of the given data is $y = 0.514706x + 1.279412$. Knowing this equation, it is now a simple procedure to determine each of the error terms. For example, we can write

$$e_1 = y_1 - [0.514706\,x_1 + 1.279412]$$

$$e_1 = 2.0 - [(0.514706)\,(2.0) + 1.279412] = -0.308824$$

Similarly,

$$e_2 = y_2 - [0.514706x_2 + 1.279412]$$

$$e_2 = 3.5 - [(0.514706)(4.0) + 1.279412] = 0.161764$$

and so on. The results are summarized in the table below. Note that the value of y determined from the least squares equation (column 4) is expressed as $y(x_i)$.

i	x_i	y_i	$y(x_i)$	e_i	e_i^2
1	2.0	2.0	2.308824	-0.308824	0.095372
2	4.0	3.5	3.338236	0.161764	0.026168
3	7.0	4.5	4.882354	-0.382354	0.146194
4	11.0	8.0	6.941178	1.058822	1.121104
5	17.0	9.5	10.029414	-0.529414	0.280279
	Sum:	27.5		Sum:	1.669117

If we sum the values in the last column, we determine that the sum of the squares of the errors (SSE) is 1.669117. This value is not particularly meaningful by itself, but, if we were to fit several different curves to the same set of data, the sum of the squares of the errors could be compared, thus providing a measure of the quality of each fit.

From the sum of the y-values, we can also determine the mean y-value as

$$\bar{y} = 27.5/5 = 5.5$$

Hence, we can determine the values of SST and r^2 as

$$\text{SST} = [(2.0 - 5.5)^2 + (3.5 - 5.5)^2 + (4.5 - 5.5)^2 + (8.0 - 5.5)^2 + (9.5 - 5.5)^2] = 39.5$$

$$r^2 = 1 - 1.669117 / 39.5 = 0.957744$$

Note that the r^2 value is close to 1, which suggests a reasonably good fit.

9.4 LEAST SQUARES CURVE FITTING IN EXCEL

If all of this seems rather complicated, don't despair. Excel will do most of the work for you. In fact, in order to fit a straight line to a data set using Excel, you need only enter the data into a worksheet, plot the data in the usual manner, and then generate a *trendline* through the data. The method of least squares will automatically be applied to the data set, and the results displayed both graphically and algebraically. Thus, you need not concern yourself with constructing extended tables, solving simultaneous equations, or evaluating error terms.

To fit a straight line to a set of data in Excel, proceed as follows:

1. Open a new worksheet and enter the *x*-data (the independent variable) in the leftmost column.

2. Enter the *y*-data (the dependent variable) in the next column.

3. Plot the data as an *x-y* graph (i.e., an *XY Chart*) with arithmetic coordinates. Do not interconnect the individual data points.

4. Click on one of the plotted data points, thus selecting the data set as the active editing object. (The data points will appear highlighted if this step is carried out correctly.) Then choose Add Trendline... from the Chart menu. Or, you may right-click on one of the data points and then select Add Trendline... from the resulting menu.

5. Specify the type of curve and request any pertinent options from the resulting Add Trendline dialog box. Generally, you should request that the equation of the curve and its associated r^2 value be displayed. You may also wish to force the curve through a specified intercept or extrapolate the curve fit forward, i.e., beyond the rightmost data point, or backward, beyond the leftmost data point.

6. Press the OK button. The curve fitting will then be carried out and the results displayed automatically.

The procedure is illustrated in the following example.

Example 9.5 Fitting a Straight Line to a Set of Data in Excel

Use Excel's Trendline feature to fit a straight line through the data given in Example 9.3. Extend the line backward, to the point $x = 0$. Display the equation of the line passing through the data and its associated r^2 value. Then use the equation to determine the force corresponding to a distance of 8.5 cm.

Following the procedure outlined above, we begin by entering and plotting the data within a worksheet, as shown in Fig. 9.5. The data are plotted as an *x-y* graph against a background of horizontal and vertical grid lines. Note that the individual data points are not interconnected.

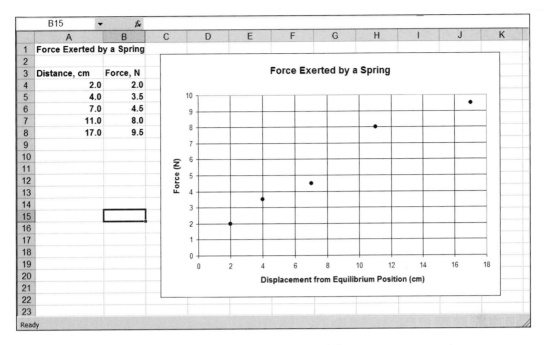

Figure 9.5 – Plotting a set of measured data as an *x-y* graph

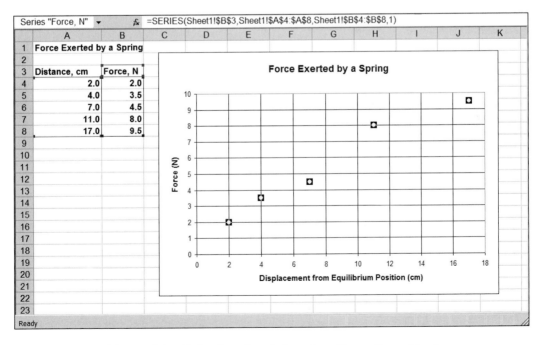

Figure 9.6 – Selecting the data set as the active object

Figure 9.6 illustrates the appearance of the graph after the data set has been selected as the active editing object. We then either right-click on one of the data points and select Add Trendline... from the resulting menu, or select Add Trendline... directly from the Chart menu, resulting in the dialog box shown in Fig. 9.7.

Notice that the Add Trendline dialog box shown in Fig. 9.7 contains two tabs. Initially, the dialog box associated with the Type tab is active, allowing us to select a linear (straight line) trendline. (Several of the remaining types of curves will be discussed in Sec. 9.6.)

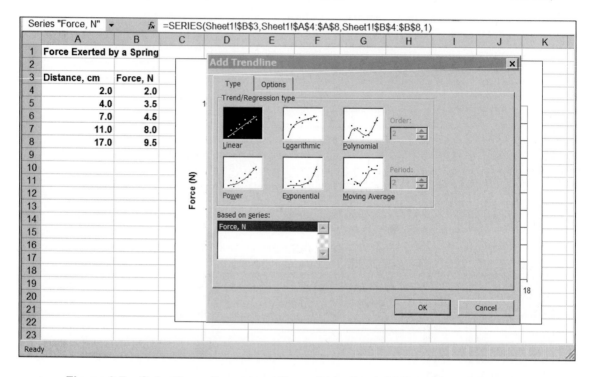

Figure 9.7 – Selecting a linear trendline within the Add Trendline dialog box

We then select the Options tab. This results in another dialog box, from which we can specify that the trendline equation and the r^2 value will be displayed on the graph. The selection of these features is indicated at the bottom of Fig. 9.8. Also, note that 2 units are indicated in the backward direction under Forecast. This causes the trendline to be extended back two units beyond the leftmost data point ($x = 2$), to the point $x = 0$. We could also have forced the trendline to intercept the y-axis at a specified location if we had wished, by selecting the Set Intercept feature and entering a y-value in the data entry area.

Clicking the OK button results in the graph shown in Fig. 9.9. Notice the straight line that has been drawn through the data set, ranging from the y-axis ($x = 0$) to the last data point ($x = 17$). This is the desired trendline. The equation of the trendline is shown as

$$y = 0.5147x + 1.2794$$

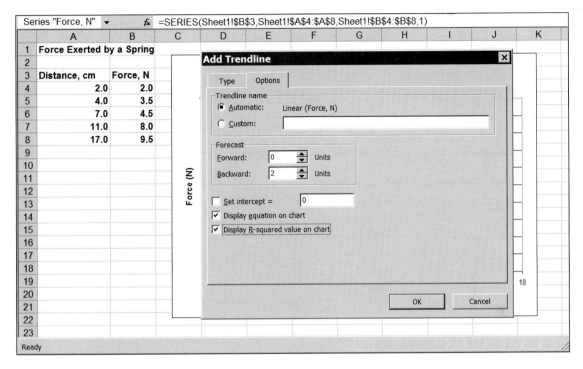

Figure 9.8 – Specifying options within the Add Trendline dialog box

as determined manually in Example 9.3. (The resulting equation for the trendline and the r^2 value have been moved above the graph in order to improve their legibility.) Moreover, we see that the calculated r^2 value is

$$r^2 = 0.9577$$

as determined in Example 9.4. *Note that the determination of the trendline equation and the r^2 value is entirely automatic.*

We can now determine the force corresponding to a distance of = 8.5 cm as follows:

$$y = (0.5147)(8.5) + 1.2794 = 5.6544$$

Hence, the force is approximately 5.7 N.

Unfortunately, the Trendline feature in Excel does not display the sum of the squares of the errors (SSE) associated with the curve fit. This value can, of course, be obtained manually within the worksheet, by determining the square of each of the error terms (once the trendline equation is known) and then summing their values.

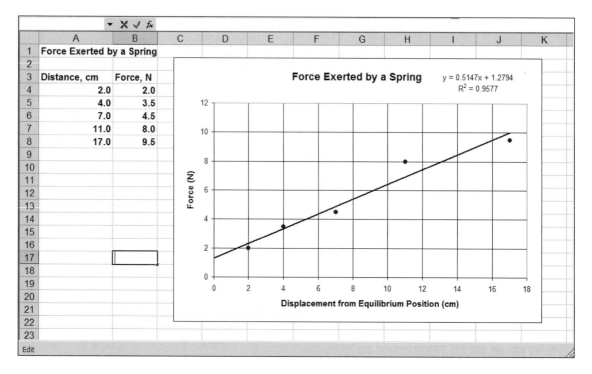

Figure 9.9 – A linear trendline passing through a set of measured data

Sometimes the equation for the desired trendline does not include enough significant figures. This situation can be remedied by right-clicking on the equation and then selecting Format Data Labels... from the resulting dialog box. The nature of the display can then be altered by clicking on the Number tab and then selecting either the Number or the Scientific category. Or, you may select the equation and then click on the Increase Decimal button in the Formatting Toolbar.

Excel also provides a Regression feature within the Analysis Toolpak. This feature will fit a straight line (called a *regression line*) to a set of data using the method of least squares. The resulting output includes the coefficients of the regression line equation, the sum of the squares of the errors, and the r^2 value. In addition, optional output can be requested, which includes a listing of the individual error terms (called *residuals*) and a plot of the data. From an engineer's perspective, however, the Regression feature may provide too much statistical information in the output summary. Much of this information is extraneous and may be confusing, as seen in the example presented below. Thus, the trendline feature discussed earlier provides a more straightforward approach to the problem of fitting a straight line through a set of data.

Example 9.6 Use of the Regression Feature in Excel

Fit a straight line through the data given in Example 9.3 using the Regression feature found in the Excel Analysis Toolpak. Include a list of the residuals in the output and generate a plot of the data.

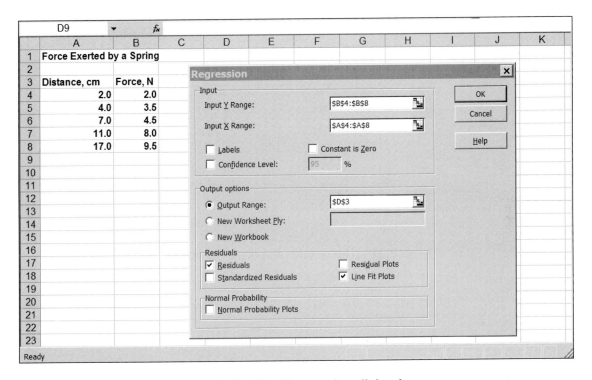

Figure 9.10 – The Regression dialog box

We begin by entering the data into a worksheet, as shown in the earlier figures (e.g., Fig. 9.5). We then select Data Analysis/Regression from the Tools menu. This results in the Regression dialog box shown in Fig. 9.10. Within this dialog box, the y-values are designated as the contents of cells B4 through B8, and the x-values as the contents of cells A4 through A8. The upper left corner of the output will appear in cell D3, as indicated in the address space corresponding to Output Range. Notice also that Residuals and Line Fit Plots have been selected as options.

Figure 9.11 shows the resulting output, once OK has been selected in the Regression dialog box. Much of this information is extraneous, of interest only to a trained statistician. Within the computed output, however, we see that the r^2 value (0.957744) is shown in cell E7, under the heading *Regression Statistics*. Similarly, the sum of the squares of the errors (1.669118) is shown in cell F15, under the heading *SS*. This value is labeled *Residual*.

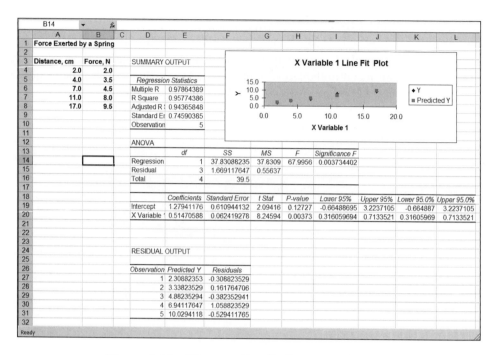

Figure 9.11 – The results of a linear regression

The coefficients in the regression equation are given in cells E19 and E20, under the heading *Coefficients*. Thus, the *y*-intercept (1.279412) is shown in cell E19, and the slope (0.514706) in cell E20. Also, the predicted *y*-values are shown in cells E27 through E31, and the corresponding error terms (residuals) in cells F27 through F31.

A plot of the given data points and the corresponding points generated by the regression line are shown in the graph at the top of the worksheet. The graph was dragged to this location to improve its legibility. It is still difficult to read, however, not only because of its size (which can easily be increased), but also because the resulting curve fit is not shown as an interconnected set of points on a straight line (compare with the results shown in Fig. 9.9).

Problems

9.6   Reconstruct the worksheet shown in Fig. 9.9, including the graph. Reposition the graph, if necessary, so that columns C, D, and E are accessible. Use the trendline formula to generate a set of *y*-values that correspond to the *x*-values given in column A. Place these *y*-values in column C. Generate the values of the corresponding errors (i.e., the residuals) in column D, and the squares of the errors in column E. Then determine the sum of the values in column E. Compare with the values obtained in Examples 9.4 and 9.6.

9.7 Reconstruct the worksheet shown in Fig. 9.9. Extend the trendline across the entire graph (i.e., from $x = 0$ to $x = 20$). Then delete the trendline and create a new one that passes through the origin. Notice the r^2 value that is now obtained and compare it with the original r^2 value. What do you conclude about the relationship between the r^2 value and the quality of the curve fit?

9.8 A polymeric material contains a solvent that dissolves as a function of time. The concentration of the solvent, expressed as a percentage of the total weight of the polymer, is shown in the following table as a function of time (repeated from Prob. 4.2).

Solvent Concentration (weight percent)	Time (sec)
55.5	0
44.7	2
38.0	4
34.7	6
30.6	8
27.2	10
22.0	12
15.9	14
8.1	16
2.9	18
1.5	20

Enter the data into an Excel worksheet, plot the data, and fit a straight line through the data. Determine the equation of the line and the corresponding r^2 value.

9.9 Repeat Prob. 9.8 using the following set of data:

Solvent Concentration (weight percent)	Time (sec)
30.2	0
44.7	2
22.5	4
41.3	6
28.8	8
14.0	10
26.2	12
11.0	14
23.4	16
14.5	18
4.2	20

Compare the results with those obtained for Prob. 9.8. Which data set results in a better fit, and why?

9.10 A popular consumer magazine tabulated the following list of weight versus overall gasoline mileage for several different sizes and types of cars:

Weight (lb)	Mileage (mpg)	Weight (lb)	Mileage (mpg)
2775	33	3325	20
2495	27	3200	21
2405	29	3450	19
2545	28	3515	21
2270	34	3495	19
2560	24	4010	19
3050	23	4205	17
3710	24	2900	24
3085	23	2555	28
2940	21	2790	21
2395	26	2190	34

From these data, develop a straight-line correlation (i.e., an equation) for gasoline mileage as a function of weight. Based upon your results, how well are the given data represented by the straight-line relationship? How might a better relationship be obtained?

9.5 FITTING OTHER FUNCTIONS TO A SET OF DATA

The method of least squares can be used to fit many different types of functions through a set of data points. Thus, we need not be confined to straight-line relationships when fitting a curve through a data set. In fact, the data obtained in many engineering applications may be better represented by an exponential function, a power function, or a polynomial rather than a straight line.

Exponential Functions

We have already discussed the fact that the exponential function

$$y = ae^{bx} \tag{9.12}$$

governs many different phenomena in engineering (see Sec. 4.5). If we take the natural log of each side of Equation (9.12), we obtain

$$\ln y = \ln a + bx \tag{9.13}$$

If we let $u = \ln y$ and $c = \ln a$, we can rewrite Equation (9.13) as

$$u = bx + c \tag{9.14}$$

which is the equation for a straight line (but *not* the straight line $y = ax + b$ that we considered earlier in this chapter). Thus, we can fit an exponential function to a set of data by proceeding as we did for a straight line, substituting $\ln y$ for y, b for a, and c for b in Equations (9.7) and (9.8); that is,

$$b\sum_{i=1}^{n} x_i + cn = \sum_{i=1}^{n} \ln y_i \tag{9.15}$$

$$b\sum_{i=1}^{n} x_i^2 + c\sum_{i=1}^{n} x_i = \sum_{i=1}^{n} x_i \ln y_i \tag{9.16}$$

After solving Equations (9.15) and (9.16) for b and c, the original coefficient a is then obtained by taking the antilog of c; that is,

$$a = e^c \tag{9.17}$$

The entire procedure is illustrated in the following example.

Example 9.7 Fitting an Exponential Function to a Set of Data

The transient behavior of a capacitor has been studied by measuring the voltage drop across the device as a function of time. The following data have been obtained.

Time (sec)	Voltage	Time (sec)	Voltage
0	10	6	0.5
1	6.1	7	0.3
2	3.7	8	0.2
3	2.2	9	0.1
4	1.4	10	0.07
5	0.8	12	0.03

With electronic devices of this type, the voltage generally varies exponentially with time. We will therefore fit an exponential function to the current set of data, using the method of least squares.

To apply the method of least squares to this data set, let us represent the time as x and the voltage as y. We then form the following table:

i	x_i	x_i^2	y_i	$\ln y_i$	$x_i \ln y_i$
1	0	0	10	2.302585	0
2	1	1	6.1	1.808289	1.808289
3	2	4	3.7	1.308333	2.616666
4	3	9	2.2	0.788457	2.365371
5	4	16	1.4	0.336472	1.345888
6	5	25	0.8	−0.223144	−1.115720
7	6	36	0.5	−0.693147	−4.158882
8	7	49	0.3	−1.203973	−8.427811
9	8	64	0.2	−1.609438	−12.875504
10	9	81	0.1	−2.302585	−20.723265
11	10	100	0.07	−2.659260	−26.592600
12	12	144	0.03	−3.506558	−42.078695
Sums:	67	529		−5.653969	−107.836263

Substituting the appropriate values into Equations (9.15) and (9.16), we obtain the following two least squares equations:

$$67b + 12c = -5.653969$$

$$529b + 67c = -107.836263$$

The solution to these equations is $b = -0.492318$ and $c = 2.277612$. Hence,

$$a = e^c = e^{2.277612} = 9.753358$$

and the desired exponential function is

$$y = ae^{bx} = 9.753358e^{-0.492318x}$$

Though the setup in the previous example enhances your understanding of the least squares process using an exponential function, it is not necessary when using Excel. In fact, Excel does all of the work for you, as was the case when we fit a straight line to a data set.

The procedure for fitting an exponential function to a set of data in Excel is the same as the procedure outlined in Sec. 9.3 for a straight line. Now, however, we select an exponential function when fitting the trendline to the data. The procedure is illustrated in the following example.

Example 9.8 Fitting an Exponential Function to a Set of Data in Excel

Fit an exponential function through the data given in Example 9.7 by plotting the data within an Excel worksheet and then fitting an exponential trendline through the graph. Display the equation of the exponential function and its corresponding r^2 value. Use the resulting equation to determine the voltage after 11 seconds.

The solution proceeds in the same manner as described in Example 9.5, where we fit a straight line to a set of data. Thus, we enter the given data into a worksheet, plot the data as an x-y graph, and pass a trendline through the graph. Now, however, we select an exponential trendline rather than a linear (straight line) trendline.

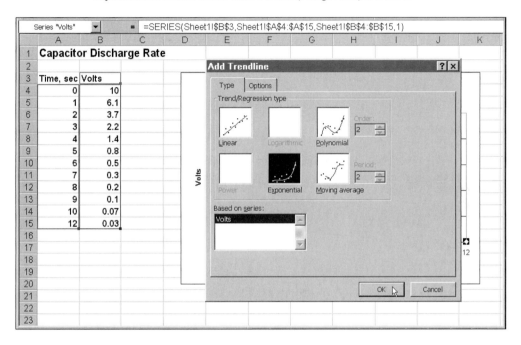

Figure 9.12 – Selecting an exponential trendline within the Add Trendline dialog box

Figure 9.12 shows the worksheet containing the data and the line graph, with the Trendline dialog box superimposed over the line graph. Notice that the Exponential icon is highlighted, indicating that an exponential function will be passed through the data.

The graph containing the trendline is shown in Fig. 9.13. Note that the trendline appears to fit the data very well. The equation of the trendline is shown as

$$y = 9.7534e^{-0.4923x}$$

which agrees with the result obtained in Example 9.7. In addition, we see that the r^2 value is 0.9988, which indicates a good fit.

To determine the voltage after 11 seconds, we write

$$y = 9.7534\ e^{-(0.4923)\,(11)} = 0.04338$$

Thus, the voltage is approximately 0.04 volt after 11 seconds.

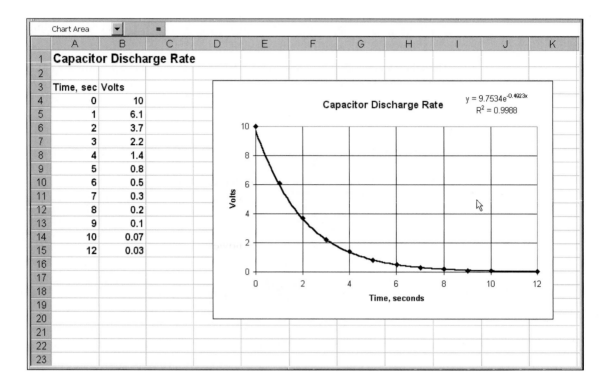

Figure 9.13 – An exponential trendline passing through a set of data

It should be mentioned that the Analysis Toolpak Regression feature, described at the end of Sec. 9.3, can also be used to fit an exponential function to a data set. When using this feature, however, the y-values are not automatically transformed into logarithmic form. Thus, the natural logarithms of the y-values must be entered into the worksheet explicitly. (They can, of course, be calculated directly from the given y-values, using the LN function.) The cell addresses of these ln y values are then specified as the Input Y Range within the Regression dialog box.

Logarithmic Functions

A *logarithmic function* is an equation of the form

$$y = a \ln x + b \tag{9.18}$$

This is clearly a linear function between y and ln x. Hence, the applicable least squares equations in this case are

$$a \sum_{i=1}^{n} \ln x_i + bn = \sum_{i=1}^{n} y_i \tag{9.19}$$

$$a \sum_{i=1}^{n} (\ln x_i)^2 + b \sum_{i=1}^{n} \ln x_i = \sum_{i=1}^{n} (\ln x_i) y_i \tag{9.20}$$

Solving these two equations for a and b will determine the particular logarithmic function that best fits the data.

In Excel, the procedure for passing a logarithmic trendline through a given data set is identical to that for a straight line or an exponential function, except that the logarithmic function is selected when specifying the type of trendline (as shown in Fig. 9.12 for an exponential trendline).

The logarithmic function can be defined in terms of base-10 logarithms as well as natural logarithms. In either case, however, the function can only be used with data sets in which the values of the independent variable (the x-values) are positive since logarithms are undefined for zero or negative values. (In Fig. 9.12, one of the x-values is equal to zero. Hence, the logarithmic function is unavailable for that particular data set.)

It is easy to show that the logarithmic function is equivalent to the exponential function

$$x = ce^{dy} \tag{9.21}$$

This is an ordinary exponential function with the independent and dependent variables interchanged. Note that the constants c and d in Equation (9.21) are related to the constants a and b in Equation (9.18). In particular,

$$c = e^{-b/a} \tag{9.22}$$

and

$$d = 1/a \tag{9.23}$$

Use of the logarithmic function in engineering applications is relatively uncommon. Hence, we will not discuss it further (see, however, Prob. 9.12 at the end of this section).

Power Functions

In Sec. 4.6 we considered the power function, which is an equation of the form

$$y = ax^b \tag{9.24}$$

This equation, like the exponential equation, describes many phenomena that occur in science and engineering.

To see how the method of least squares applies to a power function, we take the natural log of each side of Equation (9.24), resulting in

$$\ln y = \ln a + b \ln x \tag{9.25}$$

Now if we let $u = \ln x$, $v = \ln y$, and $c = \ln a$, we obtain

$$v = bu + c \tag{9.26}$$

which is the equation for a straight line. (Remember, however, that this straight line is different from those encountered earlier in this chapter.) Therefore, we can apply the method of least squares to the logs of the variables.

The least squares equations for a power function are obtained by substituting $\ln x$ for x, $\ln y$ for y, b for a, and c for b in Equations (9.7) and (9.8), resulting in

$$b\sum_{i=1}^{n} \ln x_i + cn = \sum_{i=1}^{n} \ln y_i \tag{9.27}$$

$$b\sum_{i=1}^{n} (\ln x_i)^2 + c\sum_{i=1}^{n} \ln x_i = \sum_{i=1}^{n} (\ln x_i)(\ln y_i) \tag{9.28}$$

Once these equations have been solved for b and c, the original coefficient a is obtained as

$$a = e^c \tag{9.29}$$

The use of Equations (9.27) through (9.29) to fit a power function to a set of data parallels the use of Equations (9.15) through (9.17) to fit an exponential function to a set of data, as illustrated in Example 9.7 (see Prob. 9.14 at the end of this section).

In Excel, all of this is handled automatically, by passing a trendline through a plot of the data. The procedure is the same as that described earlier. Now, however, we specify a power function when selecting the type of trendline, as described in the next example.

Example 9.9 Fitting a Power Function to a Set of Data

A chemical engineer is studying the rate at which a reactant is consumed in a chemical reaction involving the manufacture of a polymer. The following data have been obtained, showing the reaction rate (in moles per second) as a function of the concentration of the reactant (moles per cubic foot).

Concentration (moles/cu ft)	Reaction Rate (moles/second)
100	2.85
80	2.00
60	1.25
40	0.67
20	0.22
10	0.072
5	0.024
1	0.0018

With reactions of this type, the reaction rate is generally proportional to the concentration of the reactant raised to some power. Therefore, we will use Excel to determine the values of the constant of proportionality and the power to which the reactant concentration is raised by fitting a power function to the data. That is, we will determine the values for the parameters a and b in the expression

$$RR = aC^b$$

where RR is the reaction rate and C is the concentration of the reactant.

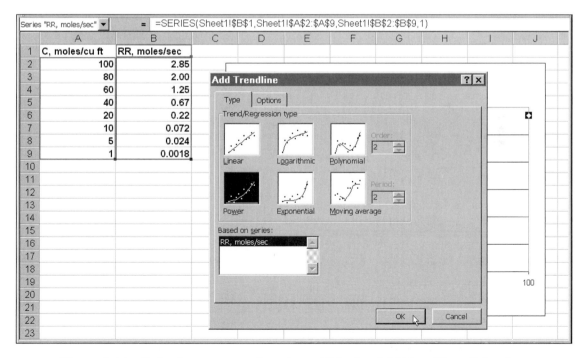

Figure 9.14 – Selecting a power function within the Add Trendline dialog box

To solve this problem, we again enter the data into an Excel worksheet and plot the data as an x-y graph. We then pass a power trendline through the graph and determine the parameters in the resulting equation. Figure 9.14 shows the worksheet containing the data, with the Trendline dialog box superimposed over the graph. The Power icon will be activated, indicating the selection of a power function, when we click on the OK button.

Figure 9.15 shows the final form of the worksheet, containing the data and the graph with the power function passed through the plotted data points. Careful inspection of the graph reveals that the equation of the power function is

$$y = 0.0018x^{1.599}$$

and the resulting r^2 value is 1. Thus, we conclude that the reaction rate can indeed be represented as a function of the reactant concentration using the power function

$$RR = 0.0018C^{1.599}$$

Remember that the method of least squares relates the *logarithm* of *y* to the *logarithm* of *x* when applied to a power function. Therefore, all of the *x*- and *y*-values in the data set must be positive, since the logarithm of a nonpositive value is undefined. If one or more data points are nonpositive, the Power icon will be inactive when selecting a trendline type (as, for example, in Fig. 9.12).

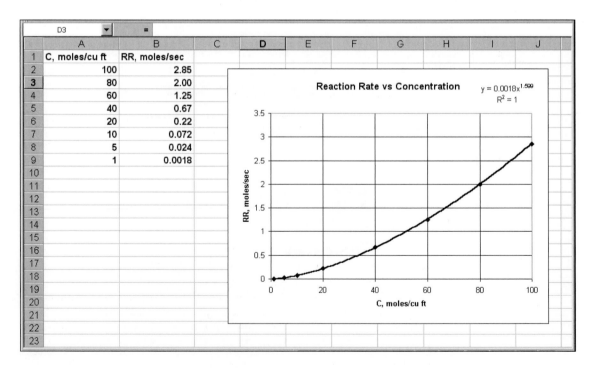

Figure 9.15 – A power function passing through a set of data

The Analysis Toolpak Regression feature can also be used to fit a power function to a data set using the method of least squares. When doing so, however, remember that the logarithms of the data must first be determined manually and placed in separate columns (or separate rows) within the worksheet. Either natural logarithms or base-10 logarithms can be used, though you must be consistent within a given data set.

Polynomials

The method of least squares can also be used to fit polynomials to a data set. The good news is that logarithmic transformations are not required. The bad news, however, is that a kth-degree polynomial requires the solution of $k+1$ simultaneous equations in $k+1$ unknowns. Thus, fitting a cubic equation involves solving four equations in four unknowns, and so on.

Let us write a kth-degree polynomial as

$$y = c_1 + c_2 x + c_3 x^2 + \cdots + c_{k+1} x^k \tag{9.30}$$

If we were to fit this polynomial to a set of data using the method of least squares, the resulting simultaneous equations would be

$$nc_1 + c_2 \sum_{i=1}^{n} x_i + c_3 \sum_{i=1}^{n} x_i^2 + \cdots + c_{k+1} \sum_{i=1}^{n} x_i^k = \sum_{i=1}^{n} y_i \tag{9.31}$$

$$c_1 \sum_{i=1}^{n} x_i + c_2 \sum_{i=1}^{n} x_i^2 + c_3 \sum_{i=1}^{n} x_i^3 + \cdots + c_{k+1} \sum_{i=1}^{n} x_i^{k+1} = \sum_{i=1}^{n} x_i y_i \tag{9.32}$$

$$\cdots \cdots$$

$$c_1 \sum_{i=1}^{n} x_i^k + c_2 \sum_{i=1}^{n} x_i^{k+1} + c_3 \sum_{i=1}^{n} x_i^{k+2} + \cdots + c_{k+1} \sum_{i=1}^{n} x_i^{2k} = \sum_{i=1}^{n} x_i^k y_i \tag{9.33}$$

Thus, we have $(k+1)$ equations in $(k+1)$ unknowns. The unknowns are $c_1, c_2, \ldots, c_{k+1}$. The straight-line curve fit discussed in Sec. 9.3 is a special case, resulting in two equations in two unknowns. If we wanted to fit a quadratic equation to the data, we would solve three equations in three unknowns, and so on.

There are several different techniques available for solving simultaneous algebraic equations such as those expressed by Equations (9.31) through (9.33). We will not present any of these techniques at this time because Excel solves the simultaneous equations for us when fitting a polynomial to the data. Note, however, that Chap. 11 discusses techniques for solving systems of simultaneous algebraic equations in Excel (see also Prob. 9.15 at the end of this section).

Example 9.10 Fitting a Polynomial to a Set of Data

The following table presents the time for a high-performance sports car to reach various speeds. The times are given in seconds and the speeds in miles per hour. (The top speed is considered to be the independent variable in this application, and time is the *dependent* variable.) Fit an equation to the data, thus obtaining an accurate mathematical relationship between acceleration time and top speed.

Top Speed (mph)	Time (sec)
30	1.9
40	2.8
50	3.8
60	5.2
70	6.5
80	8.3
90	10.4
100	12.7
110	15.6
120	19.0
130	23.2
140	31.2
150	45.1

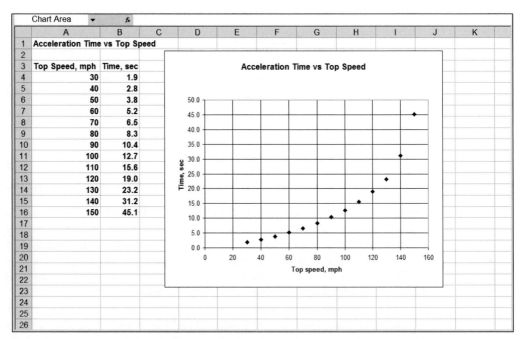

Figure 9.16 – Plotting a set of data with an unfamiliar appearance

Figure 9.16 shows an Excel worksheet containing the data and a corresponding *x-y* graph. From the shape of the graph, it is obvious that a straight-line curve fit is not satisfactory. Moreover, some experimentation with the power function and the logarithmic function shows that neither of these functions provides a satisfactory fit, either. The fit resulting from the use of an exponential function is not bad, though it appears that a better fit might be obtained using a polynomial.

Let us pass a polynomial-type trendline through the data, using the same general approach as with the other functions discussed in this chapter. Thus, Fig. 9.17 shows the Trendline dialog box superimposed over the line graph, with a polynomial of order 5 (i.e., a fifth-degree polynomial) being selected. The decision to use this particular polynomial is entirely arbitrary. (Note that the order of the polynomial can be increased or decreased by clicking on the appropriate arrow shown next to the data entry box labeled Order.)

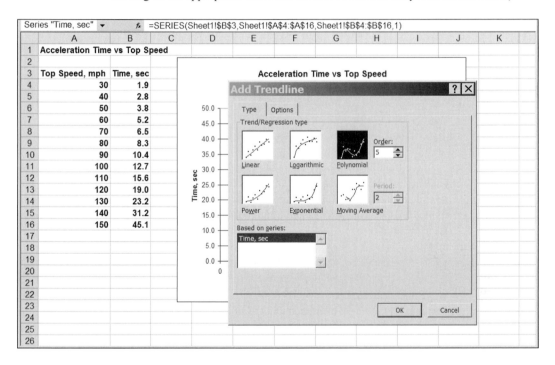

Figure 9.17 – Selecting a fifth-degree polynomial within the Add Trendline dialog box

The resulting trendline (rearranged somewhat to show the equation more clearly) is shown in Fig. 9.18(*a*). From visual inspection, we can see that the fifth-degree polynomial

$$y = 1 \times 10^{-8}x^5 - 5 \times 10^{-6}x^4 + 8 \times 10^{-4}x^3 - 0.0525x^2 + 1.777x - 20.958$$

fits the data very well. (Note that *y* represents the time to reach top speed, in seconds, and *x* represents the top speed, in mph.) The corresponding value of $r^2 = 0.9998$ further supports this argument.

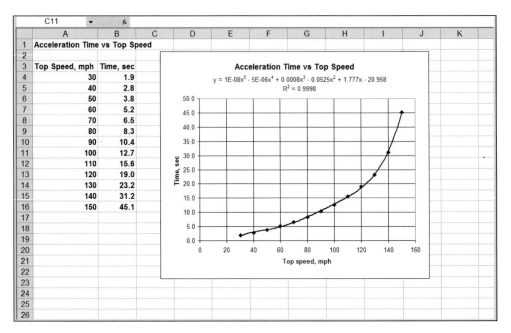

Figure 9.18(a) – A polynomial passing through a set of data

The trendline can be shown with greater precision by right-clicking on the equation, and then selecting Format Data Labels.../Number/Scientific from the resulting dialog box, as explained in Sec. 9.3. If we choose to display three decimals, we obtain the result shown in Fig. 9.18(*b*). Thus, the desired trendline (in scientific notation, with 3-decimal accuracy) is

$$y = 1.324 \times 10^{-8} x^5 - 5.124 \times 10^{-6} x^4 + 7.617 \times 10^{-4} x^3 - 5.255 \times 10^{-2} x^2 + 1.777x - 2.096 \times 10^{-1}$$

Since the first two coefficients are very small (1.324×10^{-8} and -5.124×10^{-6}, respectively), it might be argued that the fifth-degree polynomial is actually not required in order to obtain a satisfactory fit – a third-degree polynomial (i.e., a cubic equation) might be sufficient. Keep in mind, however, that these small coefficients will be multiplied by large numbers (the values of x^5 and x^4) when x takes on modestly large values. Thus, the magnitude of the first two terms in the polynomial might be significant. (We will pursue this issue further in Example 9.10, where we fit several different functions to the current set of data.)

Some warnings about higher-degree polynomials: First, the results may be inaccurate due to numerical errors in the curvefitting procedure. Second, even if the curvefitting is done accurately, use of the resulting curve fit may involve the product of very large and very small numbers, resulting in inaccurate results. And finally, the trendlines shown in Excel involve rounded coefficients, whose use may produce sizable errors with relatively large or relatively small values of x.

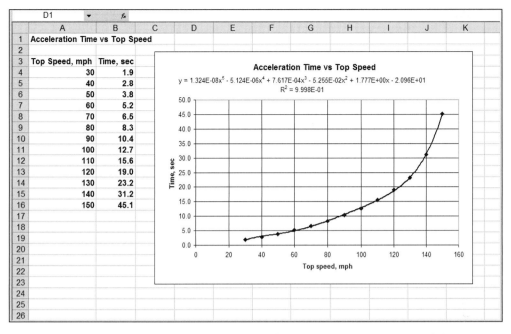

Figure 9.18(b) – Showing the power function with greater precision

Problems

9.11 For the data given in Example 9.7, carry out the following calculations:

(a) Calculate the sum of the squares of the errors using the (correct) least squares error criterion

$$e_i = \ln y_i - (\ln a + bx_i)$$

(b) Calculate the sum of the squares of the errors using the error criterion

$$e_i = y_i - ae^{bx_i}$$

(c) Use Equation (9.9) to determine a value of SST for this data set.

(d) Solve Equation (9.8) for SSE, using the value of SST obtained in part (c) and the value of $r^2 = 0.9988$ obtained in Example 9.8. Compare with the values for the sum of the squares of the errors obtained in parts (a) and (b). What do you conclude from this comparison?

9.12 The following data represent the temperature as a function of vertical depth within a chemically active settling pond. Fit the data to a logarithmic function using the method of least squares, as expressed by Equations (9.19) and (9.20).

Distance (cm)	Temp (°C)	Distance (cm)	Temp (°C)
0.1	21.2	390	45.9
0.8	27.3	710	47.7
3.6	31.8	1200	49.2
12	35.6	1800	50.5
120	42.3	2400	51.4

9.13 Enter the data given in Prob. 9.12 into an Excel worksheet and fit an appropriate trendline to the aggregate of the data. Compare the results obtained using an exponential function, a logarithmic function, a power function, and a fifth-degree polynomial.

9.14 Use Equations (9.27) through (9.29) to fit a power function to the data given in Example 9.9. Compare with the results obtained in Example 9.9, using Excel.

9.15 Verify that the curve fit obtained in Example 9.10 is correct by carrying out the following calculations:

(*a*)   Write the least squares equations required for a fifth-degree polynomial, using Equations (9.31) through (9.33) as a guide. (This requires writing six equations in six unknowns.)

(*b*)   Enter the data given in Example 9.10 into an Excel worksheet.

(*c*)   Create additional columns within the worksheet to represent the components of the various summations appearing in the six simultaneous equations. Calculate the appropriate sum at the bottom of each column.

(*d*)   Verify that the least squares equations are satisfied by substituting the summation terms into the six simultaneous equations.

9.16 Enter the data given in Example 9.9 into an Excel worksheet and carry out the following calculations:

(*a*)   Verify that the results obtained in Example 9.9 are correct by fitting a power function to the data.

(*b*)   Fit a straight line, an exponential function, and a logarithmic function to the data. Compare the results with those obtained in part (*a*).

(*c*)   Fit several different polynomials of varying degrees through the data. Compare the results with each other and with the results obtained in parts (*a*) and (*b*) above.

Additional curvefitting problems are presented at the end of this chapter.

9.6 SELECTING THE BEST FUNCTION FOR A GIVEN DATA SET

Now that we know how to fit various functions to a data set, one question remains: How do we determine which function to fit to a particular set of data? There is no simple answer to this question – the issue is usually decided by trial and error. However, the following guidelines may prove helpful:

1. Select an equation based upon the underlying theory governing the process used to collect the data.

2. Try to plot the data as a straight line. If this is successful, it will suggest the function that should be used to represent the data.

3. If a straight-line relationship cannot be obtained, try fitting different types of curves to the data. Use visual assessment to recognize a good fit, aided by the resulting values for the sum of the squares of the errors (SSE) and the r^2 parameter.

4. If a satisfactory curve fit cannot be obtained using the first two guidelines, try plotting the data differently (for example, try plotting y versus $1/x$, $1/y$ versus x, etc.).

5. In some cases, a better fit may be obtained by *scaling* the data so that the magnitude of the x-values is more or less the same as the y-values.

Each step is discussed separately below.

Selecting an Equation Based upon Theory

Measured data are often obtained from a process that is known to be governed by a particular equation, based upon underlying engineering principles. For example, chemical reaction rates are known to vary exponentially with the inverse of the absolute temperature, as illustrated in Example 9.13. In such situations, the equation reflecting the underlying theory should always be the first choice.

Plotting the Data as a Straight Line

If a data set can be plotted as a straight line on arithmetic coordinates, you should clearly fit a straight line to the data set. Recall, however, that other equations plot as straight lines with different coordinate systems, as explained in Secs. 4.5, 4.6, and 9.5. The results are summarized in the following table:

Equation Type	*Equation*	*Coordinate System*
Exponential	$y = ae^{bx}$	$\log y$ versus x (semi-log)
Logarithmic	$y = a \ln x + b$	y versus $\log x$
Power	$y = ax^b$	$\log y$ versus $\log x$ (log-log)

We have already seen that a given data set can easily be plotted with various coordinate systems in Excel by first plotting the data as an *x-y* graph (i.e., an XY Chart) and then selecting logarithmic coordinates for one or both axes. Thus, *if the data appear as a straight line when plotted in one of these coordinate systems, we will have a clear indication of what type of trendline to fit to the data.*

Example 9.11 Obtaining a Straight-Line Plot Using Different Coordinate Systems

In Example 9.9, a power function was used to represent a set of data. The selection of the power function was based upon previous familiarity with the process from which the data had been obtained. Verify the desirability of the power function by plotting the data on arithmetic coordinates, semi-log coordinates (both log *y* versus *x* and *x* versus log *y*), and log-log coordinates. Seek a coordinate system that will result in a straight-line relationship.

We begin by plotting the data on ordinary arithmetic coordinates. This is the same plot that was obtained in Example 9.9, when fitting a trendline to the data (see Fig. 9.15). From the arithmetic graph shown in Fig. 9.19, we see that this data set does not plot as a straight line on arithmetic coordinates.

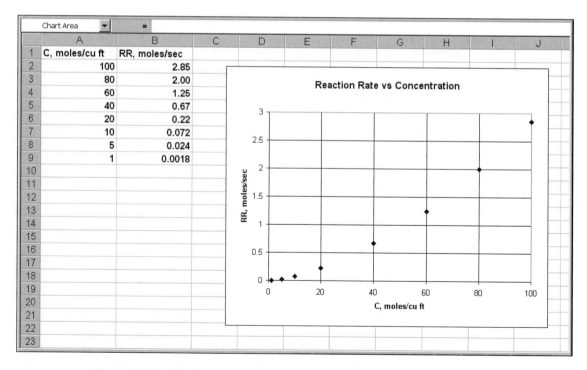

Figure 9.19 – An Excel worksheet with an accompanying *x-y* graph

We then click on the y-axis and choose Selected Axis/Scale/Logarithmic from the Format menu, as described in Chapter 4 (see Sec. 4.5). We will then obtain the semi-log plot shown in Fig. 9.20. Clearly, the data do not plot as a straight-line on semi-logarithmic coordinates. In fact, the departure from a straight-line plot is now much worse than it was before.

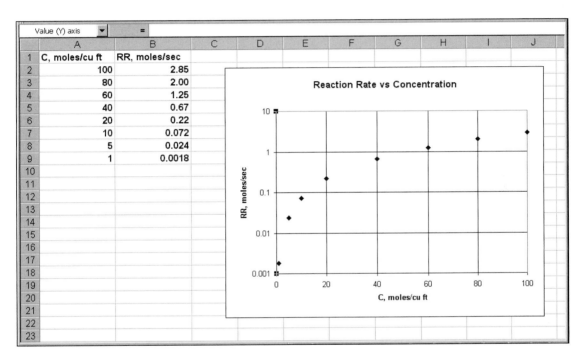

Figure 9.20 – Plotting the data as a semi-log graph

We then restore the arithmetic coordinates along the y-axis, and select a logarithmic scale along the x-axis. This results in the plot shown in Fig. 9.21. The departure from a straight line is now about as bad as in Fig. 9.20, though the curvature of the data is in the opposite direction (i.e., convex rather than concave, as viewed from the upper left).

Finally, we restore the logarithmic coordinates along the y-axis, while retaining the logarithmic coordinates along the x-axis. This results in a log-log graph, as shown in Fig. 9.22. Now the data are represented very accurately as a straight line. (Remember that we are really looking at a plot of log y versus log x, not a plot of y versus x.)

We conclude that the data cannot be represented accurately by the equation for a straight line, an exponential function, or a logarithmic function. The data can, however, be represented accurately by a power function, as originally shown in Example 9.9.

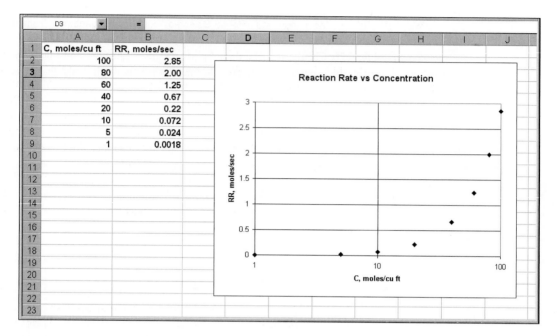

Figure 9.21 – Another type of semi-log graph (logarithmic scale along the *x*-axis)

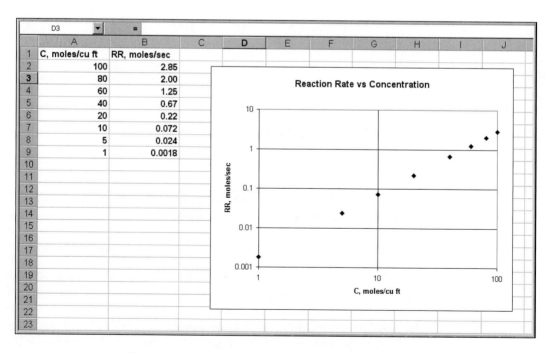

Figure 9.22 – Plotting the data as a log-log graph

Fitting Multiple Functions to a Given Data Set

It sometimes happens that a given data set cannot be plotted as a straight line in any of the commonly used coordinate systems. We may still be able to fit an equation to the data accurately, however, using a polynomial. Moreover, we may also be able to obtain an acceptable, if not excellent, fit using an exponential function, a logarithmic function, or a power function, even though we were not able to plot the data as a straight line using semi-log or log-log coordinates. Thus, we may find that several different functions will represent the data reasonably well. The question then arises as to which function best fits the data.

This question can generally be answered by visual inspection. If several different functions appear to fit the data more or less equally well, however, the sums of the squares of the errors (SSE) and the r^2 values may be helpful in deciding which function fits best.

Example 9.12 Fitting Multiple Functions to a Set of Data

In Example 9.10, we represented some performance data for several popular sports cars with a fifth-degree polynomial. Expand this study by fitting several other commonly used functions to the data. Determine which function best fits the data by comparing their respective SSE and r^2 values.

For comparative purposes, several different functions were passed through the data. The results are summarized in the following table (you will be asked to reproduce these results in Problem 9.21):

Function	Visual Assessment	r^2	SSE
Straight line	Poor	0.8384	312.0
2nd-degree polynomial	Poor	0.965	66.237
Power	Poor	0.9715	55.03
3rd-degree polynomial	Fair	0.9909	17.57
Exponential	Good	0.9922	15.06
4th-degree polynomial	Good	0.9981	3.669
5th-degree polynomial	Excellent	0.9998	0.3862
6th-degree polynomial	Excellent	0.9999	0.1931

In each case, the value of SSE was obtained from Equation (9.8), using a calculated value of 1930.9 for SST. (In the case of the exponential function, it is not actually SSE that is minimized by the method of least squares but a sum of the squares of the errors based upon the logs of y. Similarly, with a power function, the method of least squares minimizes a sum of the squares of the errors based upon the logs of y and x.) Notice that the second-degree polynomial and the power function result in poor fits, despite the fact that their r^2 values are relatively close to 1. Thus, the r^2 values are not always a reliable indicator of the quality of a curve fit.

The functions are listed in the order of increasing quality. Notice that the sums of the squares of the errors (SSE) show much greater variation than the r^2 values, which always remain between 0.8 and 1. Thus, the SSE values are a better predictor of curve-fit quality than the r^2 values. Notice also that the subjective visual assessments largely reinforce the comparisons based upon the SSE values.

Finally, it is interesting to compare the curve fits obtained with a third-degree and a fifth-degree polynomial, since this was a topic of some conjecture in Example 9.10. We see that the fifth-degree polynomial provides a much better fit than the third-degree polynomial. Thus, the argument that the last two terms in the fifth-degree polynomial are insignificant because of their small coefficients is invalid.

Substituting Other Variables for *y* and *x*

Sometimes we seek a linear relationship (i.e., a straight line) when plotting the data. This may be desirable because of some special significance that may be associated with the slope of the line or its intercept. In such situations, it may be helpful to replace one or both of the variables with a related function. For example, it may be helpful to substitute $1/y$ for y, $1/x$ for x, $\sqrt{x}$ for x, and so on, and then fit a straight line to the resulting plot.

Usually, some knowledge of the process being studied will suggest the type of substitution. For example, it is known that the logarithms of chemical reaction rates are proportional to the reciprocal of the absolute temperature. Thus, chemical engineers customarily plot the log of the reaction rate against $1/T$ (actually, they plot the reaction rate versus $1/T$ using semi-log coordinates) and expect to obtain a straight line. Similarly, it is known that the discharge velocity of a fluid leaving a tank is proportional to the square root of the height of the fluid above the discharge point. Thus, we would expect to obtain a straight line by plotting v versus $\sqrt{h}$ or v^2 versus h. A straight-line curve fit could be obtained in either of these situations by plotting the data with the appropriate substitutions.

If nothing is known about the expected behavior of the data based upon underlying principles, it may be helpful to try various variable substitutions using trial and error. Experienced data analysts refer to this as "playing" with the data. The results obtained can be surprisingly effective.

Example 9.13 Variable Substitution

The following data represent the rate at which an oxygenation reaction occurs within a water purification chamber as a function of temperature. The data were obtained by a civil engineer, who must now plot the data and fit an appropriate equation through the data.

Temperature (°K)	Reaction Rate (moles/second)
253	0.12
258	0.17
263	0.24
268	0.34
273	0.48
278	0.66
283	0.91
288	1.22
293	1.64
298	2.17
303	2.84
308	3.70

Figure 9.23 shows a plot of reaction rate (RR) versus temperature. A cubic equation (i.e., a third-degree polynomial) has been passed through the data, resulting in the equation

$$RR = 2.2056 \times 10^{-5} T^3 - 1.7064 T^2 + 4.4148T - 381.80$$

where RR represents the reaction rate, in moles per second, and T represents the absolute temperature, in degrees K. The plot indicates a good fit, as verified by the corresponding r^2 value of 0.99987.

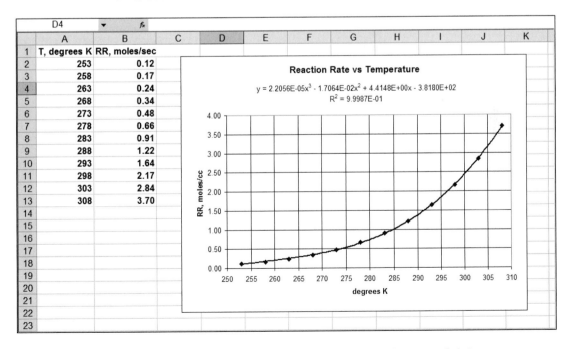

Figure 9.23 – Passing a cubic polynomial through a set of data

Though the cubic equation results in a good fit, it is known that chemical reaction rates generally vary with absolute temperature in accordance with the formula

$$RR = ke^{-E/RT}$$

where E is the *activation energy* and R is a known physical constant. Therefore, it is preferable to plot the reaction rate against the reciprocal of the absolute temperature. Moreover, if a straight line can be obtained on a semi-logarithmic plot, we will know that the exponential form of the equation is valid, and the slope of the line can be used to determine the activation energy.

Figure 9.24 shows a plot of RR versus $1/T$ on semi-logarithmic coordinates. The data clearly plot as a straight line on semi-log coordinates. Again, we see an excellent curve fit, resulting in the equation

$$RR = 2.8667 \times 10^{7} e^{-4887.2/T}$$

with a corresponding r^2 value of 0.99991.

We can now determine the activation energy from the quotient $E/R = 4887.2$ °K. Thus, if the universal constant R has a value of 1.987 (cal)/[(mole)(°K)], the activation energy for this reaction is $4887.2 \times 1.987 = 9710.9$ cal/mole.

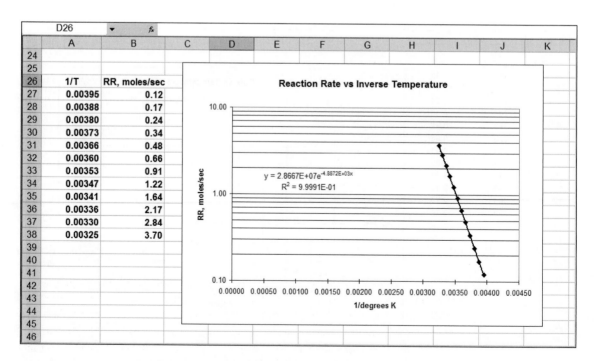

Figure 9.24 – A semi-log plot of reaction rate vs reciprocal of the absolute temperature

Scaling the Data

When fitting a curve to a set of data, it is sometimes helpful to *scale* the data so that the magnitudes of the *y*-values do not differ significantly from the magnitudes of the *x*-values. In such situations, the recommended procedure is to redefine one of the variables so that it is expressed in terms of some multiple of the original unit of measurement. For example, when plotting the velocity of a rocket as a function of time, it might be better to plot the velocity in *thousands of miles* per hour versus time in *seconds*, rather than in *miles* per hour (large numbers) versus time in *hours* (small numbers). This technique may result in a considerably better curve fit, particularly if the original *x*-values differ substantially (i.e., by several orders of magnitude) from the original *y*-values.

Example 9.14 Scaling a Data Set

A group of environmental engineers and scientists have made a prediction of the damage to the ozone layer in the Northern Hemisphere as a function of time, based upon trends related to the unrestricted use of fluorocarbon compounds. The following scenario has been suggested. (The thickness of the ozone layer is expressed in terms of an arbitrary scale ranging from 0 to 1.)

Year	Ozone Layer
1995	1.00
1996	0.97
1997	0.88
1998	0.76
1999	0.63
2000	0.50
2001	0.39
2002	0.30
2003	0.22
2004	0.17
2005	0.13
2006	0.11
2007	0.10

Fit an equation to the data, expressing the ozone layer value as a function of time.

Figures 9.25 and 9.26 show two different Excel worksheets, each containing an *x-y* graph of the data. The first worksheet (Fig. 9.25) contains the data as given, whereas the second worksheet (Fig. 9.26) contains redefined *x*-values, obtained by subtracting 1995 from each of the given values. In each graph, the data points define a distinctive S-shaped curve. A fifth-degree polynomial has been passed through each data set, since S-shaped curves are generally best represented by higher-order polynomials.

We see that both data sets can be represented accurately with a fifth-degree polynomial. In particular, the first data set results in the equation (to three decimals)

$$y = 8.767 \times 10^{-6}x^5 - 8.785 \times 10^{-2}x^4 + 3.521 \times 10^2x^3 - 7.058 \times 10^5x^2 + 7.072 \times 10^8x - 2.835 \times 10^{11}$$

with a value of $r^2 = 1.000$, whereas the second data set results in

$$y = 8.767 \times 10^{-6}x^5 - 4.023 \times 10^{-4}x^4 + 6.975 \times 10^{-3}x^3 - 4.814 \times 10^{-2}x^2 + 1.166 \times 10^{-2}x + 1.000$$

with $r^2 = 1.000$. Each of these r^2 values indicates an excellent fit, though the quality of each fit is readily apparent simply by inspection.

Some spreadsheet programs, including earlier versions of Excel, are unable to obtain a satisfactory fit by passing a polynomial through the first data set. However, the second data set consistently results in an excellent fit. Thus, in some situations, the second data set is much easier to fit.

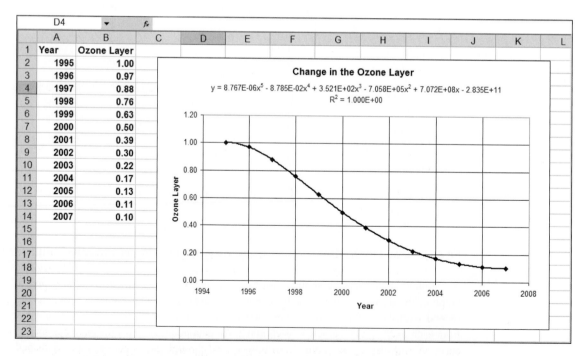

Figure 9.25 – Plotting the data with the given independent variables

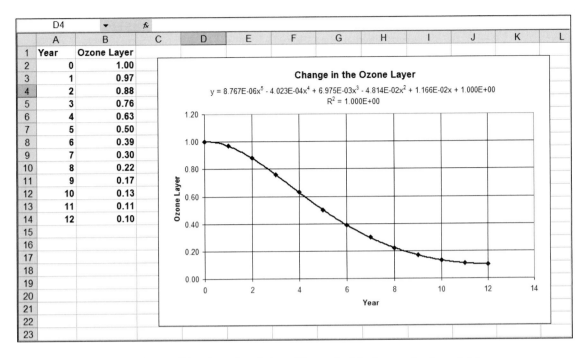

Figure 9.26 – Plotting the data with scaled independent variables

Problems

9.17 Try fitting a quadratic equation (i.e., a second-degree polynomial) and a power function to the data set given in Example 9.13. Compare the results with those given in Example 9.13.

9.18 Plot the data given in Example 9.14 two different ways within an Excel worksheet:

(a) Plot y versus $x/1000$, where y represents the ozone level and x represents the year.

(b) Plot $1000\,y$ versus x.

Fit a fifth-degree polynomial to each data set. Compare each curve fit with the results obtained in the example. Does one method for modifying the data seem better than the other? If so, can you suggest a reason for this behavior?

9.19 Try fitting some other polynomials through the modified data given in Example 9.14 (see Fig. 9.26). What is the lowest-degree polynomial that will fit the data well?

9.20 Use Excel's Regression feature to fit a fifth-degree polynomial to the sports car performance data given in Example 9.10. To do so, enter the x-values in one column, the x^2 values in the next column, the x^3 values in the next column, and so on, up to and including the values of x^5. Finally, enter the y-values in the last column. When you are prompted for the range of x-values within the Regression dialog box (see Fig. 9.10), enter a range that includes the first five columns (the values of x, x^2, x^3, etc.). Compare the outcome with the results obtained in Example 9.10.

9.21 Verify the results in Example 9.12 by fitting each type of function to the original data set. Assess each curve visually and compare your assessment with that given in the example. In addition, verify the value of SSE for each function, using the method described in the example.

9.22 A computer magazine has evaluated several popular desktop computers for price and performance. The cost and performance ratings (in arbitrary units) are listed below. Plot the cost versus performance in an Excel worksheet. Fit the best possible equation through the data.

Unit No.	Cost	Performance	Unit No.	Cost	Performance
1	$1280	72	6	$1370	112
2	1060	86	7	550	94
3	1300	95	8	800	109
4	780	89	9	1000	113
5	1010	99	10	1040	122

Use the resulting curve fit to estimate the following:

(*a*) How much would you expect to pay for a computer with a performance rating of 105?

(*b*) What performance rating would you expect from a computer that cost $2400?

(*c*) What performance rating would you expect from a computer that cost $1500?

9.23 Repeat Prob. 9.22, forcing the resulting trendline through the origin. Does this make a significant difference in the answers to the three questions in Prob. 9.22?

9.24 Several engineering students have built a wind-driven device that generates electricity. The following data have been obtained with the device (repeated from Prob. 4.5):

Wind velocity	Power	Wind velocity	Power
0 mph	0 watts	35 mph	64.5 watts
5	0.26	40	80.2
10	2.8	45	86.8
15	7.0	50	88.0
20	15.8	55	89.2
25	28.2	60	90.3
30	46.7		

Fit an appropriate equation to the data with the intercept set to zero. Use the equation to answer the following questions:

(*a*) How much power will be generated if the wind velocity is 32 mph?

(*b*) What wind velocity will generate 75 watts of power?

9.25 Fit an appropriate equation to the voltage versus time data presented in Prob. 4.1 (the data are repeated here for your convenience). Use the resulting equation to answer the following questions:

(*a*) What voltage would you expect after 1.5 seconds?

(*b*) What voltage would you expect after 15 seconds?

(*c*) How long will it take for the voltage to decline to 1.5 volts?

(*d*) How long will it take for the voltage to reach 0.17 volts?

Time (sec)	Voltage (V)	Time (sec)	Voltage (V)
0	9.8	6	0.6
1	5.9	7	0.4
2	3.9	8	0.3
3	2.1	9	0.2
4	1.0	10	0.1
5	0.8		

9.26 The following data describe the current, in milliamps, passing through an electronic device as a function of time (repeated from Prob. 4.6).

Time (sec)	Current (mA)	Time (sec)	Current (mA)
0	0	9	0.77
1	1.06	10	0.64
2	1.51	12	0.44
3	1.63	14	0.30
4	1.57	16	0.20
5	1.43	18	0.14
6	1.26	20	0.091
7	1.08	25	0.034
8	0.92	30	0.012

(a) Fit the best possible equation to the entire data set.

(b) Try breaking the data set into two segments and fit an equation through each segment. Compare with the results obtained in part (a).

(c) Using the results of parts (a) and (b), estimate the current passing through the device at 0.5 seconds and at 22.8 seconds.

(d) Using the results of parts (a) and (b), estimate the times when the current passing through the device will be exactly 1 milliampere.

9.27 A recent car magazine contains fuel economy and vehicle weight data for a number of new cars. The following table contains the weight, in pounds force, and the corresponding EPA highway mileage figures for several of these vehicles.

Vehicle	Weight (lb)	mpg
1	2684	28
2	2505	31
3	4123	19
4	3410	25
5	2647	29
6	3274	24
7	4407	14
8	3010	24
9	3055	25
10	2860	27
11	4130	18
12	3675	22
13	2375	33
14	3885	20
15	3180	26
16	3780	21
17	2778	32
18	3150	14
19	3425	23
20	2480	41

(a) Enter the data into an Excel worksheet. Then prepare an x-y graph of mpg versus weight. Identify any "outliers" (i.e., any data points that fall outside of the general cluster).

(b) Fit the best possible equation to the entire data set.

(c) Remove the outliers and then fit the best possible equation to the remaining data. Does the removal of the outliers significantly affect the outcome of the curve fit?

(d)   What mileage would you expect from a car weighing 3000 pounds, based upon the results of part (b)? What mileage would you expect based upon the results of part (c)?

9.28   An engineer has determined the shear forces acting upon two beams within a structure. The following data represent the force (in newtons) as a function of distance (in cm) from the left end of each beam.

Distance (cm)	Member A Force (N)	Member B Force (N)
0	0	0
1	3	0.03
2	6	0.20
3	8	0.57
4	9	0.79
5	11	1.15
6	12	1.29
7	14	1.36
8	15	1.60
9	16	1.62
10	18	1.68
12	20	1.93
14	22	2.10
16	24	2.08
18	26	2.17
20	28	2.28
25	33	2.25
30	38	2.39
35	42	2.42
40	46	2.50
50	54	2.47
60	61	2.54
70	68	2.57
80	75	2.62
90	82	2.60
100	88	2.65

(a)   Enter the data into an Excel worksheet in tabular form, as shown above. Leave some extra space so that you can "play" with the data, as described below.

(b)   Plot each data set (i.e., force versus distance) in a manner that will result in a linear (straight-line) relationship. You may need to use some ingenuity to do this. (*Hint*: Try plotting force (F) versus distance (x) on arithmetic coordinates, semi-log coordinates, and log-

log coordinates. You might also try plotting $1/F$ versus x, F versus $1/x$, or $1/F$ versus $1/x$.)

(c) From the graph that yields the best straight-line relationship for each data set, determine an equation for the force as a function of distance. Use the resulting two equations to estimate the shear force at a distance of 45 cm from the left end of each beam.

9.29 The following data describe the sequence of chemical reactions A→B→C. The concentrations of A, B, and C (in moles per liter) are tabulated as a function of time (in seconds).

Time	Conc A	Conc B	Conc C
0	5.0	0.0	0.0
1	4.5	0.46	0.02
2	4.1	0.84	0.06
3	3.7	1.2	0.13
4	3.4	1.4	0.22
5	3.0	1.6	0.33
6	2.7	1.8	0.45
7	2.5	1.9	0.58
8	2.3	2.0	0.72
9	2.0	2.1	0.87
10	1.8	2.2	1.0
12	1.5	2.2	1.3
14	1.2	2.2	1.6
16	1.0	2.1	1.9
18	0.83	2.0	2.2
20	0.68	1.85	2.5
25	0.41	1.5	3.1
30	0.25	1.2	3.5
35	0.15	0.93	3.9
40	0.09	0.71	4.2

(a) Enter the data into an Excel worksheet and determine an equation that will accurately represent concentration as a function of time for each of the substances A, B, and C. Note that a separate equation will be required for each data set.

(b) Using the equations determined in part (a), estimate the concentration of A, B, and C after 23 seconds.

(c) Using the equation obtained in part (a) for concentration of B versus time, determine when the concentration of B will be maximized.

9.30 An engineering student has carried out a series of measurements of tensile force versus elongation for a structural steel cylindrical sample with a diameter of 0.5 inch and a length of four inches. The data cover only the elastic (linear) region. From these measurements, the student has obtained the following table of stress versus strain, where the stress is defined as the force per unit area (in pounds force per square inch, or psi), and the strain is defined as the elongation per unit of original length (dimensionless). The student wishes to determine the *modulus of elasticity* (also known as *Young's modulus*) by measuring the slope of the line representing a plot of stress versus strain. (The modulus of elasticity is a property of the material being tested. It is independent of the dimensions of the cylinder or the magnitude of the forces applied, as long as the forces fall within the elastic region.)

Strain	Stress (psi)	Strain	Stress (psi)
0.1×10^{-3}	3016	0.7×10^{-3}	21200
0.2×10^{-3}	5983	0.8×10^{-3}	24228
0.3×10^{-3}	9191	0.9×10^{-3}	24261
0.4×10^{-3}	12178	1.0×10^{-3}	32205
0.5×10^{-3}	14908	1.1×10^{-3}	30966
0.6×10^{-3}	18292	1.2×10^{-3}	33392

(*a*) Enter the data into an Excel spreadsheet and generate a line graph of stress versus strain.

(*b*) Fit a straight line to the data, with the intercept passing through the origin.

(*c*) Determine the modulus of elasticity for this material from the equation of the straight line.

9.31 The following table presents stress–strain data for the same structural steel sample described in Prob. 9.30. Now, however, the data extend beyond the elastic region to the point of failure (i.e., the point at which the cylindrical sample will pull apart into two separate pieces).

Strain	Stress (psi)	Strain	Stress (psi)
0.02	29737	0.14	57046
0.04	37166	0.16	56593
0.06	44820	0.18	53448
0.08	44074	0.20	52103
0.10	49161	0.22	49185
0.12	53002	0.24	45386

(*a*) Enter the data into an Excel spreadsheet and generate an *x-y* graph of stress versus strain.

(*b*) Fit an appropriate curve to the data.

(*c*) Use the resulting equation to estimate the stress at a strain of 0.115. What is the corresponding elongation?

CHAPTER 10

SOLVING
SINGLE EQUATIONS

Engineers are often required to solve complicated algebraic equations. These equations may represent cause-and-effect relationships between system variables, or they may be the result of applying a physical principle to a specific problem situation. In either case, the result generally provides a detailed understanding of the problem at hand.

For example, the relationship between pressure, volume, and temperature for many real gases can be determined by the *van der Waals equation of state*, which is written as

$$\left(P + \frac{a}{V^2}\right)(V - b) = RT \tag{10.1}$$

where P is the absolute pressure, V is the volume per mole, T is the absolute temperature, R is the ideal gas constant (0.082054 liter atm/mole °K), and a and b are constants that are unique to each particular gas. (Note that Equation (10.1) reduces to the well-known ideal gas equation if a and b both equal zero. Hence, the parameters a and b represent the departure from ideal gas behavior.)

Suppose we are working with a real gas for which a and b are known. If the pressure and volume are specified, it is very easy to solve Equation (10.1) for the temperature. Similarly, it is very easy to solve Equation (10.1) for the pressure if the temperature and volume are specified. It is much more difficult, however, to solve Equation (10.1) for the volume if the temperature and pressure are specified since the equation is *nonlinear* in terms of volume (because of the V^2 term in the denominator on the left-hand side). In fact, the solution involves finding a solution to a cubic equation. Moreover, the procedure may be complicated by the presence of multiple solutions. This may require some judgment in the selection of an appropriate value, based upon the physical interpretation of the problem.

In this chapter, we will see several different approaches to solving nonlinear algebraic equations. We begin by discussing the characteristics of nonlinear equations and showing how approximate solutions can be obtained using graphical techniques. We will then see two well-known mathematical techniques for obtaining detailed numerical solutions. The chapter concludes by showing how to obtain solutions using Excel's automated Goal Seek and Solver features.

10.1 CHARACTERISTICS OF NONLINEAR ALGEBRAIC EQUATIONS

Let's begin by reviewing what is known about algebraic equations and their solutions. The *root* of an algebraic equation is the value of the independent variable that satisfies the equation (i.e., the root of the equation is the solution to the equation). For example, $x = 3$ is a solution to the simple equation $x^2 + x = 12$. Thus, the root of this equation is 3. (Note that this equation also has a second root, whose value is −4.)

An equation is said to be *linear* if it can be rearranged so that the variable representing the unknown quantity appears only to the first power. For example, Ohm's law, written as $V = iR$, is linear with respect to all three variables (V, i, and R). Linear equations have one and only one root, and it is always real. Hence, linear equations are very easy to solve using the techniques of elementary algebra.

On the other hand, an equation is said to be *nonlinear* if it *cannot* be rearranged so that the variable representing the unknown quantity appears only to the first power. Thus, if x represents the unknown quantity, a nonlinear equation might include x raised to some power other than one, or it might include the log of x, the sine of x, etc. Nonlinear equations generally cannot be solved using the techniques of elementary algebra (though there are exceptions, such as the well-known formula for obtaining the roots of a quadratic equation). Hence, they must be solved either graphically or numerically. Moreover, there may be multiple real roots, or there may not be any real roots at all (they may be expressed as complex variables; i.e., $x = u + iv$, where i is an imaginary number, defined as the square root of −1). Clearly, nonlinear equations are much more difficult to solve than linear equations.

A *polynomial* equation is a special case of a nonlinear equation. A polynomial equation can include powers of x but not terms such as log x, sin x, e^x, etc. The highest power of x is called the *degree* of the polynomial (it is also called the *order* of the polynomial). Thus, a quadratic equation is a second-degree polynomial because the highest power of x is 2; a cubic equation is a third-degree polynomial because the highest power of x is 3; and so on.

The following information is known about polynomial equations:

1. An nth-degree polynomial can have no more than n real roots.
2. If the degree of a polynomial is *odd*, there will *always* be *at least one* real root.
3. Complex roots always exist in *pairs*, if they exist at all. Each pair of complex roots consists of *complex conjugates*; that is, $x_1 = u + iv$ and $x_2 = u - iv$.

We will confine our attention to real roots of equations within this book.

Example 10.1 Identifying the Real Roots of a Polynomial

The following equation determines the location of the maximum shear force along a beam, where x represents the distance from the left end of the beam.

$$2x^3 - \frac{5}{x^2} = 3$$

How many real roots might there be for this equation?

To answer this question, let us first rearrange the equation into the form

$$2x^5 - 3x^2 - 5 = 0$$

When written in this form, it is easy to see that the equation is a fifth-degree polynomial. Therefore, there will be at least one real root. If there is *only* one real root, there will be two pairs of complex roots. There may also be three real roots and one pair of complex roots. Or there may be five real roots.

When real roots do exist, we have no assurance that any of them will be positive. They can be negative, or they can equal zero. It is not unusual for multiple real roots to be mixed in sign (some positive, some negative, etc.). In such situations, the physical situation described by the equation generally suggests which root to choose.

Problems

10.1 What can you conclude about the number of real roots for each of the following equations?

(a) $3x + 10 = 0$
(b) $3x^2 + 10 = 0$
(c) $3x^3 + 10 = 0$
(d) $x + \cos x = 1 + \sin x$

10.2 Rearrange the van der Waals equation of state into the form $f(x) = 0$. Once you have completed the rearrangement, what can you conclude about the number of real roots for the van der Waals equation if the volume per mole (V) is considered the unknown quantity?

10.3 In order to determine the temperature distribution within a one-dimensional solid, engineers must often solve the equation

$$x \tan x = c$$

where c is a known positive constant. What can you conclude about the number of real roots for this equation?

10.2 SOLVING EQUATIONS GRAPHICALLY

Approximate graphical solutions are easy to obtain, particularly with a spreadsheet. Basically, the procedure is to write the equation in the form $f(x) = 0$ and then plot $f(x)$ versus x. The points where $f(x)$ crosses the x-axis (i.e., the values of x that cause $f(x)$ to equal 0) are real roots of the equation. Generally, these points can be read directly from the graph. You can also interpolate between the tabulated values to find the point where $f(x) = 0$.

Example 10.2 Solving a Polynomial Equation Graphically

In Example 10.1 we established that the equation

$$f(x) = 2x^5 - 3x^2 - 5 = 0$$

has at least one real root and possibly more. Prepare an Excel worksheet in which $f(x)$ is tabulated as a function of x over the interval $-10 \le x \le 10$. Plot the data as an x-y graph and determine the real roots that fall within this interval.

The worksheet containing the tabulated data and the corresponding x-y graph is shown in Fig. 10.1. Notice the Excel formula that is used to generate the contents of cell B2; that is, =2*A2^5–3*A2^2–5. This formula (with relative cell addressing) was used to generate all of the remaining values in column B. From the tabulated values, we see that the function crosses the x-axis at a point between $x = 1$ and $x = 2$. (Using linear interpolation, we obtain the crossover point as $x = 1.113$.) Hence, the equation appears to have one real root within the specified interval, at approximately $x = 1.113$.

The x-y graph in Fig. 10.1 is not very helpful because the range of the y-values is too large. Therefore, let us retabulate the data over the interval $0 \le x \le 2$ using a finer set of x-values. Figure 10.2 shows a more detailed plot of $f(x)$ vs x over the interval $0 \le x \le 1.6$. Now we can see quite clearly, from both the tabulated data and the graph, that $f(x)$ crosses the x-axis at a point slightly beyond $x = 1.4$.

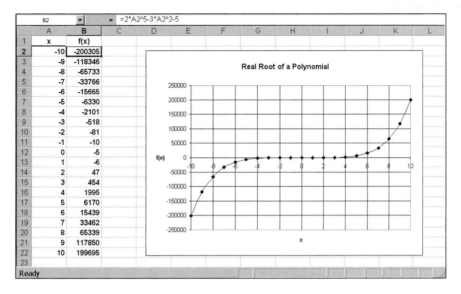

Figure 10.1 – Solving an equation by plotting f(x) vs x

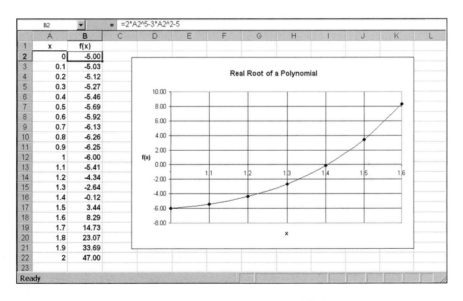

Figure 10.2 – A localized plot of f(x) vs x

We conclude that the given equation has only one real root within the original interval, and its value is approximately $x = 1.4$. If this approximate value (or the value of 1.403, based upon linear interpolation) is sufficiently accurate, the problem is solved. However, we will see how a more accurate value can be obtained later in this chapter, using several different numerical methods.

Problems

10.4 Determine a real root for the equation given in Prob. 10.1(c) using the graphical method described above.

10.5 Determine the positive real roots for the equation given in Prob. 10.1(d) using the graphical method described above.

10.6 The van der Waals constants for carbon dioxide have been determined to be $a = 3.592$ liter^2 atm/mole^2 and $b = 0.04267$ liter/mole. Solve the van der Waals equation of state [Equation (10.1)] for the volume per mole (V) of carbon dioxide under the following conditions:

(a) $P = 1$ atm, $T = 300$ °K

(b) $P = 10$ atm, $T = 400$ °K

(c) $P = 0.1$ atm, $T = 300$ °K

In each case, compare your solution with the volume obtained using the ideal gas law ($PV = RT$). Remember that $R = 0.082054$ liter atm / mole °K.

10.7 The van der Waals constants for an organic compound have been determined as $a = 40.0$ liter^2 atm/mole^2 and $b = 0.2$ liter/mole. Solve the van der Waals equation of state [Equation (10.1)] for the volume per mole (V) of this compound under the following conditions:

(a) $P = 1$ atm, $T = 300$ °K

(b) $P = 10$ atm, $T = 400$ °K

(c) $P = 0.1$ atm, $T = 300$ °K

In each case, compare your solution with the results obtained for carbon dioxide in Prob. 10.6 and the results obtained using the ideal gas law.

10.8 For the equation given in Prob. 10.3, determine

(a)   The three smallest positive roots.

(b)   The two largest negative roots (i.e., the two negative roots closest to the origin).

Assume a value of $c = 2$ in all calculations.

10.3 SOLVING EQUATIONS NUMERICALLY

We now turn our attention to two different mathematical methods that are used to solve nonlinear algebraic equations by successsive approximations. Methods such as these form the basis for the procedures used in Excel's *Goal Seek* and *Solver* features. Thus, these methods provide insight into how Excel goes about solving algebraic equations.

Interval-Reduction Techniques

When using an interval-reduction technique, the basic idea is to determine the approximate location of a root within a specified interval and then eliminate a part of the interval, retaining the portion of the original interval containing the root. This process is repeated several times, until the interval containing the root is very small. The desired root will lie somewhere within this final small interval. (The exact location of the root will not be required if the final interval is sufficiently small.)

One of the most widely used interval reduction techniques is the *method of bisection* (also called the *method of interval halving*). To use this method, *we must begin with an interval containing one and only one real root.* (Thus, a plot of the function $f(x)$ versus x will cross the x-axis once and only once.) An appropriate interval can usually be determined by graphical or tabular analysis, as described in Sec. 10.2. The interval is then divided in half, and the half-interval containing the root is retained. The process is then repeated with the remaining half-interval. The detailed procedure is outlined below.

1. Begin with an interval $a \leq x \leq b$, which is known to contain one real root. Since $f(x)$ will cross the x-axis once within this interval, the value of $f(x)$ at each of the end points will have a different sign; that is, either $f(a) < 0$ and $f(b) > 0$, or $f(a) > 0$ and $f(b) < 0$.

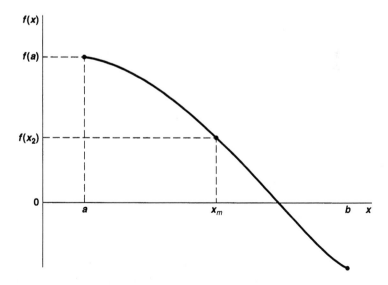

Figure 10.3 – The method of bisection (root falls within right half-interval)

2. Determine the interval midpoint, x_m, as

$$x_m = \frac{a+b}{2} \tag{10.2}$$

3. Evaluate $f(x_m)$ and compare the sign with that of $f(a)$.

(a) If $f(a)$ and $f(x_m)$ both have the same sign, the root must lie within the right half-interval, as illustrated in Fig. 10.3. Then assign the value of x_m to a and repeat steps 2 and 3 using the new interval $a \leq x \leq b$.

(b) If $f(a)$ and $f(x_m)$ differ in sign, the root must lie within the left half-interval, as illustrated in Fig. 10.4. Then assign the value of x_m to b and repeat steps 2 and 3 using the new interval $a \leq x \leq b$.

4. Repeat steps 2 and 3 until a and b are very close together. As a rule, the computation is continued until $(b - a)/(b_0 - a_0) \leq \varepsilon$, where $(b_0 - a_0)$ represents the original interval and ε represents some specified small number.

5. If ε is sufficiently small, the exact location of the root within the final interval is not critical. The root is often assumed to be located at the midpoint. Or, if a more precise result is desired, linear interpolation can be used to determine where the line connecting $f(a)$ and $f(b)$ crosses the x-axis.

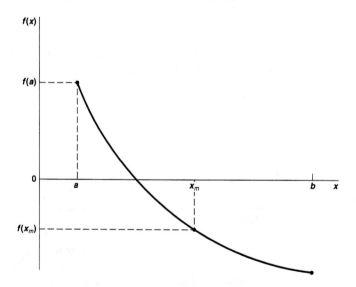

Figure 10.4 – The method of bisection (root falls within left half-interval)

Interval-reduction techniques always result in a valid solution, provided the initial interval is chosen properly. Their disadvantages are the relatively inaccurate final answer and the relatively great amount of computational effort required to obtain a solution.

Example 10.3 Solving a Polynomial Equation Using the Method of Bisection

In Example 10.2, we determined that the equation

$$f(x) = 2x^5 - 3x^2 - 5 = 0$$

has a real root within the interval $1.3 \le x \le 1.5$. Let us now determine a more accurate value for the root, using the method of bisection. We will choose a value of $\varepsilon = 0.01$ as a maximum value for the size of the final interval.

We begin with $a_0 = 1.3$, $b_0 = 1.5$, and $x_m = 1.4$. The corresponding values for $f(x)$ are $f(a_0) = -2.644141$, $f(x_m) = -0.123521$, and $f(b_0) = 3.4375$. Since $f(a_0)$ and $f(x_m)$ are both negative but $f(b_0)$ is positive, we conclude that the root lies in the right half-interval. Therefore, we set $a = 1.4$ and repeat the process within the interval $1.4 \le x \le 1.5$.

The series of eliminations is summarized below. At the end of each calculation, we test the stopping condition; that is, we determine whether or not the ratio of the final to the initial interval, $(b - a) / (b_0 - a_0)$, exceeds the value of $\varepsilon = 0.01$. If so, another elimination is required. The half-interval to be retained for the next calculation is indicated in parentheses.

Interval No.				$(b-a)/(b_0-a_0)$
1	$a = 1.3$	$x_m = 1.4$	$b = 1.5$	1
	$f(a) = -2.644141$	$f(x_m) = -0.123521$	$f(b) = 3.4375$	(*Retain R*)
2	$a = 1.4$	$x_m = 1.45$	$b = 1.5$	0.5
	$f(a) = -0.123521$	$f(x_m) = 1.51197$	$f(b) = 3.4375$	(*Retain L*)
3	$a = 1.4$	$x_m = 1.425$	$b = 1.45$	0.25
	$f(a) = -0.123521$	$f(x_m) = 0.659921$	$f(b) = 1.51197$	(*Retain L*)
4	$a = 1.4$	$x_m = 1.4125$	$b = 1.425$	0.125
	$f(a) = -0.123521$	$f(x_m) = 0.25986$	$f(b) = 0.659921$	(*Retain L*)
5	$a = 1.4$	$x_m = 1.40625$	$b = 1.4125$	0.0625
	$f(a) = -0.123521$	$f(x_m) = 0.066116$	$f(b) = 0.25986$	(*Retain L*)
6	$a = 1.4$	$x_m = 1.403125$	$b = 1.40625$	0.03125
	$f(a) = -0.123521$	$f(x_m) = -0.029211$	$f(b) = 0.066116$	(*Retain R*)
7	$a = 1.403125$	$x_m = 1.4046875$	$b = 1.40625$	0.015625
	$f(a) = -0.029211$	$f(x_m) = 0.018325$	$f(b) = 0.066116$	(*Retain L*)
8	$a = 1.403125$	$x_m = 1.4039062$	$b = 1.4046875$	0.0078125
	$f(a) = -0.029211$	$f(x_m) = -0.005473$	$f(b) = 0.018325$	

Within the last interval, $1.403125 \le x \le 1.4046875$, we see that the interval ratio has been reduced to 0.0078125, which is less than the value of $\varepsilon = 0.01$. Hence, the elimination procedure stops, and we conclude that the desired root falls within this last interval. If we interpolate linearly across the interval, we determine that the desired root is $x = 1.404084$. This value is close to the answer obtained graphically in Example 10.2, though it is more accurate and more precise.

Iterative Techniques

Iterative techniques are another widely used class of methods for solving nonlinear equations. These methods are based upon a trial-and-error strategy that begins with an initial guess and then uses this value to calculate a more refined (i.e., a more accurate) value. The newly calculated value is then used as a guess to obtain another value that is still more refined, and so on. The process continues until the last calculated value is very close to its corresponding guess.

The most widely used iterative technique is the *Newton-Raphson method,* which is based upon the formula

$$x_{i+1} = x_i - \frac{f(x_i)}{f'(x_i)} \tag{10.3}$$

In this expression, x_i represents the assumed value (i.e., the guess) for the ith iteration, x_{i+1} represents the calculated value, $f(x_i)$ represents the given equation evaluated at $x = x_i$, and $f'(x_i)$ represents the first derivative of the equation with respect to x, evaluated at x_i. The process continues until the magnitude of the ratio $(x_i - x_{i-1}) / x_i$ does not exceed some specified value; that is, until

$$\left| (x_i - x_{i-1}) / x_i \right| \le \varepsilon \tag{10.4}$$

This last condition is known as the *stopping condition* or the *convergence criterion.*

To understand the basis for the Newton-Raphson method, let x^* represent the desired root of the equation and let x represent some other point, so that $f(x^*) = 0$ but $f(x) \ne 0$ when $x \ne x^*$. Now let us express $f(x^*)$ in terms of $f(x)$ using the first two terms of a Taylor series:

$$f(x^*) = f(x) + f'(x)(x^* - x) \tag{10.5}$$

We now solve for x^*, making use of the fact that $f(x^*) = 0$. This results in

$$x^* = x - \frac{f(x)}{f'(x)} \tag{10.6}$$

The value of x^* obtained from Equation (10.6) will be only approximate because we have retained only the first two terms of the Taylor series expansion. If we assume, however, that the value of x^* obtained from Equation (10.6) is closer to the desired root than the original value of x, then Equation (10.6) suggests the procedure given by Equation (10.3).

Example 10.4 Solving a Polynomial Equation Using the Newton-Raphson Method

Use the Newton-Raphson method to solve the equation

$$f(x) = 2x^5 - 3x^2 - 5 = 0$$

for a real root within the interval $1.3 \leq x \leq 1.5$. Use the value $x = 1.4$ as a first guess and $\varepsilon = 0.00001$ for the convergence criterion. Compare your answer with the result obtained in Example 10.3.

In order to use the Newton-Raphson method, we must first determine $f'(x)$, the first derivative of the given equation. Applying the rules of calculus, we obtain

$$f'(x) = 10x^4 - 6x$$

Hence, we can write the Newton-Raphson equation as

$$x_{i+1} = x_i - \frac{f(x_i)}{f'(x_i)} = x_i - \frac{2x_i^5 - 3x_i^2 - 5}{10x_i^4 - 6x_i} = \frac{8x_i^5 - 3x_i^2 + 5}{10x_i^4 - 6x_i}$$

Let us refer to the initial guess, $x = 1.4$, as x_0. Then we can write

$$x_1 = \frac{8(1.4)^5 - 3(1.4)^2 + 5}{10(1.4)^4 - 6(1.4)} = 1.404115$$

$$|(x_1 - x_0) / x_1| = |(1.404115 - 1.4) / 1.404115| = 0.002931$$

$$x_2 = \frac{8(1.404115)^5 - 3(1.404115)^2 + 5}{10(1.404115)^4 - 6(1.404115)} = 1.404086$$

$$|(x_2 - x_1) / x_2| = |(1.404115 - 1.404086) / 1.404086| = 0.000021$$

$$x_3 = \frac{8(1.404086)^5 - 3(1.404086)^2 + 5}{10(1.404086) - 6(1.404086)} = 1.404086$$

$$|(x_3 - x_2) / x_3| = |(1.404086 - 1.404086) / 1.404086| = 0$$

Thus, we conclude that the desired root (rounded to seven significant figures) is $x = 1.404086$. This value agrees very closely with the final interpolated value $x = 1.404084$, obtained in Example 10.3 using the method of bisection. Notice, however, that the current solution converged after only three iterations, though seven elimination steps were required to obtain a similar result with the method of bisection.

Although the Newton-Raphson method converged very quickly in the last example, this is not always the case. In fact, in some situations the Newton-Raphson method (like all iterative methods) may actually *diverge*; that is, the successive values of x may move *further away* from the desired root. (Notice what happens to Equation (10.3) when $f'(x_i) = 0$.) In fact, it can be shown that the Newton-Raphson method will converge only if

$$\left| \frac{f(x)f''(x)}{f'(x)^2} \right| < 1 \tag{10.7}$$

in the vicinity of the root. As a rule, the method is much more likely to converge if the initial guess is close to the actual root.

Comparing Methods

It is interesting to compare the characteristics of iterative techniques with those of interval-reduction techniques. Iterative techniques frequently converge to a solution rapidly, though convergence is not guaranteed. Interval-reduction techniques are generally less accurate, though they will always produce a solution provided the initial interval is selected properly (spanning one and only one root).

The Newton-Raphson method and the method of bisection are both implemented frequently on computers. In fact, both methods are often used as exercises in programming courses. As a practical matter, however, it is much easier to use Excel's **Goal Seek** or **Solver** features than to program either method from scratch. We will see how they are used in the following sections of this chapter.

Problems

10.9   For the polynomial equation solved in Example 10.4, evaluate Equation (10.7) at the root ($x = 1.404086$). Is the magnitude less than 1? Might this be related to the rapid convergence experienced in Example 10.4?

10.10 Determine a real root for the equation given in Problem 10.1(c) using

   (a)   the method of bisection.

   (b)   the Newton-Raphson method.

   Compare the results and the computational effort required to obtain each solution. Compare your answers with those obtained in Prob. 10.4.

10.11 Determine the two smallest positive real roots for the equation given in Prob. 10.1(*d*) using

(*a*) the method of bisection.

(*b*) the Newton-Raphson method.

Compare the results and the computational effort required to obtain each solution. Compare your answers with those obtained in Prob. 10.5.

10.12 Solve the van der Waals equation [Equation (10.1)] for the volume per mole (*V*) of an organic compound at 10 atm pressure and 400 °K. The van der Waals constants for this particular compound are $a = 40.0$ liter^2 atm/mole^2 and $b = 0.2$ liter/mole. Solve the equation using

(*a*) the method of bisection.

(*b*) the Newton-Raphson method.

Compare the answers obtained using the two methods and compare the answers with those obtained in Prob. 10.7(*b*).

10.13 Determine the smallest positive root and the largest negative root (the negative root closest to the origin) for the following equation, originally given in Prob. 10.8:

$$x \tan x = 2$$

Solve the equation using

(*a*) the method of bisection.

(*b*) the Newton-Raphson method.

Compare the answers obtained using the two methods, and compare the answers with the corresponding answers obtained in Prob. 10.8.

10.4 SOLVING EQUATIONS IN EXCEL USING GOAL SEEK

The material in the previous three sections was intended to provide you with a general understanding of the characteristics of single algebraic equations, and to introduce you to some classical methods for determining numerical solutions. Now let us see how numerical solutions can be obtained quickly and easily using Excel's Goal Seek feature. This feature permits rapid solutions of algebraic equations using numerical procedures similar to those described in Sec. 10.3.

Remember that these methods are based upon a series of successive approximations derived from an initial guess. The accuracy of each successive approximation is refined until the values *converge* to a stable solution. You should understand, however, that iterative calculations do not always converge to a stable solution. Moreover, in the case of equations that possess multiple roots, the calculations may converge to the wrong root. Thus, there is no assurance that a converged solution provides a meaningful interpretation of a given physical problem.

To solve a single algebraic equation with Goal Seek, proceed as follows:

1. Enter an initial guess in one of the cells on the worksheet.

2. Enter a formula for the equation, in the form $f(x) = 0$, in another cell. Within this formula, express the unknown quantity x as the cell address containing the initial guess.

3. Select Goal Seek from the Tools menu.

4. When the Goal Seek dialog box appears, enter the following information:

    (*a*) The address of the cell containing the formula in the Set cell entry location.

    (*b*) The value 0 in the To value entry location.

    (*c*) The address of the cell containing the initial value in the entry location labeled By changing cell. Then select OK.

A new dialog box labeled Goal Seek Status will then appear, telling you whether or not Excel has been able to solve the problem (i.e., whether convergence has been obtained). If a solution has been obtained, the value of the root will appear in the cell originally containing the initial guess. The value in the cell containing the formula will show a value that is close to zero (but usually not exactly zero). This last value will also appear within the Goal Seek Status dialog box.

The likelihood of obtaining a converged solution will be enhanced if the initial guess is as close as possible to the desired root. Thus, we would like to use an approximate value for the root as the initial guess. This value can often be established using the graphical or tabular techniques described in Sec. 10.2.

A warning message will be generated if the computation does not converge or if an inappropriate value (such as the square root of a negative number) is about to be generated during the course of the computation. You may also receive an error message indicating a *circular reference* if the formula for the given equation is not entered into Excel correctly, as outlined in steps 1-4 above (see Sec. 2.6 for more information about circular references). The following example illustrates the correct use of the Goal Seek feature.

Example 10.5 Solving a Polynomial Equation in Excel Using Goal Seek

In Example 10.2, we found that the equation

$$f(x) = 2x^5 - 3x^2 - 5 = 0$$

has a real root at approximately $x = 1.4$. Let us now make use of Excel's Goal Seek feature to obtain a more accurate solution. We will select the value $x = 1.4$ as a first guess, since we already know that this is the approximate value of the root.

Figure 10.5 shows an Excel worksheet with the initial guess (1.4) entered into cell B3, and the formula for $f(x)$ entered into cell B5. (Notice the Excel formula for $f(x)$ shown in the formula bar.) The numerical value -0.12352 appearing in cell B5 is the value of $f(x)$ evaluated at $x = 1.4$.

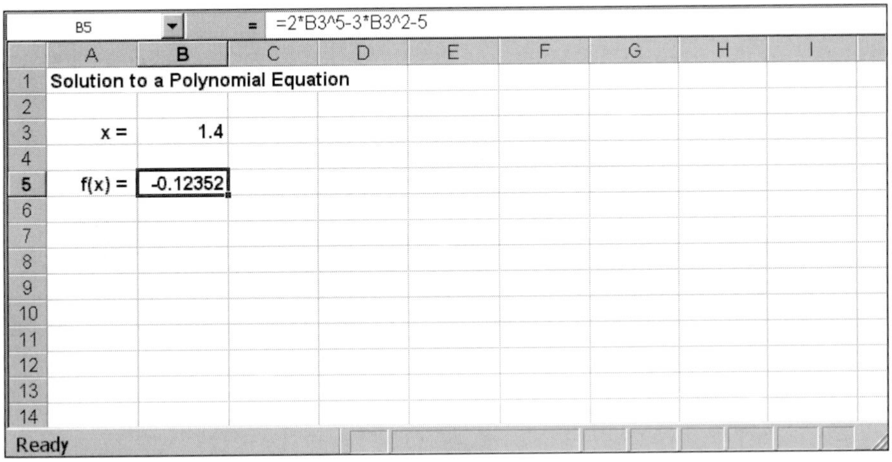

Figure 10.5 – Preparing to solve an algebraic equation with Goal Seek

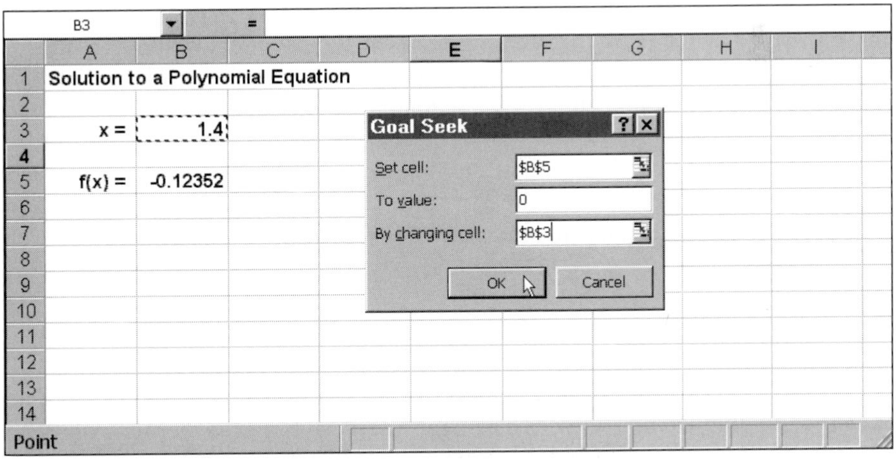

Figure 10.6 – The Goal Seek dialog box

Our goal, of course, is to determine the value of x that causes $f(x)$ to equal zero. To do so, we select Goal Seek from the Tools menu. Figure 10.6 shows the resulting Goal Seek dialog box. The information that has been entered into the dialog box indicates that the value of $f(x)$ in cell B5 will be driven to zero (or approximately zero) by changing the value of x in cell B3.

The resulting solution is shown in Fig. 10.7. The Goal Seek Status dialog box, which has replaced the previous dialog box, indicates that a solution has been found that drives the value of the formula to -3.5625×10^{-5} (approximately zero). The corresponding value of x, which is the desired solution, now appears in cell B3. Thus, Excel has found a solution of $x = 1.404085$. This value is consistent with the approximate solution obtained earlier. The corresponding value of $f(x)$, $-3.6E-5$, appears in cell B5. This is simply a duplication (rounded) of the value shown in the Goal Seek Status dialog box.

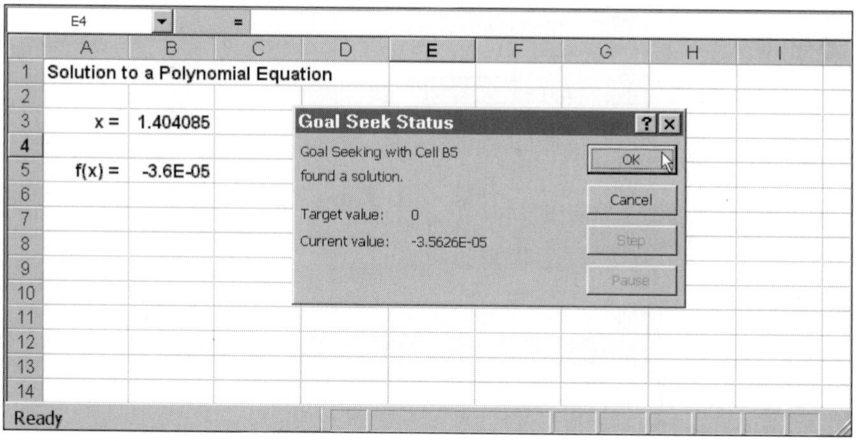

Figure 10.7 – The final solution obtained using Goal Seek

Microsoft does not clearly explain how Goal Seek goes about finding a solution. However, a technical memorandum states that it uses a "simple linear search" to explore beyond the initial guess. The search direction is based upon the outcome of explorations in the neighborhood of the initial guess. If the linear search is not successful, Goal Seek utilizes an unspecified iterative process to find a solution. An error message is generated if a converged solution is not obtained after an unspecified number of iterations.

Example 10.6 Convergence Considerations

Sometimes a great deal can be learned by studying a problem whose solution is well known. Thus, let us use Goal Seek to determine the roots of the equation

$$\tan(x) = 0$$

within the interval $0 \leq x \leq \pi$. Anyone who has ever taken a course in trigonometry should recognize that this equation has roots at $x = 0$ and $x = \pi$, and a discontinuity at $x = \pi/2$.

Figure 10.8 shows an Excel worksheet in which tan(x) is tabulated and plotted as a function of x within the specified interval. The graph verifies that the equation has two roots, one at the origin and the other at $x = \pi$, with a discontinuity at $x = \pi/2$ (approximately $x = 1.5708$). The graph does not show the location of the roots accurately, however, because of the relatively large interval spanned by the independent variable x.

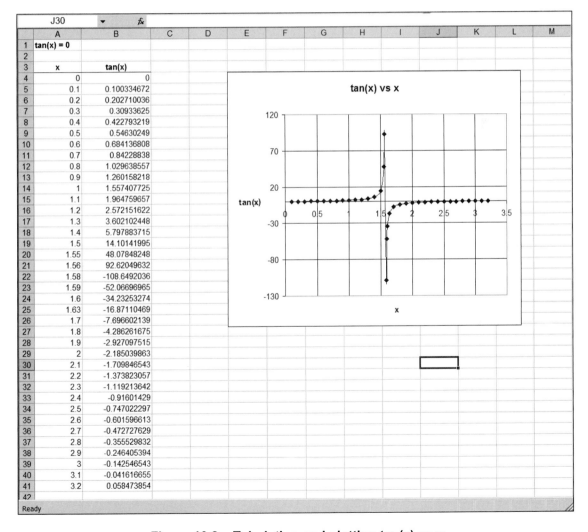

Figure 10.8 – Tabulating and plotting tan(*x*) vs *x*

Let us first use Goal Seek to find a root of the equation using $x = 1$ as a starting point, as shown in Fig 10.9(a). Goal Seek finds a root at approximately $x = 0$, as shown in Fig. 10.9(b). (Goal Seek reports the root at $x = 3.0112 \times 10^{-6}$ rather than $x = 0$ because of numerical roundoff.) Intuitively this is the root we would expect to find, since the starting point is closer $x = 0$ than $x = \pi$. There is no guarantee, however, that Goal Seek will converge to this particular root.

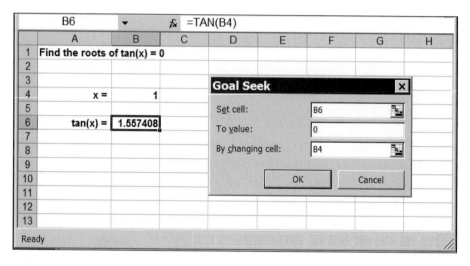

Figure 10.9(a) – Seeking a root of tan(x) = 0, beginning at x = 1

Now let us repeat the process, beginning at $x = 2$. To do so, we enter the value 2 in cell B4, as shown in Fig 10.10(a). The solution found by Goal Seek is approximately $x = 3.1408$, as shown in Fig. 10.10(b). This value is very close to the exact solution at $x = \pi$.

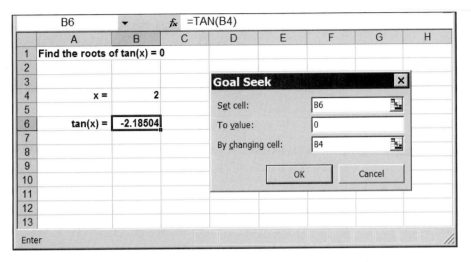

Figure 10.10(a) – Seeking a root of tan(x) = 0, beginning at x = 2

Figure 10.10(b) – Goal Seek finds a root at (approximately) x = π

Finally, let us attempt a solution with Goal Seek beginning at $x = 1.57$, which is slightly below the location of the discontinuity. (This is admittedly a bad choice, though it does illustrate a point.) Figure 10.11(a) shows the initial setup. Goal Seek is unable to find a solution from this starting point. Figure 10.11(b) shows the resulting error message, warning that Goal Seek "may not have found a solution." Note that the value of tan(x) shown in cell B6 is not close to zero, indicating a lack of convergence.

Excel includes another feature, called Solver, which may be able to obtain a converged solution when Goal Seek cannot. Solver is entirely independent of Goal Seek. The use of Solver is discussed in the next section.

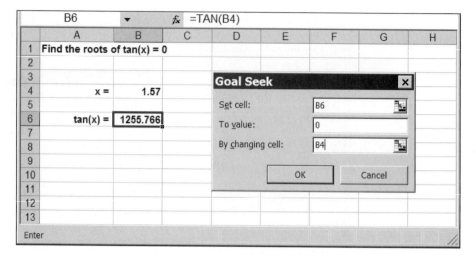

Figure 10.11(a) – Seeking a root of tan(x) = 0, beginning at x = 1.57

Figure 10.11(b) – Goal Seek does not converge from this starting point

10.5 SOLVING EQUATIONS IN EXCEL USING SOLVER

Single algebraic equations can also be solved using Excel's Solver feature, though Solver is intended primarily for solving constrained optimization problems (see Chapter 16). The advantage to using Solver is that you can specify constraints (i.e., auxiliary conditions) on the independent variable. For example, you can restrict the independent variable to nonnegative values or require that the independent variable falls within prescribed bounds. You can also specify a convergence criterion, rather than accept the convergence criterion that is built into Goal Seek. Otherwise, however, the procedure for using Solver is very similar to that used with Goal Seek.

Solver is an Excel add-in, like the Analysis Toolpak that we used to generate histograms (see Chapter 8). To install Solver, choose Add-Ins from the Tools menu. Then select Solver Add-In from the resulting Add-Ins dialog box. Once the Solver feature has been installed, it will remain installed unless it is removed, by reversing the above procedure.

To solve a single equation with Solver, proceed as follows:

1. Enter an initial guess in one of the cells on the worksheet.

2. Enter an equation for $f(x)$ in another cell, expressed as an Excel formula. Within this formula, express the unknown quantity x as the address of the cell containing the initial guess.

3. Select Solver from the Tools menu.

4. When the Solver Parameters dialog box appears, enter the following information:

    (*a*) The address of the cell containing the formula for $f(x)$ in the Set Target Cell location.

    (*b*) Select Value of in the Equal to line. Then enter 0 within the associated data area. (In other words, set the target cell equal to a value of zero.)

    (*c*) Enter the address of the cell containing the initial value for x in the area labeled By Changing Cells.

    (*d*) If you wish to restrict the range of x, click on the Add button under the heading Subject to the Constraints. Then provide the following information within the Add Constraint dialog box:

        (*i*) The address of the cell containing the initial value for x in the Cell Reference location.

        (*ii*) Specify the type of constraint (i.e., $\leq$ or $\geq$) from the pull-down menu.

        (*iii*) Enter the bounding value in the Constraint data area. Then select OK to return to the Solver Parameters dialog box.

        Note that you can change or delete a constraint after it has been added.

    (*e*) To change the convergence criterion, click on Options within the Solver Parameters dialog box. Then enter a numerical value for the convergence criterion in the designated location.

5. When all of the required information has been entered correctly, select Solve. This will initiate the actual solution procedure. When the computation is finished, a new dialog box labeled Solver Results will appear, telling you

whether or not Solver has been able to solve the problem (i.e., whether convergence has been obtained). If a solution has been obtained, the value of the root will appear in the cell originally containing the initial guess. The cell containing the formula will show a value that is zero or very close to zero.

Like Goal Seek, convergence is not guaranteed. Hence, Solver will generate an error message if the computation does not converge, or if a mathematical error (e.g., an attempt to calculate the square root of a negative number) is detected during the course of the computation. Solver will also generate an error message if it detects a circular reference in the formula for $f(x)$.

Since Solver and Goal Seek use different mathematical solution procedures, the values obtained with Solver may not be identical to those obtained using Goal Seek, though they should be very close. Remember that both features arrive at their respective solutions through a series of refined approximations rather than by providing exact answers.

Example 10.7 Solving a Polynomial Equation in Excel Using Solver

In several previous examples we solved the polynomial equation

$$f(x) = 2x^5 - 3x^2 - 5 = 0$$

and found a positive real root in the vicinity of $x = 1.4$. Let us now use Solver to find a root of this equation, with the added restriction that $x \geq 0$. Use a convergence criterion of 10^{-5} (i.e., 0.00001).

Figure 10.12 shows an Excel worksheet containing an initial guess in cell B3 and a formula for the given equation in cell B5, as before. We also see the Solver Parameters dialog box, which was obtained by selecting Solver from the Tools menu. The cell address of the equation (B5) has been entered into the Set Target Cell area. Notice the selection requiring that the value of the target cell be set equal to zero. The cell address containing the initial guess (B3) has also been entered into the By Changing Cells area.

To add the nonnegativity restriction, we select Add near the bottom of the Solver Parameters dialog box, resulting in the Add Constraint dialog box shown in Fig. 10.13. We then enter the cell address of the independent variable (B3) in the left data area, we select the type of constraint ($\geq$) in the center data area, and we provide the bounding value (0) in the right data area. Once the constraint has been entered correctly, we select OK at the bottom of the dialog box, which returns us to the Solver Parameters dialog box, as shown in Fig. 10.14. The entire problem, including the constraint specification, now appears at the bottom of the Solver Parameters dialog box.

Before initiating the solution we must first set the specified convergence criterion, overriding the default value of 0.001. Thus, we click on the Options button in the Solver Parameters dialog box and enter the desired value (0.00001) in the Convergence area of the Options dialog box, as shown in Fig. 10.15.

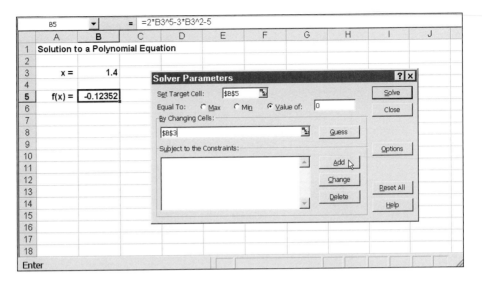

Figure 10.12 – The Solver Parameters dialog box

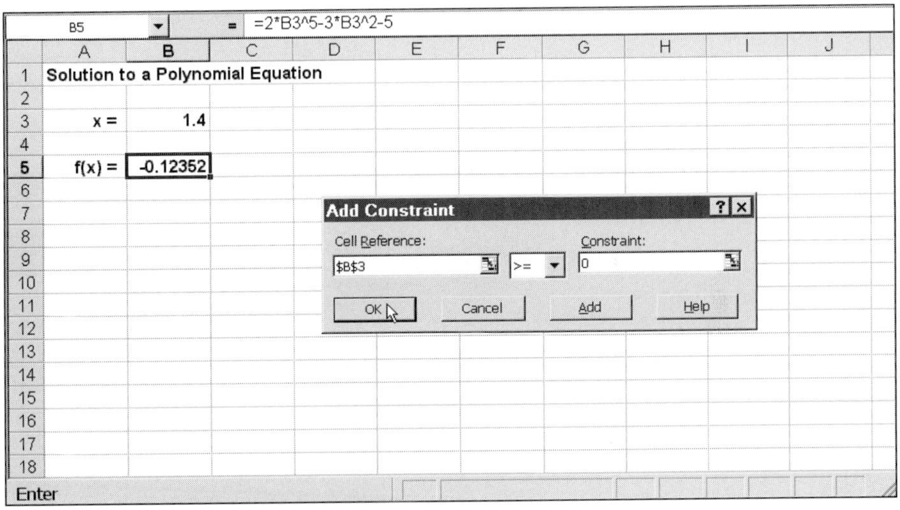

Figure 10.13 – Adding a constraint

We are now ready to proceed with the solution. To do so, we select Solve from the Solver Parameters dialog box. Once a solution has been obtained, the resulting values appear in cells B3 and B5, as shown in Fig. 10.16. Thus, we see that the desired solution is $x = 1.404086$, resulting in a value of $f(x) = 3.59 \times 10^{-8}$. This is very close to the values obtained in the earlier examples.

Note that the Solver Results dialog box replaces the Solver Parameters dialog box once the solution has been obtined. Our response to the Solver Results dialog box is Keep Solver Solution followed by OK. Neither the Reports nor the Save Scenario feature should be selected, since we already have the information being sought.

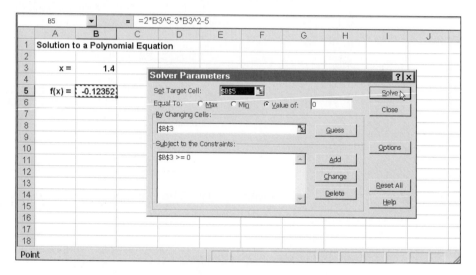

Figure 10.14 – The completed Solver Parameters dialog box

Solver Options

Max Time:	100 seconds	OK
Iterations:	100	Cancel
Precision:	0.000001	Load Model...
Tolerance:	5 %	Save Model...
Convergence:	0.00001	Help

☐ Assume Linear Model   ☐ Use Automatic Scaling
☐ Assume Non-Negative   ☐ Show Iteration Results

Estimates
  ◉ Tangent
  ○ Quadratic

Derivatives
  ◉ Forward
  ○ Central

Search
  ◉ Newton
  ○ Conjugate

Figure 10.15 –The Solver Options dialog box

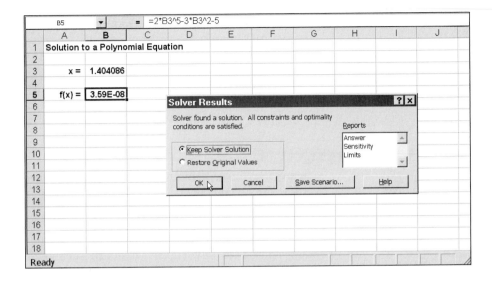

Figure 10.16 – The final solution obtained using Solver

Solver uses a sophisticated strategy known as the *generalized reduced gradient method* for minimizing mathematical functions. Its use to drive $f(x)$ to zero in order to solve a nonlinear algebraic equation is a special case of this powerful computational procedure. We will explain how Solver is used to solve a more complicated class of problems in Chapter 16.

Problems

Solve each of the following problems in Excel using either Goal Seek or Solver.

10.14 Determine a real root for the equation given in Prob. 10.1(*c*):

$$3x^3 + 10 = 0$$

10.15 Determine the smallest positive real root for the equation given in Prob. 10.1(*d*):

$$x + \cos x = 1 + \sin x$$

10.16 Solve the van der Waals equation [Equation (10.1)] for the volume per mole (*V*) of an organic compound at 10 atm pressure and 400 °K. The van der Waals constants for this particular compound are $a = 40.0$ liter^2 atm/mole^2 and $b = 0.2$ liter/mole. Compare your answer with the results obtained in Prob. 10.7(*b*).

10.17 Determine the smallest positive root and the largest negative root (the negative root closest to the origin) for the equation originally given in Prob. 10.3, with $c = 2$:

$$x \tan x = 2$$

Compare your answers with the results obtained in Prob. 10.8.

10.18 An angular support member within a bridge is subject to a coaxial compressive force. If the horizontal force component is 120 lb_f and the vertical force component is 175 lb_f, what angle does the member make with the horizontal?

10.19 Many consumers buy expensive items, such as cars and houses, by borrowing the purchase cost from a bank and then repaying the loan on a constant monthly basis. If P is the total amount of money that is borrowed initially, the amount of the monthly payment, A, can be determined from the formula

$$A = P \left[\frac{i(1+i)^n}{(1+i)^n - 1} \right]$$

where i is the monthly interest rate, expressed as a *fraction* (not a percentage), and n is the total number of payments.
    Suppose you borrow $10,000 to buy a car.

(*a*)   If the nominal interest rate is 8 percent API (which is equivalent to a fractional monthly interest rate of $0.08/12 = 0.006667$) and you borrow the money for 36 months, how much money will you have to repay each month?

(*b*)   If you are required to repay $350 each month for 36 months, what is the corresponding monthly interest rate?

(*c*)   If you choose to repay $350 each month at the 0.006667 monthly interest rate, how many payments (i.e., how many months) will be required to repay the loan?

(*d*)   Do all of the above questions require the use of the Goal Seek feature in Excel?

10.20 An equation that sometimes arises in the design of a structure is

$$\tanh x = \tan x + 1$$

where tanh x is the hyperbolic tangent of x and x represents the angle (in radians, measured from the horizontal) at which an external force must be applied in order to distribute the force properly. Determine the first positive root of this equation; that is, the value of x within the interval $0 < x < \pi$, that satisfies this equation. Use Excel's TAN and TANH functions to obtain your solution.

10.21 A horizontal beam of length L is supported at each end ($x = 0$ and $x = L$), as shown in Fig. 10.17. Suppose an external vertical force F is applied to the beam at location a, where a is measured from the left end of the beam (from Fig. 10.17, note that $a + b = L$). Then the vertical deflection of the beam at some point x, $x \le a$, is determined by the expression

$$y = \frac{Fbx}{6EIL}(L^2 - x^2 - b^2), \qquad 0 \le x \le a$$

where y is the vertical deflection, E is the *modulus of elasticity* (also known as *Young's modulus*), and I is the *moment of inertia* of the cross-sectional area of the beam. (Note that E depends only on the beam material, and I depends only on the beam geometry.)

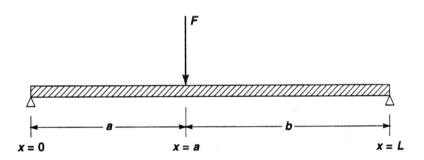

Figure 10.17 – A horizontal beam with a point load

Suppose a 10-ft beam is made of steel ($E = 30 \times 10^6$ psi), and the moment of inertia is $I = 5$ in^4. If a vertical force of 15,000 lb$_f$ is applied at $a = 4$ ft, determine

(a) The maximum deflection, which occurs at $x = \sqrt{(L^2 - b^2)/3}$.

(b) The location to the left of the vertical force (solve for $x < a$) at which the deflection is 0.75 times the maximum deflection.

Remember to use consistent units (convert ft to inches).

10.22 Four-bar linkages are commonly used to transmit mechanical movement in various ways. A typical four-bar linkage is shown in Fig. 10.18. Member AD is anchored to its supporting base and does not move. Member AB rotates about point A, resulting in a reciprocating movement of members BC and CD. Therefore, the angle θ_2 changes in response to the change in θ_1.

$$AD = AB\cos\theta_1 - CD\cos\theta_2 + \sqrt{(BC)^2 - (CD\sin\theta_2 - AB\sin\theta_1)^2}$$

Suppose $AB = 1.5$ in, $BC = 6$ in, $CD = 3.5$ in, and $AD = 5$ in.

(a)   Determine the value of θ_2 corresponding to $\theta_1 = 45°$.

(b)   By plotting several additional values of θ_2 against θ_1, determine the maximum value of θ_2 for this configuration.

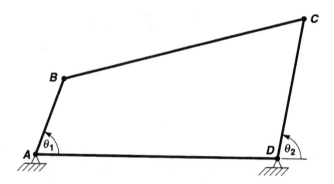

Figure 10.18 – A four-bar linkage

10.23 Chemical, civil, and mechanical engineers must often determine the power required to pump a fluid through a series of pipes. To do so, we must first determine a *friction factor*. If the fluid flow is turbulent and the pipes are smooth, the friction factor can be determined as

$$1.1513 / \sqrt{f} = \ln(\mathrm{Re}\ \sqrt{f} / 2.51)$$

In this equation f represents the friction factor, a positive value that does not exceed 0.1, and Re represents the *Reynolds number*, which is a function of the fluid properties, the pipe diameter, and the nominal fluid velocity. The Reynolds number increases with increasing fluid velocity. The flow becomes turbulent at higher velocities, when the Reynolds number exceeds approximately 3000. Hence the above equation is valid only when Re exceeds 3000.

Determine the values of the friction factor corresponding to Reynolds numbers of 5000, 12,000, and 20,000. Does the friction factor increase or decrease as the Reynolds number increases?

10.24 Mechanical engineers often study the movement of damped oscillating objects. Such studies play an important role in the design of automobile suspensions, among other things.

The horizontal displacement of an object is given as a function of time by the equation

$$x = x_0 e^{-\beta t}\left[\cos(\omega t) + (\beta/\omega)\sin(\omega t)\right]$$

The parameters β and ω depend upon the mass of the object and the dynamic characteristics of the system (i.e., the spring constant and the damping constant).

For a given system, suppose $x_0 = 8$ in, $\beta = 0.1$ s^{-1}, and $\omega = 0.5$ s^{-1}. Carry out the following calculations for this system:

(a) Plot x versus t over the interval $0 \leq t \leq 30$ seconds.

(b) Determine, as accurately as possible, the point where x first crosses the t-axis (i.e., determine the value of t corresponding to $x = 0$).

(c) Determine, as accurately as possible, the point where x crosses the t-axis for the second time.

(d) If x_0 and β retain their original values, what value of ω will cause x to cross the t-axis for the first time at $t = 2.5$ seconds?

10.25 Electrical engineers often study the transient behavior of RLC circuits (containing resistors, inductors, and capacitors). Such studies play an important role in the design of various electronic components.

Suppose an open RLC circuit has an initial charge of q_0 stored within a capacitor. When the circuit is closed (by throwing a switch), the charge flows out of the capacitor and through the circuit in accordance with the equation

$$q = q_0 e^{-Rt/(2L)} \cos\left[\sqrt{\frac{1}{LC} - \left(\frac{R}{2L}\right)^2}\, t\right]$$

In this equation, R is the resistive value of the circuit, L is the value of the inductance, and C is the value of the capacitance.

For a particular circuit, suppose $q_0 = 10^3$ coulombs, $R = 0.5 \times 10^3$ ohms, $L = 10$ henries, and $C = 10^{-4}$ farads. Carry out the following calculations for this system:

(a) Plot q versus t over the interval $0 \le t \le 0.5$ seconds.

(b) Determine, as accurately as possible, the point where q first crosses the t-axis (i.e., determine the value of t corresponding to $q = 0$).

(c) Determine, as accurately as possible, the point where q crosses the t-axis for the second time.

(d) If q_0, L, and C retain their original values, what value of R will cause q to cross the t-axis for the first time at $t = 0.06$ second?

Notice the similarity between this problem and the vibrating mass described in Prob. 10.24.

CHAPTER 11

SOLVING SIMULTANEOUS EQUATIONS

Simultaneous algebraic equations arise in many different engineering applications. We have already seen one such application, when fitting a curve to a set of data using the method of least squares (see Sec. 9.3). There are many other applications in such diverse fields as heat conduction, solid mechanics, electrical circuits, molecular diffusion, and fluid mechanics. In fact, the solution of simultaneous algebraic equations is one of the most commonly used mathematical solution techniques.

The systems of simultaneous equations that arise in engineering can be either *linear* or *nonlinear*. In a system of *linear* equations, none of the unknown quantities are raised to a power or appear as arguments within a trigonometric function, a logarithmic function, a square root, and so on; thus, the effect of each unknown quantity is one of direct proportionality. Therefore, linear equations are easier to solve than nonlinear equations. Moreover, the techniques used to solve systems of linear equations are different than the techniques used for nonlinear equations.

We will examine both linear and nonlinear systems of equations within this chapter. We begin by describing matrix notation and its application to simultaneous linear equations. We then explain how these concepts can be used to solve systems of linear algebraic equations in Excel. Finally, we will see another approach to the solution of simultaneous algebraic equations, based upon the use of Excel's Solver feature, that can be used with either linear or nonlinear systems of equations.

Before launching into matrix notation and its associated solution techniques, however, let us see how a system of simultaneous equations can be written in general terms using traditional algebraic notation. Suppose we have a system of n simultaneous linear algebraic equations in n unknowns. Using traditional notation, we would typically write the equations in the following manner:

$$a_{11}x_1 + a_{12}x_2 + a_{13}x_3 + \cdots + a_{1n}x_n = b_1$$
$$a_{21}x_1 + a_{22}x_2 + a_{23}x_3 + \cdots + a_{2n}x_n = b_2 \qquad (11.1)$$
$$\cdots\cdots$$
$$a_{n1}x_1 + a_{n2}x_2 + a_{n3}x_3 + \cdots + a_{nn}x_n = b_n$$

Within these equations, the a_{ij} and b_i terms represent known values, whereas the x_js represent the unknown quantities. Notice that the coefficients a_{ij} have double subscripts. The first subscript (i) represents the equation number (i.e., the row number), whereas the second subscript (j) represents the number of the unknown quantity (i.e., the column number).

11.1 MATRIX NOTATION

A *matrix* is simply a two-dimensional array of numbers. The elements of a matrix are characterized by a row number and a column number. A matrix that contains only one column is called a *vector*. Thus, the system of equations given by (11.1) becomes, in matrix notation,

$$AX = B \qquad (11.2)$$

where the matrix A and the vectors X and B are written as

$$A = \begin{bmatrix} a_{11} & a_{12} & \cdots & a_{1n} \\ a_{21} & a_{22} & \cdots & a_{2n} \\ \cdots & \cdots & \cdots & \cdots \\ a_{n1} & a_{n2} & \cdots & a_{nn} \end{bmatrix} \qquad X = \begin{bmatrix} x_1 \\ x_2 \\ \cdots \\ x_n \end{bmatrix} \qquad B = \begin{bmatrix} b_1 \\ b_2 \\ \cdots \\ b_n \end{bmatrix} \qquad (11.3)$$

The advantage to matrix notation is that entire matrices or vectors can be manipulated collectively using a single symbolic notation. We shall see that this simplifies the way we view the process for solving a system of simultaneous equations.

Example 11.1 Writing a System of Simultaneous Equations in Matrix Form

Write the following system of equations in matrix form:

$$2x_1 + 3x_2 = 8$$
$$4x_1 - 3x_2 = -2$$

From Equation (11.3), we can write the above equations in matrix form as

$$\begin{bmatrix} 2 & 3 \\ 4 & -3 \end{bmatrix} \begin{bmatrix} x_1 \\ x_2 \end{bmatrix} = \begin{bmatrix} 8 \\ -2 \end{bmatrix}$$

or simply

$$AX = B$$

where

$$A = \begin{bmatrix} 2 & 3 \\ 4 & -3 \end{bmatrix} \qquad X = \begin{bmatrix} x_1 \\ x_2 \end{bmatrix} \qquad B = \begin{bmatrix} 8 \\ -2 \end{bmatrix}$$

Matrix Multiplication

Two matrices can be multiplied by one another, resulting in a product matrix having the same number of rows as the first matrix multiplicand and the same number of columns as the second. The first multiplicand must have the same number of columns as the second multiplicand has rows.

Suppose A is an $m \times n$ matrix (m rows and n columns) and B is an $n \times p$ matrix. Note that A has n columns and B has n rows, as required. Then the matrix product $AB = C$ is defined on an element-by-element basis in the following manner:

$$\sum_{j=1}^{n} a_{ij} b_{jk} = c_{ik} \qquad i = 1, 2, \ldots, m \qquad k = 1, 2, \ldots, p \qquad (11.4)$$

The resulting C matrix will contain m rows and p columns.

Matrix multiplication is not commutative; that is, the product BA is generally not the same as the product AB. In fact, *the product BA will not even be defined unless the number of columns of B is the same as the number of rows of A*. If A and B are both *square $n \times n$ matrices*, however, we can form either product $C1 = AB$ or $C2 = BA$, though $C1$ will generally differ from $C2$.

Matrix multiplication is really not as complicated as it appears. The rule is work *across* (by columns) in A and *down* (by rows) in X.

Example 11.2 Matrix Multiplication

Suppose A is a 2×3 matrix and B is a 3×2 matrix, where each matrix is defined as

$$A = \begin{bmatrix} 1 & 2 & 3 \\ 4 & 5 & 6 \end{bmatrix} \qquad B = \begin{bmatrix} 7 & 8 \\ 9 & 10 \\ 11 & 12 \end{bmatrix}$$

Determine the matrix product $C = AB$.

We first observe that A has the same number of columns as B has rows (namely, three). Hence, the desired matrix multiplication can be carried out. The resulting product matrix, C, will have two rows (because A has two rows) and two columns (because B has two columns).

From Equation (11.4), we can write

$$c_{11} = a_{11}\,b_{11} + a_{12}\,b_{21} + a_{13}\,b_{31} = (1)(7) + (2)(9) + (3)(11) = 58$$

Similarly,

$$c_{12} = a_{11}\,b_{12} + a_{12}\,b_{22} + a_{13}\,b_{32} = (1)(8) + (2)(10) + (3)(12) = 64$$

$$c_{21} = a_{21}\,b_{11} + a_{22}\,b_{21} + a_{23}\,b_{31} = (4)(7) + (5)(9) + (6)(11) = 139$$

$$c_{22} = a_{21}\,b_{12} + a_{22}\,b_{22} + a_{23}\,b_{32} = (4)(8) + (5)(10) + (6)(12) = 154$$

Hence, we can write the desired product as

$$C = \begin{bmatrix} 58 & 64 \\ 139 & 154 \end{bmatrix}$$

Equation (11.4) undergoes some simplification if the second multiplicand is a vector. Thus, if A is an $m \times n$ matrix and X is a vector having n elements (that is, n rows), the matrix product $AX = B$ can be written on an element-by-element basis in the following manner:

$$\sum_{j=1}^{n} a_{ij} x_j = b_i \qquad i = 1, 2, \ldots, m \qquad (11.5)$$

The resulting B vector will contain m elements (m rows) since the A matrix has m rows.

Example 11.3 Reconstructing a System of Simultaneous Equations

Beginning with the matrix form of the equations given in Example 11.1 and using Equation (11.5), reconstruct the original two equations.

From Equation (11.5), we can write the first equation as

$$\sum_{j=1}^{2} a_{1j}x_j = b_1$$

Substituting the values of the coefficients, we obtain

$$2x_1 + 3x_2 = 8$$

which is the first of the original two equations.

Similarly, we can write the second equation as

$$\sum_{j=1}^{2} a_{2j}x_j = b_2$$

or

$$4x_1 - 3x_2 = -2$$

This is the second of the original two equations.

Other Matrix Operations

For completeness, three other matrix operations should be discussed, though they are not required to solve simultaneous equations. These operations are *matrix addition*, *matrix subtraction*, and *scalar multiplication*.

Two matrices can be added if they both contain the same number of rows and the same number of columns. Thus, if A and B are each $m \times n$ matrices, the matrix sum is written as $A + B = C$. On an element-by-element basis, the matrix sum is determined simply by adding the elements of A to the corresponding elements of B. In other words,

$$a_{ij} + b_{ij} = c_{ij} \tag{11.6}$$

Note that C will also be an $m \times n$ matrix.

Matrix subtraction is directly analogous to matrix addition. Thus, if A and B are each $m \times n$ matrices, the matrix difference is written as $A - B = C$, where C is also an $m \times n$ matrix. On an element-by-element basis, the matrix difference is determined by subtracting the elements of B from the corresponding elements of A. That is,

$$a_{ij} - b_{ij} = c_{ij} \tag{11.7}$$

Scalar multiplication involves multiplying each element of an $m \times n$ matrix by a constant k. Thus, we can write scalar multiplication as $kA = C$. On an element-by-element basis, we can write

$$ka_{ij} = c_{ij} \tag{11.8}$$

Again, note that C will be an $m \times n$ matrix.

These three operations are less complicated than matrix multiplication. We mention them here in order to provide a balanced overview of elementary matrix operations.

Example 11.4 Matrix Addition, Matrix Subtraction, and Scalar Multiplication

Suppose we are given the matrices

$$A = \begin{bmatrix} 4 & 7 \\ -2 & 5 \end{bmatrix} \qquad B = \begin{bmatrix} 1 & -8 \\ 4 & 4 \end{bmatrix}$$

Determine the matrix sum $A + B$, the matrix difference $A - B$, and the product $3A$.

Using Equations (11.6), (11.7), and (11.8), we obtain

$$A + B = \begin{bmatrix} (4+1) & (7-8) \\ (-2+4) & (5+4) \end{bmatrix} = \begin{bmatrix} 5 & -1 \\ 2 & 9 \end{bmatrix}$$

$$A - B = \begin{bmatrix} (4-1) & (7+8) \\ (-2-4) & (5-4) \end{bmatrix} = \begin{bmatrix} 3 & 15 \\ -6 & 1 \end{bmatrix}$$

$$3A = \begin{bmatrix} 3(4) & 3(7) \\ 3(-2) & 3(5) \end{bmatrix} = \begin{bmatrix} 12 & 21 \\ -6 & 15 \end{bmatrix}$$

Special Matrices

Before proceeding further, we must discuss two special matrices, the *identity matrix*, I, and the *inverse matrix*, A^{-1}.

The *identity matrix* is analogous to the constant 1 in ordinary algebra. It is a square matrix, containing n rows and n columns. For example (for a 5×5 matrix),

$$I = \begin{bmatrix} 1 & 0 & 0 & 0 & 0 \\ 0 & 1 & 0 & 0 & 0 \\ 0 & 0 & 1 & 0 & 0 \\ 0 & 0 & 0 & 1 & 0 \\ 0 & 0 & 0 & 0 & 1 \end{bmatrix} \tag{11.9}$$

The identity matrix has the important property that

$$IA = AI = A \tag{11.10}$$

provided A is also a square matrix having n rows and n columns. Note that this is analogous to writing

$$1 \times a = a \times 1 = a \tag{11.11}$$

in ordinary algebra.

The *inverse matrix* is analogous to a reciprocal in ordinary algebra. It is a square matrix, containing n rows and n columns, which has the important property that

$$A^{-1}A = AA^{-1} = I \tag{11.12}$$

Note that this is analogous to writing

$$a^{-1}a = aa^{-1} = 1 \tag{11.13}$$

in ordinary algebra. You should understand, however, that the elements of A^{-1} are *not* the reciprocals of the elements of A. In fact, calculating A^{-1} is quite complicated, as we will see in Sec. 11.3. For now, we will let Excel compute the inverse matrix for us when it is needed.

Example 11.5 Properties of the Inverse Matrix

The inverse of the A matrix (i.e., the coefficient matrix) given in Example 11.1 is

$$A^{-1} = \begin{bmatrix} 1/6 & 1/6 \\ 2/9 & -1/9 \end{bmatrix}$$

The classical method for obtaining the inverse of a matrix is obtained is relatively complicated and beyond the scope of this book. Note, however, that *the elements of the inverse matrix are not the reciprocals of the corresponding elements of A*.

Let us use the rules of matrix multiplication to show that $AA^{-1} = I$.

Writing out the components of AA^{-1}, we obtain

$$AA^{-1} = \begin{bmatrix} 2 & 3 \\ 4 & -3 \end{bmatrix} \begin{bmatrix} 1/6 & 1/6 \\ 2/9 & -1/9 \end{bmatrix}$$

$$AA^{-1} = \begin{bmatrix} (2)(1/6)+(3)(2/9) & (2)(1/6)+(3)(-1/9) \\ (4)(1/6)+(-3)(2/9) & (4)(1/6)+(-3)(-1/9) \end{bmatrix}$$

$$AA^{-1} = \begin{bmatrix} (1/3+2/3) & (1/3-1/3) \\ (2/3-2/3) & (2/3+1/3) \end{bmatrix} = \begin{bmatrix} 1 & 0 \\ 0 & 1 \end{bmatrix} = I$$

as expected. You may wish to work out the product $A^{-1}A$ and show that it is also equal to the identity matrix I (see Prob. 11.6).

Not all square matrices have a corresponding inverse. Square matrices that do not have an inverse are referred to as *singular matrices*. A singular matrix can be identified by the fact that its *determinant* is equal to zero.

Simultaneous Linear Equations

Now we will make use of the identity matrix and the inverse matrix to solve a system of simultaneous linear equations. Let us begin by writing the system of simultaneous equations as

$$AX = B \tag{11.14}$$

If we premultiply by A^{-1}, we obtain

$$A^{-1}AX = A^{-1}B \tag{11.15}$$

But $A^{-1}AX$ can be written as $IX = X$. Hence, we can write

$$X = A^{-1}B \tag{11.16}$$

Therefore, we conclude that *a system of simultaneous linear algebraic equations* $AX = B$ *can be solved by determining the inverse,* A^{-1}, *of the coefficient matrix A and then forming the product* $A^{-1}B$. The resulting product will be a vector, X, whose components are the desired unknown quantities.

Example 11.6 Solving Simultaneous Linear Equations Using Matrix Inversion

Solve the system of equations given in Example 11.1 using the method described above. For your convenience, the given system of equations is repeated below.

$$2x_1 + 3x_2 = 8$$
$$4x_1 - 3x_2 = -2$$

In matrix form, we can write the system of equations as

$$AX = B$$

where

$$A = \begin{bmatrix} 2 & 3 \\ 4 & -3 \end{bmatrix} \qquad X = \begin{bmatrix} x_1 \\ x_2 \end{bmatrix} \qquad B = \begin{bmatrix} 8 \\ -2 \end{bmatrix}$$

Furthermore, in the last example we established that

$$A^{-1} = \begin{bmatrix} 1/6 & 1/6 \\ 2/9 & -1/9 \end{bmatrix}$$

Hence, the solution can be obtained from Equation (11.16) as

$$X = A^{-1}B = \begin{bmatrix} 1/6 & 1/6 \\ 2/9 & -1/9 \end{bmatrix} \begin{bmatrix} 8 \\ -2 \end{bmatrix} = \begin{bmatrix} 8/6 - 2/6 \\ 16/9 + 2/9 \end{bmatrix} = \begin{bmatrix} 1 \\ 2 \end{bmatrix}$$

Therefore, the desired solution is $x_1 = 1$, $x_2 = 2$.

Problems

11.1 Write each of the following systems of simultaneous equations in matrix form:

(a)
$$x_1 - 2x_2 + 3x_3 = 17$$
$$3x_1 + x_2 - 2x_3 = 0$$
$$2x_1 + 3x_2 + x_3 = 7$$

(b)
$$0.1x_1 - 0.5x_2 + x_4 = 2.7$$
$$0.5x_1 - 2.5x_2 + x_3 - 0.4x_4 = -4.7$$
$$x_1 + 0.2x_2 - 0.1x_3 + 0.4x_4 = 3.6$$
$$0.2x_1 + 0.4x_2 - 0.2x_3 = 1.2$$

(c)
$$11x_1 + 3x_2 \qquad + \quad x_4 + 2x_5 = 51$$
$$4x_2 + 2x_3 \qquad + \quad x_5 = 15$$
$$3x_1 + 2x_2 + 7x_3 + \quad x_4 \qquad = 15$$
$$4x_1 \qquad + 4x_3 + 10x_4 + \quad x_5 = 20$$
$$2x_1 + 5x_2 + \quad x_3 + 3x_4 + 13x_5 = 92$$

11.2 Determine the matrix product $AB = C$ for each of the following pairs of matrices:

(a) $\quad A = \begin{bmatrix} 1 & 2 \\ 3 & 4 \\ 5 & 6 \end{bmatrix} \qquad B = \begin{bmatrix} 7 & 8 & 9 & 10 \\ 11 & 12 & 13 & 14 \end{bmatrix}$

(b) $\quad A = \begin{bmatrix} 1 & 2 & 3 \end{bmatrix} \qquad B = \begin{bmatrix} 4 \\ 5 \\ 6 \end{bmatrix}$

(c) $\quad A = \begin{bmatrix} 1 \\ 2 \\ 3 \end{bmatrix} \qquad B = \begin{bmatrix} 4 & 5 & 6 \end{bmatrix}$

(d) $\quad A = \begin{bmatrix} 1 & 2 & 3 \\ 4 & 5 & 6 \\ 7 & 8 & 9 \end{bmatrix} \qquad B = \begin{bmatrix} 5 \\ -3 \\ 2 \end{bmatrix}$

11.3 Reconstruct a system of simultaneous algebraic equations from the following matrix equation:

$$\begin{bmatrix} 4 & -2 & 1 \\ 1 & 3 & -7 \\ -2 & 0 & -5 \end{bmatrix} \begin{bmatrix} x_1 \\ x_2 \\ x_3 \end{bmatrix} = \begin{bmatrix} 6 \\ 34 \\ 14 \end{bmatrix}$$

11.4 Suppose we are given the following square matrices:

$$A = \begin{bmatrix} 6 & 0 & -3 \\ -1 & 4 & 9 \\ 8 & -5 & 2 \end{bmatrix} \qquad B = \begin{bmatrix} 5 & -5 & -7 \\ 0 & 9 & 2 \\ 4 & -1 & 0 \end{bmatrix}$$

Using these two matrices, carry out each of the following matrix operations:

(a)	$A + B = C$	(d)	$2A = C$
(b)	$B + A = C$	(e)	$AB = C$
(c)	$A - B = C$	(f)	$BA = C$

Are the matrix sums (a) and (b) equivalent? Are the matrix products (e) and (f) equivalent? What do you conclude from these two comparisons?

11.5 Suppose I is a 3×3 identity matrix and A is the following square matrix:

$$A = \begin{bmatrix} 6 & 0 & -3 \\ -1 & 4 & 9 \\ 8 & -5 & 2 \end{bmatrix}$$

(a) Determine the matrix product IA

(b) Determine the matrix product AI

Are (a) and (b) equivalent?

11.6 Determine the matrix product $A^{-1}A$ using the matrices given in Examples 11.1 and 11.5. Compare with the results obtained in Example 11.5. Is $A^{-1}A$ equivalent to AA^{-1}?

11.7 Determine whether or not the following matrix is the inverse of the A matrix given in Prob. 11.5.

$$B = \begin{bmatrix} 0.132832 & 0.037594 & 0.030075 \\ 0.185464 & 0.090226 & -0.127820 \\ -0.067670 & 0.075188 & 0.060150 \end{bmatrix}$$

11.8 Consider the system of three equations in three unknowns, written in matrix form as $AX = B$, where

$$A = \begin{bmatrix} 4 & -2 & 1 \\ 1 & 3 & -7 \\ -2 & 0 & -5 \end{bmatrix} \qquad X = \begin{bmatrix} x_1 \\ x_2 \\ x_3 \end{bmatrix} \qquad B = \begin{bmatrix} 6 \\ 34 \\ 14 \end{bmatrix}$$

Suppose the inverse of A is known to be

$$A^{-1} = \begin{bmatrix} 0.163043 & 0.108696 & -0.119570 \\ -0.206520 & 0.195652 & -0.315220 \\ -0.065220 & -0.043480 & -0.152170 \end{bmatrix}$$

Using this information, solve for the unknowns x_1, x_2, and x_3.

11.2 MATRIX OPERATIONS IN EXCEL

Matrix operations can easily be carried out in Excel. To do so, we need to make use of *arrays* since a matrix is represented as an array in Excel.

An array is a block of cells that is referenced collectively, in the same manner as a single cell. Any operation that is carried out on an array will produce the same outcome for all of the cells within the array. In addition, an array can be specified as a single argument to a function. This is equivalent to specifying the individual cells as multiple arguments.

An array is represented as a block of cells enclosed in braces ({}). For example, {A1:C3} refers to an array that is composed of the block of cells extending from cell A1 to cell C3. *The braces are added automatically by Excel as a part of the array specification. Thus, you should not attempt to include the braces when writing an array.*

To specify an array operation (i.e., a matrix operation) within an Excel worksheet, proceed as follows.

1. Select the block of cells that make up the array. Then move the cell marker to the upper left cell within the block.

2. Enter an array formula, as required. You may include cell ranges within the formula. However, after you have entered the formula, *do not* enclose the formula in braces.

3. Press Ctrl-Shift-Enter (on an Intel-type computer) or Command-Return (on a Macintosh) simultaneously. This will cause the formula to appear within a pair of braces, thus identifying the block of cells as an array. The indicated operations will then be carried out.

Example 11.7 Matrix Addition, Matrix Subtraction, and Scalar Multiplication in Excel

Repeat the operations shown in Example 11.4 within an Excel worksheet. Thus, we wish to determine the matrix sum $A + B$, the matrix difference $A - B$, and the product $3A$ from the following two matrices:

$$A = \begin{bmatrix} 4 & 7 \\ -2 & 5 \end{bmatrix} \qquad B = \begin{bmatrix} 1 & -8 \\ 4 & 4 \end{bmatrix}$$

The Excel worksheet is shown in Fig. 11.1. Within this worksheet, the elements of the A and B matrices were entered in the normal manner. (The vertical lines forming the matrix borders were added by selecting Cells/Border from the Format menu.)

In order to determine the matrix sum, we first select the block of cells extending from B6 to C7. The cell marker will appear over cell B6. We then enter the formula =B3:C4+F3:G4 (without enclosing braces) and press Ctrl-Shift-Enter (or Command-Return) simultaneously. The resulting matrix sum will then appear within cells B6 through C7, as shown in Fig. 11.1.

Notice the array formula {=B3:C4+F3:G4} shown in the formula bar. This formula (without the enclosing braces) was entered while the cell marker was highlighting cell B6, though the formula applies to the entire selected block of cells. The braces were added automatically, after pressing Ctrl-Shift-Enter (or Command-Return).

Similarly, the matrix difference was determined by selecting the block of cells extending from F6 to G7. The formula =B3:C4–F3:G4 was then entered, without the enclosing braces. The braces were added automatically, after pressing Ctrl-Shift-Enter (or Command-Return). The result of the matrix subtraction operation was then generated in cells F6 through G7.

Finally, the scalar product was determined by selecting the block of cells extending from B9 through C10, entering the formula =3*B3:C4, and then pressing Ctrl-Shift-Enter (or Command-Return).

	B6		= {=B3:C4+F3:G4}					
	A	**B**	C	D	E	F	G	H
1	Matrix Addition, Subtraction and Scalar Multiplication							
2								
3	A =	4	7		B =	1	-8	
4		-2	5			4	4	
5								
6	A + B =	5	-1		A - B =	3	15	
7		2	9			-6	1	
8								
9	3A =	12	21					
10		-6	15					
11								
12								
Ready								

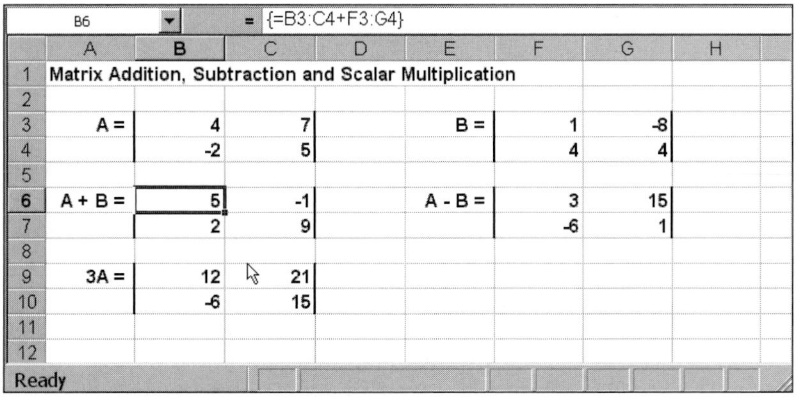

Figure 11.1 – Matrix operations in Excel

11.3 SOLVING SIMULTANEOUS EQUATIONS IN EXCEL USING MATRIX INVERSION

One way to solve a system of simultaneous linear equations in Excel is to use the matrix inversion method described in Sec. 11.1. Excel includes a library function, MINVERSE, for determining the inverse of a matrix, and another, MMULT, for carrying out matrix multiplication. Hence, it is quite simple to solve the system of equations $AX = B$ using these two functions. The general procedure is to first determine A^{-1} using the MINVERSE function and then obtain the matrix product $A^{-1}B$ using the MMULT function. The resulting product will be the vector X, which contains the desired unknown values.

The detailed procedure is as follows:

1. Enter the elements of the coefficient matrix A into an $n \times n$ block of cells.

2. Enter the right-hand side B within another block (a single column) of n cells.

3. Identify where you want the inverse A^{-1} to appear. (You must set aside another $n \times n$ block of cells.) Then highlight this block of cells, place the cell marker in the upper left cell, and enter the formula =MINVERSE( ). Place the range of the coefficient matrix inside the parentheses; for example, =MINVERSE(A1:B2). Then press Ctrl-Shift-Enter (or Command-Return) simultaneously. The elements of A^{-1} will appear within the block.

4. Now identify where you want the solution X to appear. (You must set aside a block of n cells within a single column.) Then highlight this block of cells, place the cell marker in the top cell, and enter the formula =MMULT(,). Place the range of the inverse matrix inside the parentheses, followed by a comma, followed by the range of the right-hand side; for example, =MMULT(F1:G2,D1:D2). Then press Ctrl-Shift-Enter (or Command-Return) simultaneously. The elements of X (the unknown values) will then appear within the block.

5. It is a good idea to verify your solution by calculating the matrix product AX and then comparing the resulting product with the original right-hand-side vector, B. If the computation has been carried out correctly, they will be the same (or very nearly so).

6. You may also wish to evaluate the determinant of A using the MDETERM library function. (Remember that a determinant whose value is zero indicates a singular matrix, which does not have an inverse.) To do so, enter the formula =MDETERM( ) within a cell, placing the range of A within the parentheses; for example, =MDETERM(A1:B2). MDETERM returns a single value that may be positive, negative, or zero. If the value is zero (or very close to zero, since numerical roundoff may come into play), then the system of equations does not have a solution and you need not proceed further. Thus, you might choose to begin by evaluating the determinant of A, before attempting to determine its inverse.

Example 11.8 Solving Simultaneous Equations in Excel Using Matrix Inversion

In Example 11.6 we outlined a procedure for solving the following system of simultaneous equations:

$$2x_1 + 3x_2 = 8$$
$$4x_1 - 3x_2 = -2$$

The solution procedure assumed that we knew the inverse of the coefficient matrix. Let us now use Excel to evaluate its determinant, determine its inverse, and then solve the system of equations.

Figure 11.2 shows an Excel worksheet containing the coefficient matrix (i.e., the A matrix) in cells A4:B5, and the right-hand side (i.e., the B vector) in cells G4:G5. Notice that the worksheet also contains space reserved for the inverse of the coefficient matrix (A^{-1}), the solution vector (i.e., the X vector), and the matrix product AX. These quantities will be determined from the given information. Each matrix or vector has been labeled and placed within a border in order to make the worksheet more legible.

We begin by evaluating the determinant of A in cell B7 using the MDETERM function, as shown in the formula bar. This value is -18; hence, we conclude that A is nonsingular and we can therefore determine its inverse.

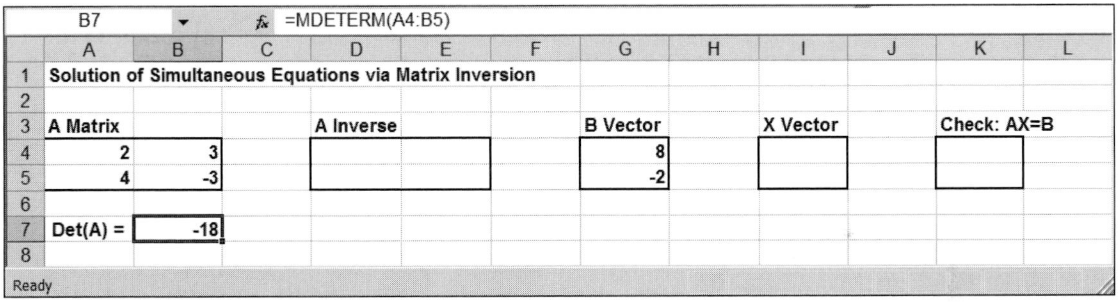

B7		▼	*fx*	=MDETERM(A4:B5)							
	A	B	C	D	E	F	G	H	I	J	K	L
1	Solution of Simultaneous Equations via Matrix Inversion											
2												
3	A Matrix			A Inverse			B Vector		X Vector		Check: AX=B	
4	2	3					8					
5	4	-3					-2					
6												
7	Det(A) =	-18										
8												

Ready

Figure 11.2 – Preparing to solve two simultaneous equations using matrix inversion

D4		▼	*fx*	{=MINVERSE(A4:B5)}							
	A	B	C	D	E	F	G	H	I	J	K	L
1	Solution of Simultaneous Equations via Matrix Inversion											
2												
3	A Matrix			A Inverse			B Vector		X Vector		Check: AX=B	
4	2	3		0.166667	0.166667		8					
5	4	-3		0.222222	-0.11111		-2					
6												
7	Det(A) =	-18										
8												

Ready   Sum=0.444444444

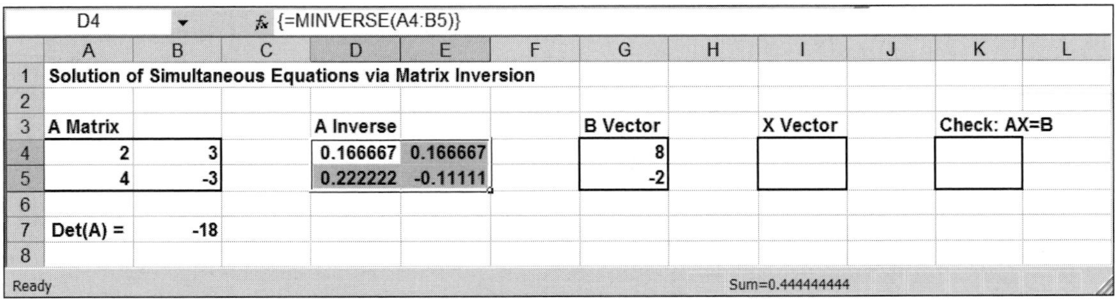

Figure 11.3 –Calculating the inverse of the A matrix

Figure 11.3 shows the elements of the inverse matrix in cells D4:E5. These values were obtained using the MINVERSE function, as shown in the formula bar.

In Fig. 11.4 we see the desired solution vector in cells I4:I5. The solution vector was obtained by multiplying the inverse of the coefficient matrix by the right-hand side; that is, by forming the matrix product $A^{-1}B$. The matrix multiplication was carried out using the MMULT function, as shown in the formula bar. Thus, we can now see that the desired solution is $x_1 = 1$, $x_2 = 2$, which is consistent with the results obtained in Example 11.6.

Finally, we can check our solution by forming the product AX and comparing it to the original right-hand-side vector, B. Figure 11.5 shows the resulting product in cells K4:K5. These values were obtained using the MMULT function, as can be seen in the formula bar. Clearly, the calculated value agrees with the original B vector, thus assuring us that the calculated solution is correct.

	I5	▼		f_x {=MMULT(D4:E5,G4:G5)}								
	A	B	C	D	E	F	G	H	I	J	K	L
1	Solution of Simultaneous Equations via Matrix Inversion											
2												
3	A Matrix			A Inverse			B Vector		X Vector		Check: AX=B	
4	2	3		0.166667	0.166667		8		1			
5	4	-3		0.222222	-0.11111		-2		2			
6												
7	Det(A) =	-18										
8												
Ready									Sum=3			

Figure 11.4 – Obtaining a solution to the simultaneous equations

	K4	▼		f_x {=MMULT(A4:B5,I4:I5)}								
	A	B	C	D	E	F	G	H	I	J	K	L
1	Solution of Simultaneous Equations via Matrix Inversion											
2												
3	A Matrix			A Inverse			B Vector		X Vector		Check: AX=B	
4	2	3		0.166667	0.166667		8		1		8	
5	4	-3		0.222222	-0.11111		-2		2		-2	
6												
7	Det(A) =	-18										
8												
Ready									Sum=6			

Figure 11.5 – Checking the solution

Problems

Use Excel to solve each of the following problems.

11.9 Suppose we are given the following square matrices (these matrices were originally given in Prob. 11.4).

$$A = \begin{bmatrix} 6 & 0 & -3 \\ -1 & 4 & 9 \\ 8 & -5 & 2 \end{bmatrix} \qquad B = \begin{bmatrix} 5 & -5 & -7 \\ 0 & 9 & 2 \\ 4 & -1 & 0 \end{bmatrix}$$

Representing each matrix as an array, carry out each of the following matrix operations.

(a) $A + B = C$ (c) $A - B = C$

(b) $B + A = C$ (d) $2A = C$

11.10 Carry out the matrix multiplication operations given in Prob. 11.2. Use the MMULT library function to carry out the matrix multiplication.

11.11 Verify that the A^{-1} matrix given in Example 11.5 is correct using the MINVERSE library function. (The original A matrix is given in Example 11.1.) Then verify that $A^{-1}A = AA^{-1} = I$ for these matrices, using the MMULT library function.

11.12 Solve Prob. 11.5 using the MMULT function.

11.13 Solve Prob. 11.7 using the MMULT function.

11.14 Solve the system of simultaneous equations given in Prob. 11.8 using Excel. Verify that the solution is correct by forming the product AX and comparing this product with the original B vector.

11.15 Solve each of the following systems of simultaneous equations, originally given in Prob. 11.1. In each case, verify that the solution is correct using the method described in the last problem.

(a) $\begin{aligned} x_1 - 2x_2 + 3x_3 &= 17 \\ 3x_1 + x_2 - 2x_3 &= 0 \\ 2x_1 + 3x_2 + x_3 &= 7 \end{aligned}$

(b) $\begin{aligned} 0.1x_1 - 0.5x_2 \qquad\quad + \quad x_4 &= 2.7 \\ 0.5x_1 - 2.5x_2 + \quad x_3 - 0.4x_4 &= -4.7 \\ x_1 + 0.2x_2 - 0.1x_3 + 0.4x_4 &= 3.6 \\ 0.2x_1 + 0.4x_2 - 0.2x_3 \qquad\quad &= 1.2 \end{aligned}$

(c) $11x_1 + 3x_2 \qquad + \quad x_4 + 2x_5 = 51$
$\qquad 4x_2 + 2x_3 \qquad + \quad x_5 = 15$
$\quad 3x_1 + 2x_2 + 7x_3 + \quad x_4 \qquad = 15$
$\quad 4x_1 \qquad + 4x_3 + 10x_4 + \quad x_5 = 20$
$\quad 2x_1 + 5x_2 + \quad x_3 + \quad 3x_4 + 13x_5 = 92$

11.16 Solve each of the following systems of simultaneous equations. In each case, verify that the solution is correct using the method described in Prob. 11.14.

(a) $15x_1 + 20x_2 = 25$
$\quad 5x_1 + 10x_2 = 12$

(b) $4x_1 - 2x_2 + x_3 = 6$
$\quad x_1 + 3x_2 - 7x_3 = 34$
$\quad -2x_1 \qquad - 5x_3 = 14$

(c) $\quad x_1 + \quad x_2 + 5x_3 - 12x_4 = 29$
$\qquad \quad x_2 \qquad + 16x_4 = 21$
$\quad 7x_1 + 2x_2 \qquad - 12x_4 = 5$
$\qquad \qquad x_3 - \quad x_4 = 6$

11.4 SOLVING SIMULTANEOUS EQUATIONS IN EXCEL USING SOLVER

Excel's Solver feature provides an altogether different approach to the solution of simultaneous equations. This method can be used with nonlinear as well as linear systems of equations.

Remember that Solver is an Excel add-in, like the Analysis Toolpak used to generate histograms (see Chapter 8). To install Solver, choose Add-Ins from the Tools menu. Then select Solver Add-In from the resulting Add-Ins dialog box. Once the Solver feature has been installed, it will remain installed unless it is removed by reversing the above procedure.

Now let us see how Solver can be used to solve a system of simultaneous equations. The equations need not necessarily be linear. Suppose the equations are represented as

$$f_1(x_1, x_2, \ldots, x_n) = 0 \tag{11.17}$$

$$f_2(x_1, x_2, \ldots, x_n) = 0 \tag{11.18}$$

$$\ldots \ldots$$

$$f_n(x_1, x_2, \ldots, x_n) = 0 \tag{11.19}$$

Thus, we have a system of n equations in n unknowns. We wish to find the values of $x_1, x_2, \ldots, x_n$ that cause each of the equations to equal zero.

One way to do this is to force the function

$$y = f_1^2 + f_2^2 + \cdots + f_n^2 \tag{11.20}$$

to zero; that is, to find the values of $x_1, x_2, \ldots, x_n$ that cause Equation (11.20) to equal zero. Since all of the terms on the right-hand side of Equation (11.20) are squares, they will all be greater than or equal to zero. Hence, the only way that y can equal zero is for each of the individual fs to also equal zero. Therefore, the values of $x_1, x_2, \ldots, x_n$ that cause y to equal zero will be the solution to the given system of equations. The general approach used with Solver is to define a *target function* consisting of the squares of the individual equations, as indicated by Equation (11.20), and to then determine the values of $x_1, x_2, \ldots, x_n$ that cause the target function to equal zero.

To solve a system of simultaneous equations with Solver, proceed as follows:

1. Enter an initial guess for each independent variable $(x_1, x_2, \ldots, x_n)$ in a separate cell on the worksheet.

2. Enter the equations for $f_1, f_2, \ldots, f_n$ and y in separate cells, expressed as Excel formulas. Within these formulas, express the unknown quantities $x_1, x_2, \ldots, x_n$ as the addresses of the cells containing the initial guess.

3. Select Solver from the Tools menu.

4. When the Solver Parameters dialog box appears, enter the following information:

    (*a*)   The address of the cell containing the formula for y in the Set Target Cell location.

    (*b*)   Select Value of in the Equal to line. Then enter 0 within the associated data area. (In other words, determine the values of $x_1, x_2, \ldots, x_n$ that will drive the target function to zero.)

    (*c*)   Enter the range of cell addresses containing the initial values of $x_1, x_2, x_3, \ldots, x_n$ in the area labeled By Changing Cells.

    (*d*)   If you wish to restrict the range of the independent variables, click on the Add button under the heading Subject to the Constraints. Then provide the following information within the Add Constraint dialog box for each of the independent variables:

      (*i*)   The cell address containing the initial value of the independent variable in the Cell Reference location.

      (*ii*)   The type of constraint (i.e., $\leq$ or $\geq$) from the pull-down menu.

      (*iii*) The limiting value in the Constraint data area.

      (*iv*)   Select OK to return to the Solver Parameters dialog box or select Add to add another constraint.

Note that you can always change a constraint or delete a constraint after it has been added.

(e)   When all of the required information has been entered correctly, select Solve. This will initiate the actual solution procedure.

A new dialog box labeled Solver Results will then appear, telling you whether or not Solver has been able to solve the problem. If a solution has been obtained, the desired values of the independent variables will appear in the cells that originally contained the initial values. The cell containing the target function will show a value that is zero or nearly zero.

Example 11.9 Solving Simultaneous Linear Equations in Excel Using Solver

Determine the solution to the following system of equations using Solver.

$$3 x_1 + 2 x_2 - x_3 = 4$$
$$2 x_1 - x_2 + x_3 = 3$$
$$x_1 + x_2 - 2 x_3 = -3$$

For clarity, let us refer to the first equation as $f(x_1, x_2, x_3)$, the second as $g(x_1, x_2, x_3)$, and the third as $h(x_1, x_2, x_3)$. Thus, we can write the given system of equations as

$$f = 3 x_1 + 2 x_2 - x_3 - 4 = 0$$
$$g = 2 x_1 - x_2 + x_3 - 3 = 0$$
$$h = x_1 + x_2 - 2 x_3 + 3 = 0$$

We wish to find the values of x_1, x_2, and x_3 that will cause $f(x_1, x_2, x_3)$, $g(x_1, x_2, x_3)$, and $h(x_1, x_2, x_3)$ to equal zero.

Now let us form the sum

$$y = f^2 + g^2 + h^2$$

If $f = 0$, $g = 0$, and $h = 0$, then y will also equal zero. For any other values of f, g, and h (whether positive or negative), however, y will be greater than zero. Hence, we can solve the given system of equations by finding the values of x_1, x_2, and x_3 that cause y to equal zero. We will use Excel's Solver feature to determine the values of x_1, x_2, and x_3 that drive the target function, y, to zero.

An Excel worksheet containing the initial setup is shown in Fig. 11.6, and the corresponding cell formulas are shown in Fig. 11.7. Note that cells B3, B4, and B5 contain the initial assumed values for x_1, x_2, and x_3. Cells B7, B8, and B9 contain the formulas for $f(x_1, x_2, x_3)$, $g(x_1, x_2, x_3)$, and $h(x_1, x_2, x_3)$; and cell B11 contains the formula for y. The values within these cells in Fig. 11.6 are based upon the initial values chosen for x_1, x_2, and x_3.

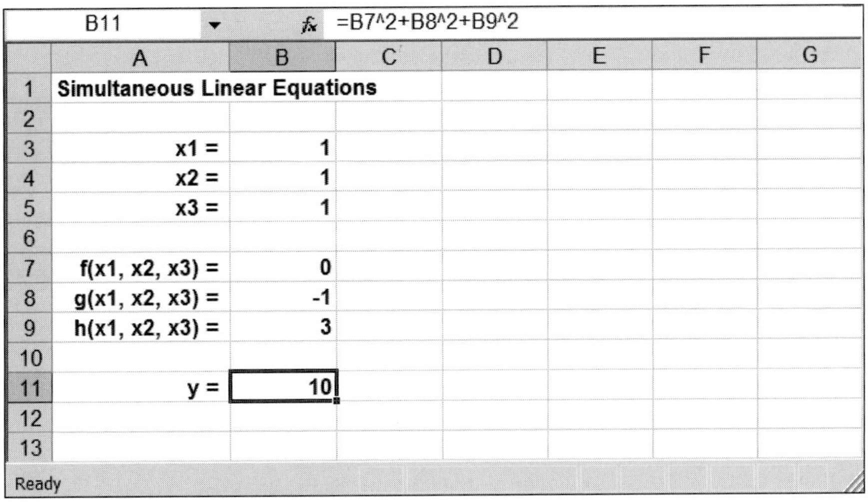

B11	▼	f_x	=B7^2+B8^2+B9^2			
	A	**B**	**C**	**D**	**E**	**F**	**G**
1	**Simultaneous Linear Equations**						
2							
3	x1 =	1					
4	x2 =	1					
5	x3 =	1					
6							
7	f(x1, x2, x3) =	0					
8	g(x1, x2, x3) =	-1					
9	h(x1, x2, x3) =	3					
10							
11	y =	10					
12							
13							

Ready

Figure 11.6 – Preparing to solve three simultaneous linear equations using Solver

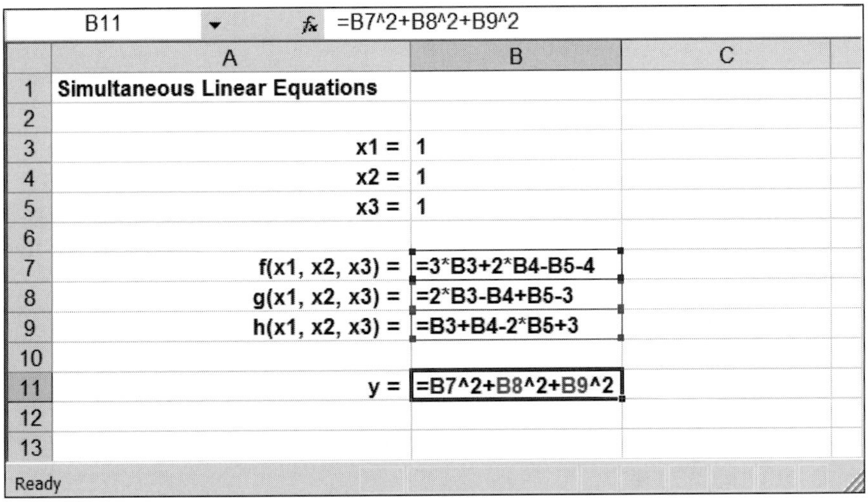

B11	▼	f_x	=B7^2+B8^2+B9^2	
	A	**B**	**C**
1	**Simultaneous Linear Equations**		
2			
3	x1 =	1	
4	x2 =	1	
5	x3 =	1	
6			
7	f(x1, x2, x3) =	=3*B3+2*B4-B5-4	
8	g(x1, x2, x3) =	=2*B3-B4+B5-3	
9	h(x1, x2, x3) =	=B3+B4-2*B5+3	
10			
11	y =	=B7^2+B8^2+B9^2	
12			
13			

Ready

Figure 11.7 – The corresponding cell formulas

Next, we select Solver from the Tools menu, and we enter the address of the target function (B11) and the range of addresses of the independent variables (B3:B5) in the Solver Parameters dialog box, as shown in Fig. 11.8. Notice the specification that the target function be set equal to zero. We then select Solve to initiate the solution procedure.

Once a solution has been found the resulting values are shown in their respective cells within the worksheet, and the Solver Results dialog box appears, as shown in Fig. 11.9. Thus, we see that the solution to the three equations is $x_1 = 1$, $x_2 = 2$, and $x_3 = 3$. The corresponding values for f, g, h, and y are each approximately zero, as expected.

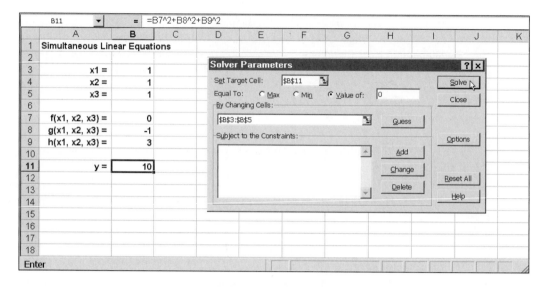

Figure 11.8 – Setting up the problem in Solver

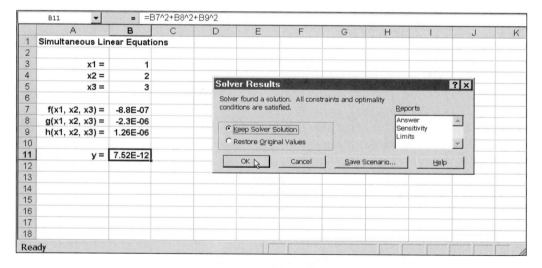

Figure 11.9 – The final solution

When solving simultaneous linear equations in Excel, be mindful of the fact that a solution obtained using matrices is (in principle) exact, whereas the solution obtained using Solver's generalized reduced gradient method is approximate. Hence, the matrix approach is generally preferred. Its accuracy can be compromised, however, by the accumulation of numerical roundoff errors in the matrix inversion process. This is especially true if there are many simultaneous equations, or if the coefficient matrix is close to being singular.

The equations given in the preceding example were all linear. Solver can also obtain the solution to *nonlinear* problems by following the same procedure as is used with linear problems. Remember, however, that nonlinear problems may have more than one solution, and the solutions need not necessarily be real. Hence, it may be desirable to restrict the range of the independent variables, as illustrated in the following example.

Example 11.10 Solving Simultaneous Nonlinear Equations in Excel Using Solver

Determine the solution to the following two simultaneous nonlinear equations using Solver. Restrict the range of the independent variables to nonnegative values.

$$x_1^2 + 2 x_2^2 - 5 x_1 + 7 x_2 = 40$$

$$3 x_1^2 - x_2^2 + 4 x_1 + 2 x_2 = 28$$

Let us rearrange each of these equations as follows:

$$f(x_1, x_2) = x_1^2 + 2 x_2^2 - 5 x_1 + 7 x_2 - 40 = 0$$

$$g(x_1, x_2) = 3 x_1^2 - x_2^2 + 4 x_1 + 2 x_2 - 28 = 0$$

Thus, we seek the values of x_1 and x_2 that cause $f(x_1, x_2)$ and $g(x_1, x_2)$ to equal zero. The procedure will be to form the target function

$$y(x_1, x_2) = f^2 + g^2$$

We will then use Solver to determine the values of x_1 and x_2 that force the target function, and hence f and g, to zero. These values will represent a solution to the given equations.

	B9	▼	f_x =B6^2+B7^2			
	A	B	C	D	E	F	G
1	Simultaneous Nonlinear Equations						
2							
3	x1 =	1					
4	x2 =	1					
5							
6	f(x1, x2) =	-35					
7	g(x1, x2) =	-20					
8							
9	y =	1625					
10							
11							
Ready							

Figure 11.10 – Preparing to solve two simultaneous nonlinear equations using Solver

Figure 11.10 shows an Excel worksheet containing the initial problem statement. Notice that x_1 and x_2 have each been assigned an initial value of 1, as shown in cells B3 and B4. The corresponding values for f, g, and y are shown in cells B6, B7, and B9, respectively. These values were obtained by evaluating the formulas shown in Fig. 11.11, based upon the initial values for x_1 and x_2.

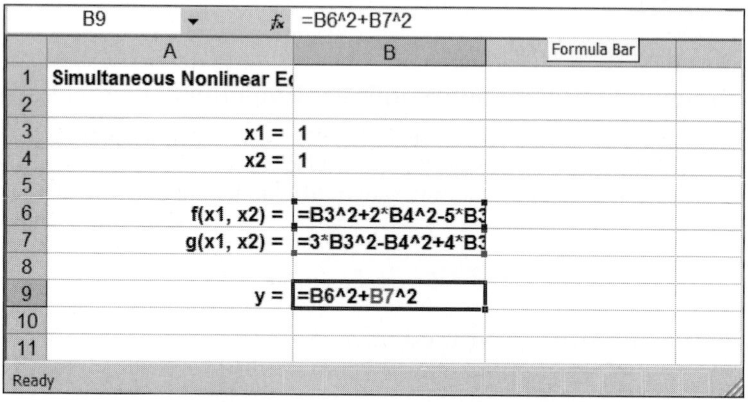

Figure 11.11 – The corresponding cell formulas

We now select Solver from the Tools menu, resulting in the Solver Parameters dialog box shown in Fig. 11.12. The address of the target function (B9) is entered near the top, followed by a specification that the target function take on a value of zero. The range of independent variables (B3:B4) is then entered in the area labeled By Changing Cells.

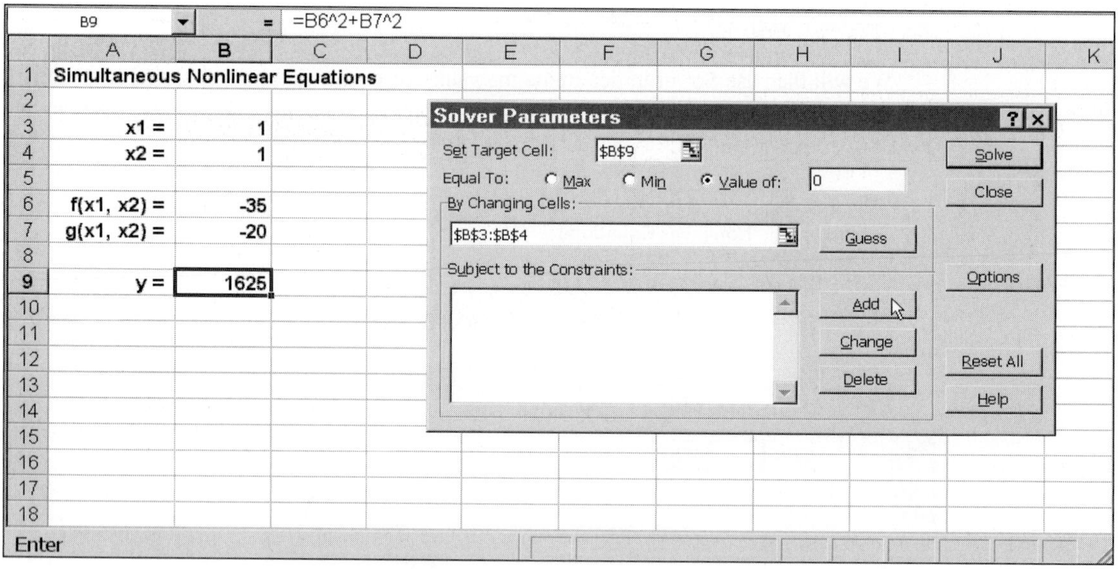

Figure 11.12 – Setting up the problem in Solver

To include the nonnegativity restrictions, we select Add near the bottom of the Solver Parameters dialog box. The Add Constraint dialog box then appears, as shown in Fig. 11.13. Within this dialog box, we enter the cell address, the type of constraint ($\leq$, $=$, or $\geq$), and the constraining value (in this case, 0) for each independent variable.

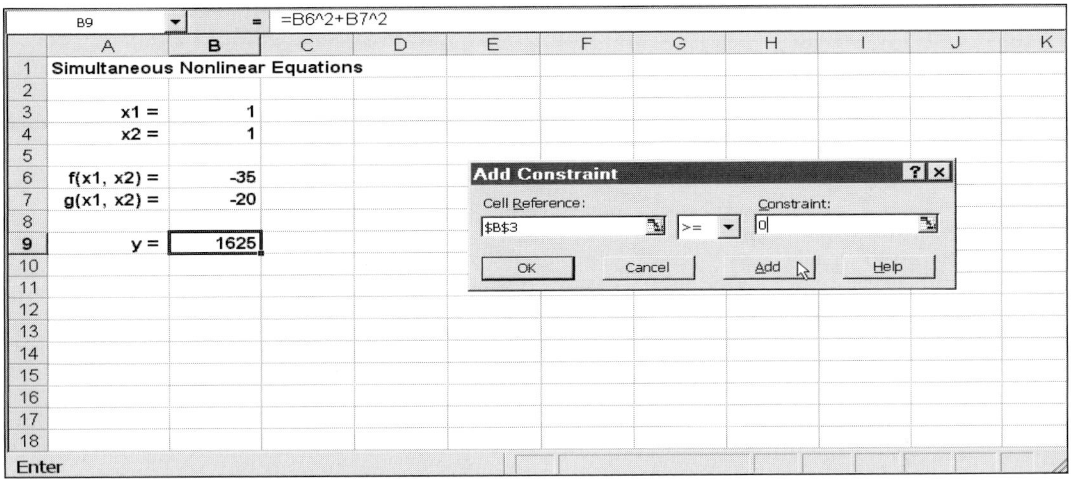

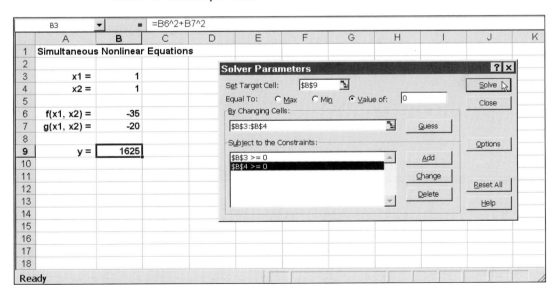

Figure 11.13 – Adding the first nonnegativity restriction

After both restrictions have been added, the Solver Parameters dialog box will reappear, with the constraints included, as shown in Fig. 11.14. We then select Solve to initiate the solution procedure.

Figure 11.14 – The final problem setup, including the nonnegativity restrictions

Once a solution has been found, the resulting values are shown in cells B3 and B4 within the worksheet, and the Solver Results dialog box appears, as shown in Fig. 11.15. Thus, we see that the solution to the given equations is approximately $x_1 = 2.70$ and $x_2 = 3.37$. In addition, we see that the corresponding values for the given equations f and g, and the target function y, are each very close to zero.

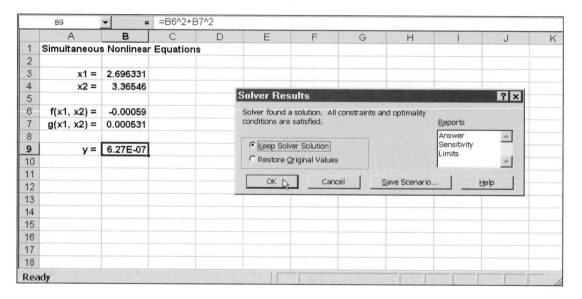

Figure 11.15 – The final solution

Problems

11.17 Solve Prob. 11.16(a) in Excel using Solver. Compare the solution with that obtained previously.

11.18 Solve Probs. 11.16(b) and (c) in Excel using Solver. Compare the effort involved in obtaining these solutions with the effort involved using the matrix-inversion method.

11.19 Solve the following systems of nonlinear algebraic equations in Excel using Solver.

(a) $4x_1^3 - \sqrt{3x_2} = 20$

$x_1 x_2^2 + 2/x_1 = 50$

(b) $\sin 2x_1 + x_1 \ln x_2 = -0.3$

$\ln (x_1^2) - \cos 3x_2 = 2$

(c) $\sin x_1 + \cos x_2 - \ln x_3 = 0$
$\cos x_1 + 2 \ln x_2 + \sin x_3 = 3$
$3 \ln x_1 - \sin x_2 + \cos x_3 = 2$

11.20 The following equations describe the current flowing across each of the resistors in the electrical circuit shown in Fig. 11.16. The equations were obtained by applying Kirchhoff's current law to each outer node, and Kirchhoff's voltage law to each closed loop within the circuit. (These equations make certain assumptions about the direction of individual current flows.)

$$i_1 + i_2 + i_3 = 2 \qquad\qquad 10i_1 - 40i_2 + 10i_4 = 0$$
$$i_1 - i_4 + i_6 = 0 \qquad\qquad 10i_4 + 2i_6 - 50i_7 = 0$$
$$i_6 + i_7 + i_{10} = 3 \qquad\qquad 50i_7 - 10i_8 - 5i_{10} = 0$$
$$i_8 - i_9 - i_{10} = 0 \qquad\qquad 20i_5 - 10i_8 - 10i_9 = 0$$
$$i_3 - i_5 - i_9 = 0 \qquad\qquad 40i_2 - 5i_3 - 20i_5 = 0$$

(a) Rearrange these equations into a more traditional form as 10 equations in 10 unknowns.

(b) Solve for the unknown current flows in Excel, using either of the methods described in this chapter.

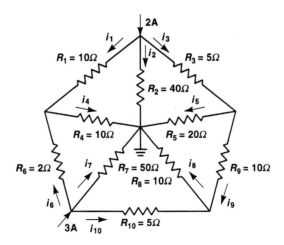

Figure 11.16 – An electrical circuit containing multiple resistors

11.21 The following equations describe the electrical circuit shown in Fig. 11.17. These equations were obtained by applying Kirchhoff's voltage law to each closed loop.

$$V_1 - i_1(R_1 + R_2) + i_2 R_2 = 0$$

$$V_2 - i_2(R_2 + R_3 + R_4) + i_1 R_2 + i_6 R_3 + i_3 R_4 = 0$$

$$V_3 - i_3(R_4 + R_9) + i_2 R_4 + i_6 R_9 = 0$$

$$-V_3 - i_4(R_5 + R_{10}) = 0$$

$$V_4 - i_6(R_3 + R_7 + R_8 + R_9) + i_2 R_3 + i_3 R_9 + i_5 R_7 = 0$$

$$-V_1 - i_5(R_6 + R_7) + i_6 R_7 = 0$$

Typically, the voltage sources (V_1 through V_4) and the resistances (R_1 through R_{10}) are known, and the resulting current flows within each loop (i_1 through i_6) must be determined.

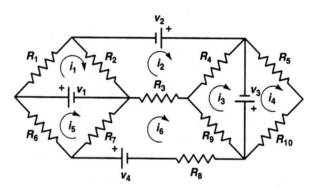

Figure 11.17 – An electrical circuit containing multiple resistors

(*a*) Rearrange these equations into a more traditional form as six equations in six unknowns.

(*b*) Solve for the unknown current flows in Excel, assuming the following values for the voltages and resistances. Use either of the methods described in this chapter to obtain a solution.

$V_1 = 12$ volts	$R_1 = 100\ \Omega$	$R_6 = 150\ \Omega$
$V_2 = 5$ volts	$R_2 = 200\ \Omega$	$R_7 = 50\ \Omega$
$V_3 = 1.5$ volts	$R_3 = 100\ \Omega$	$R_8 = 100\ \Omega$
$V_4 = 5$ volts	$R_4 = 150\ \Omega$	$R_9 = 50\ \Omega$
	$R_5 = 200\ \Omega$	$R_{10} = 100\ \Omega$

11.22 The following equations describe the truss shown in Fig. 11.18. These equations were obtained by setting the sum of the horizontal forces and the sum of the vertical forces equal to zero at each pin.

$R_1 + T_1 \cos 60° + T_2 = 0$	(lower left pin)
$R_2 + T_1 \sin 60° = 0$	(lower left pin)
$-T_2 - T_3 \cos 60° + T_4 \cos 60° + T_5 = 0$	(lower middle pin)
$T_3 \sin 60° + T_4 \sin 60° = 0$	(lower middle pin)
$-T_5 - T_6 \cos 60° = 0$	(lower right pin)
$T_6 \sin 60° + R_3 = 0$	(lower right pin)
$-T_1 \cos 60° + T_3 \cos 60° + T_7 - F_1 \cos \theta_1 = 0$	(upper left pin)
$-T_1 \sin 60° - T_3 \sin 60° - F_1 \sin \theta_1 = 0$	(upper left pin)
$-T_4 \cos 60° - T_7 + T_6 \cos 60° - F_2 \cos \theta_2 = 0$	(upper right pin)
$-T_4 \sin 60° - T_6 \sin 60° - F_2 \sin \theta_2 = 0$	(upper right pin)

The objective is to determine the internal tensile forces (T_1 through T_7) and the reactive forces (R_1, R_2, and R_3) when the external forces (F_1 and F_2) and their angles (θ_1 and θ_2) are specified.

(*a*) Rearrange these equations into a more traditional form as 10 equations in 10 unknowns.

(*b*) Use Excel to solve for the unknown forces, assuming that $F_1 = 10,000$ lb$_f$, $F_2 = 7,000$ lb$_f$, $\theta_1 = 75°$, and $\theta_2 = 45°$. (Remember to convert the angles from degrees to radians when using the SIN and COS functions in Excel.) Use either of the methods described in this chapter to obtain a solution.

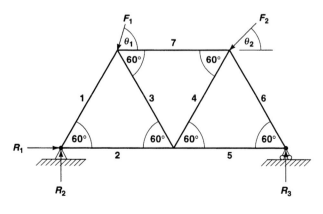

Figure 11.18 – A simple truss

11.23 A furnace wall is made up of three separate materials, as shown in Fig. 11.19. Let k_i represent the thermal conductivity of material i, let Δx_i represent the corresponding thickness (i = 1, 2, 3), and let h_1 and h_2 represent the convective heat transfer coefficients at the outer surfaces. If $T_a > T_0 > T_1 > T_2 > T_3 > T_b$, then the following steady-state heat transfer equations apply:

$$h_1(T_a - T_0) = \frac{k_1}{\Delta x_1}(T_0 - T_1)$$

$$\frac{k_1}{\Delta x_1}(T_0 - T_1) = \frac{k_2}{\Delta x_2}(T_1 - T_2)$$

$$\frac{k_2}{\Delta x_2}(T_1 - T_2) = \frac{k_3}{\Delta x_3}(T_2 - T_3)$$

$$\frac{k_3}{\Delta x_3}(T_2 - T_3) = h_2(T_3 - T_b)$$

Typically, the outer temperatures (T_a and T_b), the thermal conductivities, the material thicknesses, and the convective heat transfer coefficients are known. The objective is to determine the temperatures at the surfaces (T_0 and T_3) and the temperatures at the interfaces (T_1 and T_2).

(a) Rearrange these equations into a more traditional form as four equations in four unknowns.

(b) Using Excel, solve for the unknown temperatures, based upon the following given information (all units are consistent):

$\Delta x_1 = 0.5$ cm	$k_1 = 0.01$ cal/(cm)(s)(C°)
$\Delta x_2 = 0.3$ cm	$k_2 = 0.15$ cal/(cm)(s)(C°)
$\Delta x_3 = 0.2$ cm	$k_3 = 0.03$ cal/(cm)(s)(C°)
$T_a = 200°C$	$h_1 = 1.0$ cal/(cm²)(s)(C°)
$T_b = 20°C$	$h_2 = 0.8$ cal/(cm²)(s)(C°)

(c) Use the results of part (b) to determine the heat flux, in cal/(cm²)(s). (The heat flux is represented by either the left-hand side or the right-hand side of any of the above four equations.)

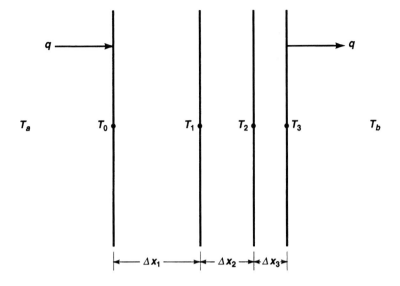

Figure 11.19 – A furnace wall consisting of three separate materials

11.24 A steel company manufactures four different types of steel alloys, called A1, A2, A3, and A4. Each alloy contains small amounts of chromium (Cr), molybdenum (Mo), titanium (Ti), and nickel (Ni). The required composition of each alloy is given below.

Alloy	*Cr*	*Mo*	*Ti*	*Ni*
A1	1.6%	0.7%	1.2%	0.3%
A2	0.6	0.3	1.0	0.8
A3	0.3	0.7	1.1	1.5
A4	1.4	0.9	0.7	2.2

Suppose the alloying materials are available in the following amounts:

Material	*Availability*
Cr	1200 kg/day
Mo	800
Ti	1000
Ni	1500

Using Excel, determine the daily production rate for each alloy in metric tons per day (1 metric ton = 1000 kg).

11.25 The *catenary* is a classical problem concerning the shape of a uniform cable hanging under its own weight. Figure 11.20 shows a drawing of the cable and its associated coordinate system. Notice that the horizontal distance x is measured from the lowest point (so that x_a is negative and x_b is positive), and that the lowest point is a distance v above the vertical datum.

Based upon this coordinate system, the length of the cable is calculated as

$$L = v\left[\sinh(x_b/v) - \sinh(x_a/v)\right]$$

and the difference in height between the end points is calculated as

$$h = v\left[\cosh(x_b/v) - \cosh(x_a/v)\right]$$

Also, the width spanned by the cable is given by

$$w = x_b - x_a$$

Suppose a 100-ft cable is suspended in such a manner that $x_a = -30$ ft and $h = 20$ ft. Solve for the corresponding values of x_b, w, and v in Excel using Solver. (Note that this problem requires solving a system of *nonlinear* algebraic equations.)

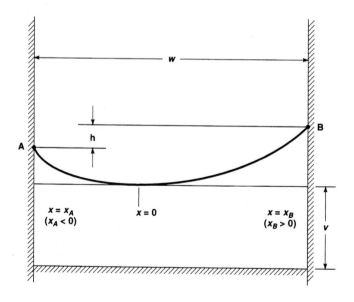

Figure 11.20 – A catenary

11.26 In Chapter 10 we learned that the horizontal displacement of a damped oscillating object as a function of time is given by the equation

$$x = x_0 e^{-\beta t} \left[\cos(\omega t) + (\beta / \omega) \sin(\omega t) \right]$$

(see Prob. 10.24). The parameters β and ω depend upon the mass of the object and the dynamic characteristics of the system (i.e., the spring constant and the damping constant).

Determine the values of x_0, β, and ω for an object that experiences the following displacement as a function of time:

t	x
1.0 sec	7.822 in
2.0	1.533
3.0	−2.979

Obtain your solution with Excel, using Solver.

CHAPTER 12

EVALUATING INTEGRALS

Many engineering problems involve the evaluation of integrals. Sometimes an integral is required in order to determine an area or the effect of some accumulation with respect to time. Or an integral may be required to evaluate a scientific formula. In any event, the need to evaluate integrals occurs frequently in many different areas of engineering.

Suppose, for example, that a gas is enclosed within a volume V. If the temperature of the gas varies with time and the gas is well mixed (so that there are no spatial variations), then the average temperature (with respect to time) is given by the following expression:

$$\overline{T} = \frac{\displaystyle\int_0^{t_{max}} T(t)\,dt}{\displaystyle\int_0^{t_{max}} dt} = \frac{1}{t_{max}} \int_0^{t_{max}} T(t)\,dt \tag{12.1}$$

where $T(t)$ represents the temperature of the gas at any given time and t represents time, where $0 \leq t \leq t_{max}$.

If the pressure-volume-temperature relationship of the gas is governed by the van der Waals equation of state, that is,

$$\left(P + \frac{a}{V^2}\right)(V - b) = RT \tag{12.2}$$

then the average pressure (with respect to time) is given by the expression

$$\overline{P} = \frac{1}{t_{max}} \int_{0}^{t_{max}} P(t)\,dt \tag{12.3}$$

Substituting Equation (12.2) into Equation (12.3), the average pressure is given by

$$\overline{P} = \frac{R \int_{0}^{t_{max}} T(t)\,dt}{(V - b)t_{max}} - \frac{a}{V^2} \tag{12.4}$$

Thus, we must evaluate an integral in order to obtain an explicit value for either $\overline{T}$ or $\overline{P}$. In order to do so, we will need some information about the exact variation of the temperature with time. That is, we will have to replace the general expression $T(t)$ with an explicit formula for temperature as a function of time.

If $T(t)$ can be represented by some simple formula, it may be possible to evaluate the integral in Equations (12.1) and (12.3) using the rules learned in calculus. However, if $T(t)$ is represented by some relatively complex formula, or by a graph or a tabulated set of data, then the integral cannot be evaluated using the classical rules of calculus. In such situations, various *numerical techniques* can be used to evaluate the integral with reasonable accuracy.

In this chapter we will discuss some simple numerical integration techniques and see how they can be implemented within Excel. We will also discuss a method for integrating measured data when significant scatter is present.

12.1 THE TRAPEZOIDAL RULE

Suppose we are given a continuously varying function $y = f(x)$, defined over the interval $a \le x \le b$, as illustrated in Fig. 12.1. Then the integral

$$I = \int_{a}^{b} f(x)\,dx = \int_{a}^{b} y\,dx \tag{12.5}$$

can be interpreted as the area under the curve, as indicated in Fig. 12.1. The function $y = f(x)$ is referred to as the *integrand*.

Now suppose we approximate this irregularly shaped area with a large number of adjacent, narrow, rectangular intervals, as illustrated in Fig. 12.2. Then we can think of the integral as the sum of the areas of these rectangular intervals.

The area of each interval can be determined as the product of the height, y, and the width, Δx. Thus,

$$I = \sum_{i=1}^{n} A_i \qquad (12.6)$$

where

$$A_i = y_i \, \Delta x_i \qquad (12.7)$$

and n is the total number of intervals.

If we combine Equations (12.6) and (12.7), we obtain

$$I = \sum_{i=1}^{n} y_i \, \Delta x_i \qquad (12.8)$$

This is the basis for the *trapezoidal rule*, which is the simplest approach to carrying out a numerical integration. The larger the number of intervals and the smaller the width of each interval, the better the approximation. The selection of a satisfactory interval size will depend upon the particular integrand.

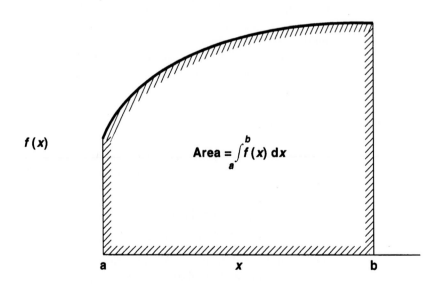

Figure 12.1 – Interpreting an integral as the area under a curve

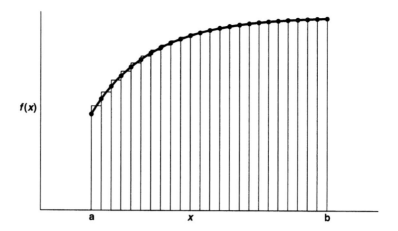

Figure 12.2 – Approximating the area under a curve with a series of rectangles

Unequally Spaced Data

Suppose we are given a set of n data points: (x_1, y_1), (x_2, y_2), . . . , (x_n, y_n), where $x_1 = a$ and $x_n = b$. These data points define $n-1$ rectangular intervals, where the width of the ith interval is given by

$$\Delta x_i = x_{i+1} - x_i \tag{12.9}$$

The height of the rectangle associated with the ith interval can be expressed as the average of the y-values at the interval boundaries; that is,

$$\bar{y}_i = \frac{y_i + y_{i+1}}{2} \tag{12.10}$$

Hence the area of the rectangular interval is

$$A_i = \bar{y}_i \Delta x_i = \frac{(y_i + y_{i+1})(x_{i+1} - x_i)}{2} \tag{12.11}$$

and the entire integral can be approximated by

$$I = \int_a^b y \, dx = \sum_{i=1}^{n-1} A_i = \frac{1}{2} \sum_{i=1}^{n-1} (y_i + y_{i+1})(x_{i+1} - x_i) \tag{12.12}$$

Equation (12.12) represents the trapezoidal rule for *unequally spaced data*.

Example 12.1 The Trapezoidal Rule for Unequally Spaced Data

The current passing through an electrical inductor can be determined from the equation

$$i = \frac{1}{L} \int_0^t v \, dt$$

where

i = current (amperes)

L = inductance (henries)

v = voltage

t = time (seconds)

Suppose a current of 2.15 amperes is induced over a period of 500 milliseconds (ms). The variation in voltage with time during this period is given below.

t (ms)	v (volts)	t	v	t	v	t	v
0	0	40	45	90	45	180	27
5	12	50	49	100	42	230	21
10	19	60	50	120	36	280	16
20	30	70	49	140	33	380	9
30	38	80	47	160	30	500	4

Determine the value of the inductance, using the above equation.

In order to determine the inductance, we must first evaluate the integral

$$I = \int_0^{0.5} v \, dt$$

If we use the trapezoidal rule, we will have 19 intervals since there are 20 data points. The calculations using the trapezoidal rule are summarized in the table below.

Note that the times have been converted to seconds within this table (column 2). The values in the fourth column of the table were obtained using Equation (12.9), and the values in the fifth column were obtained from Equation (12.10). The last column contains the products of the values in the fourth and fifth columns (the interval areas).

Point No.	t (seconds)	v (volts)	Width, Δt (seconds)	Height, $\bar{v}$ (volts)	Area, $\bar{v}\,\Delta t$ (volt) (s)
1	0	0	0.005	6.0	0.030
2	0.005	12	0.005	15.5	0.0775
3	0.01	19	0.01	24.5	0.245
4	0.02	30	0.01	34.0	0.340
5	0.03	38	0.01	41.5	0.415
6	0.04	45	0.01	47.0	0.470
7	0.05	49	0.01	49.5	0.495
8	0.06	50	0.01	49.5	0.495
9	0.07	49	0.01	48.0	0.480
10	0.08	47	0.01	46.0	0.460
11	0.09	45	0.01	43.5	0.435
12	0.10	42	0.02	39.0	0.780
13	0.12	36	0.02	34.5	0.690
14	0.14	33	0.02	31.5	0.630
15	0.16	30	0.02	28.5	0.570
16	0.18	27	0.05	24.0	1.200
17	0.23	21	0.05	18.5	0.925
18	0.28	16	0.10	12.5	1.250
19	0.38	9	0.12	6.5	0.780
20	0.50	4			
				Total:	10.7675

At the bottom of the table, beneath the last column, we see that the sum of the individual areas is 10.7675 (volt) (s). Hence, we conclude that the area under the curve (the value of the integral) is 10.7675 and the value of the inductance is

$$L = \frac{1}{i} \int_{0}^{0.5} v\, dt = \frac{10.7675}{2.15} = 5.0081 \text{ henries}$$

We conclude that the inductance is approximately 5 henries.

Excel does not include any built-in features for evaluating integrals. Since the calculations are carried out in a tabular format, however, they lend themselves very naturally to a spreadsheet solution. Thus, it is easy to evaluate an integral in Excel by following the layout given in the above example, as illustrated in the next example.

Example 12.2 The Trapezoidal Rule for Unequally Spaced Data in Excel

Repeat the problem given in Example 12.1 using Excel to carry out the numerical integration.

The procedure is the same as that used in Example 12.1. Now, however, we will determine the individual areas and the sum of the areas within an Excel worksheet, as shown in Fig. 12.3.

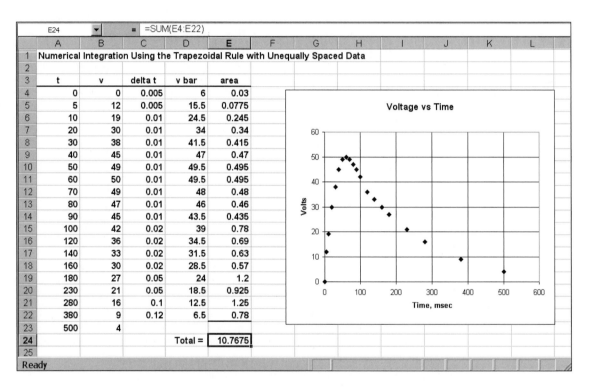

Figure 12.3 – Numerical integration using the trapezoidal rule with unequally spaced data

Within this worksheet, the first two columns contain the given data. The third column contains the calculated interval widths, in units of seconds rather than milliseconds. Thus, the value in cell C4 is obtained from the formula =0.001*(A5–A4), the value in cell C5 is obtained from =0.001*(A6–A5), and so on.

The fourth column contains the calculated height of each interval. Therefore, the value in cell D4 is obtained from the formula =0.5*(B5+B4), the value in cell D5 is obtained from =0.5*(B6+B5), and so on. Similarly, column five contains the area of each interval. Hence, the value in cell E4 is obtained as =C4*D4, the value in cell E5 is obtained as =C5*D5, and so on.

The sum of the areas is calculated in cell E24 from the formula =SUM(E4:E22). We see that this value is 10.7675. Therefore, we conclude that the value of the desired integral is 10.7675 (volt) (s). This value agrees with the value obtained in the previous example. Hence, we determine that the unknown inductance is 5.0081 henries, as in the last example.

Notice that the worksheet also contains a graph of the given data, simply to illustrate the manner in which the voltage varies with time. It is the area under this curve that is represented by the desired integral.

Equally Spaced Data

Now suppose we have n equally spaced data points, where the successive x-values are separated by a distance Δx. Then Equation (12.12) simplifies to

$$I = \int_a^b y\,dx = \frac{1}{2}(y_1 + 2y_2 + 2y_3 + \cdots + 2y_{n-2} + 2y_{n-1} + y_n)\Delta x \qquad (12.13)$$

or

$$I = \left(\frac{y_1 + y_n}{2} + \sum_{i=2}^{n-1} y_i \right) \Delta x \qquad (12.14)$$

Equation (12.14) represents the trapezoidal rule for *equally spaced data*.

Example 12.3 The Trapezoidal Rule for Equally Spaced Data

Earlier in this chapter, we were given an expression for the average pressure of a van der Waals gas when the temperature of the gas varies with time over the interval $0 \le t \le t_{max}$; that is,

$$\overline{P} = \frac{R \displaystyle\int_0^{t_{max}} T(t)\,dt}{(V - b)\,t_{max}} - \frac{a}{V^2} \qquad (12.4)$$

In this expression, P is the absolute pressure, V is the volume per mole, T is the absolute temperatue, R is the ideal gas constant (0.082054 liter atm/mole $^\circ$K), and the symbols a and b are van der Waals constants, which are unique to each particular gas.

Suppose one mole of carbon dioxide is enclosed within a one-liter container. Determine the average pressure if the temperature increases with time in accordance with the expression

$$T = 300 + 12t$$

over the interval $0 \leq t \leq 100$ seconds. The van der Waals constants for carbon dioxide are $a = 3.592$ liter^2 atm/mole^2 and $b = 0.04267$ liter/mole.

In order to determine the average pressure, we must evaluate the expression

$$\overline{P} = \frac{R \int_0^{100} (300 + 12t)\, dt}{100\,(V - b)} - \frac{a}{V^2}$$

Substituting the appropriate numerical values, this equation becomes

$$\overline{P} = \frac{0.082054 \int_0^{100} (300 + 12t)\, dt}{100\,(1 - 0.04267)} - \frac{3.592}{1^2}$$

or

$$\overline{P} = 8.571 \times 10^{-4} \int_0^{100} (300 + 12t)\, dt - 3.592$$

We must evaluate the integral in order to determine the average pressure. This particular integral can easily be evaluated using the classical rules of calculus, resulting in a value of $I = 9 \times 10^4$ and $\overline{P} = 73.547$ atm. For illustrative purposes, however, let us evaluate the integral using the trapezoidal rule. If we select 21 equally spaced points, we can generate the following table:

Point No.	t (seconds)	T (°K)	Point No	t (seconds)	T (°K)
1	0	300	12	55	960
2	5	360	13	60	1020
3	10	420	14	65	1080
4	15	480	15	70	1140
5	20	540	16	75	1200
6	25	600	17	80	1260
7	30	660	18	85	1320
8	35	720	19	90	1380
9	40	780	20	95	1440
10	45	840	21	100	1500
11	50	900			

Substituting these values into Equation (12.14), we obtain

$$I = \left[\frac{300 + 1500}{2} + (360 + 420 + 480 + \cdots + 1380 + 1440) \right] 5 = 9 \times 10^4$$

In this case, the result of the trapezoidal rule integration agrees exactly with the result obtained using calculus. This is because the integrand is a straight line, so the area determined for each interval is exact. If the integrand were a more complicated function, however, we would see that the trapezoidal rule offers only an approximation to the correct answer, with the accuracy of the approximation increasing as the number of intervals (the number of points) increases.

We can now determine the average pressure of the gas, as originally requested. The pressure can be determined as

$$\overline{P} = (8.571 \times 10^{-4})(9 \times 10^4) - 3.592 = 73.547$$

Hence, the average pressure is 73.547 atm. This is the same value obtained using calculus.

In Excel, use of the trapezoidal rule with equally spaced data points is particularly simple since formulas can be used to construct most of the tabular entries.

Example 12.4 The Trapezoidal Rule for Equally Spaced Data in Excel

Use Excel to determine the average pressure of one mole of carbon dioxide in a one-liter container when the temperature increases with time, as described in Example 12.3.

In Example 12.3, we established that the average pressure is given by the expression

$$\overline{P} = 8.571 \times 10^{-4} \int_{0}^{100} (300 + 12t)\, dt - 3.592$$

Hence, this problem requires that we evaluate the given integral using Excel. To do so, we create the Excel worksheet shown in Fig. 12.4. The first column contains the times, ranging from 0 to 100 seconds. All of these values, other than the first, are generated using cell formulas. Thus, the value shown in cell A5, for example, is determined by the cell formula =A4+5. This formula is then copied to the remaining cells in column A. Fig. 12.5 shows the cell formulas used to generate the values given in Fig. 12.4.

Similarly, all of the temperatures shown in column B, other than the first, are generated with cell formulas. For example, the value shown in cell B5 is determined by the formula =B4+12*(A5−A4). This formula is then copied to the remaining cells in column B, as shown in Fig. 12.5.

	F25	▼		=	=((B4+B24)/2+D25)*5								
	A	B	C	D	E	F	G	H	I	J	K	L	M	N
1	Numerical Integration Using the Trapezoidal Rule with Equally Spaced Data												
2													
3	Time	Temp											
4	0	300											
5	5	360		360									
6	10	420		420									
7	15	480		480									
8	20	540		540									
9	25	600		600									
10	30	660		660									
11	35	720		720									
12	40	780		780									
13	45	840		840									
14	50	900		900									
15	55	960		960									
16	60	1020		1020									
17	65	1080		1080									
18	70	1140		1140									
19	75	1200		1200									
20	80	1260		1260									
21	85	1320		1320									
22	90	1380		1380									
23	95	1440		1440									
24	100	1500											
25			Sum =	17100	I =	90000							
26													

Ready

Figure 12.4 – Numerical integration using the trapezoidal rule with equally spaced data

Column D contains all of the temperature values, other than the first and last. These values were obtained by copying the corresponding values from column B (using Copy/Paste Special/Values). The sum of these values is shown at the bottom of column D, in cell D25. This sum represents the general summation term given in Equation (12.14).

Finally, the value of the integral (I = 90,000) is shown in cell F25. This value was determined using Equation (12.14). The general strategy was to add the average of the first and last temperatures to the value in cell D25 and then multiply this sum by the interval width (5). The highlighted cell formula shown in cell F25 of Fig. 12.5 illustrates how this strategy was carried out.

Once the value of the integral is known, we can substitute it into the equation for the average pressure; that is,

$$\overline{P} = (8.571 \times 10^{-4})(9 \times 10^4) - 3.592 = 73.547$$

Thus, we obtain the value of 73.547 atmospheres for the average pressure, as in the previous example.

	F25 ▼	=	=((B4+B24)/2+D25)*5				
	A	B	C	D	E	F	G
1	Numerical Integratio						
2							
3	Time	Temp					
4	0	300					
5	=A4+5	=B4+12*(A5-A4)		360			
6	=A5+5	=B5+12*(A6-A5)		420			
7	=A6+5	=B6+12*(A7-A6)		480			
8	=A7+5	=B7+12*(A8-A7)		540			
9	=A8+5	=B8+12*(A9-A8)		600			
10	=A9+5	=B9+12*(A10-A9)		660			
11	=A10+5	=B10+12*(A11-A10)		720			
12	=A11+5	=B11+12*(A12-A11)		780			
13	=A12+5	=B12+12*(A13-A12)		840			
14	=A13+5	=B13+12*(A14-A13)		900			
15	=A14+5	=B14+12*(A15-A14)		960			
16	=A15+5	=B15+12*(A16-A15)		1020			
17	=A16+5	=B16+12*(A17-A16)		1080			
18	=A17+5	=B17+12*(A18-A17)		1140			
19	=A18+5	=B18+12*(A19-A18)		1200			
20	=A19+5	=B19+12*(A20-A19)		1260			
21	=A20+5	=B20+12*(A21-A20)		1320			
22	=A21+5	=B21+12*(A22-A21)		1380			
23	=A22+5	=B22+12*(A23-A22)		1440			
24	=A23+5	=B23+12*(A24-A23)					
25			Sum =	=SUM(D5:D23)		I =	=((B4+B24)/2+□
26							
Ready							

Figure 12.5 – The corresponding cell formulas

Effect of Interval Width

When carrying out a numerical integration, the accuracy of the answer is strongly influenced by the width of the intervals. The use of narrower intervals results in a more accurate answer. Since a decrease in the interval width results in a corresponding increase in the number of intervals, we conclude that *a greater number of intervals results in a more accurate answer*. We illustrate this in the next example.

Example 12.5 Accuracy and the Number of Intervals

Let us once again solve a problem whose answer is known in order to illustrate a point. In particular, let us use the trapezoidal rule to evaluate the integral

$$I = \int_{0}^{2} 3x^2 \, dx$$

This problem is easily solved using classical calculus, resulting in the value $I = 8$. We will solve the problem repeatedly, using the trapezoidal rule with $n = 4$, 10, 20, 50, and 100 equally spaced intervals. By comparing each result with the known solution ($I = 8$), we can see the effect of the number of intervals (interval width) on the accuracy of the numerical integration.

The numerical integrations are carried out in a manner similar to that described in the last example. Figure 12.6 shows an Excel worksheet containing the calculations for $n = 4$, 10, and 20 intervals. (The calculations are not shown for $n = 50$ and $n = 100$ in order to conserve space, though the results are similar and the final values of I are included in the overall summary below.)

	C15		f_x =(SUM(B9:B13)+SUM(B10:B12))*D6/2											
	A	B	C	D	E	F	G	H	I	J	K	L	M	N
1	Integrate 3x^2 from x=0 to x=2 using the trapezoidal rule													
2														
3	(Answer from calculus: I = 8 exactly)													
4														
5														
6	Four intervals		Delta x = 0.5			Ten intervals		Delta x = 0.2			Twenty intervals		Delta x = 0.1	
7														
8	x	y				x	y				x	y		
9	0.0	0.00				0.0	0.00				0.0	0.00		
10	0.5	0.75				0.2	0.12				0.1	0.03		
11	1.0	3.00				0.4	0.48				0.2	0.12		
12	1.5	6.75				0.6	1.08				0.3	0.27		
13	2.0	12.00				0.8	1.92				0.4	0.48		
14						1.0	3.00				0.5	0.75		
15		I =	8.25			1.2	4.32				0.6	1.08		
16						1.4	5.88				0.7	1.47		
17						1.6	7.68				0.8	1.92		
18						1.8	9.72				0.9	2.43		
19						2.0	12.00				1.0	3.00		
20											1.1	3.63		
21							I =	8.04			1.2	4.32		
22											1.3	5.07		
23											1.4	5.88		
24											1.5	6.75		
25											1.6	7.68		
26											1.7	8.67		
27											1.8	9.72		
28											1.9	10.83		
29											2.0	12.00		
30														
31												I =	8.01	
32														
	Ready													

Figure 12.6 – Accuracy of the integral vs number of intervals

The values of the integral obtained using the trapezoidal rule (I) are summarized in Fig. 12.7 as a function of the number of intervals (n) and the interval size (Δx). The results confirm our intuition, that the accuracy of the numerical integration increases as the number of intervals increases within given bounds (or, equivalently, as the width of each interval decreases). For this function, the value of the integral approaches the true answer ($I = 8$) very closely when 50 or more intervals are utilized in the calculation. (Note that the trapezoidal rule yields $I = 8.0016$ when $n = 50$.) You should understand, however, that the choice of a reasonable number of intervals will depend on the curvature of the function being integrated as well as the range of the independent variable. In general, functions that exhibit a great deal of curvature will require a smaller interval width, and hence a larger number of intervals, in order to obtain a relatively accurate solution.

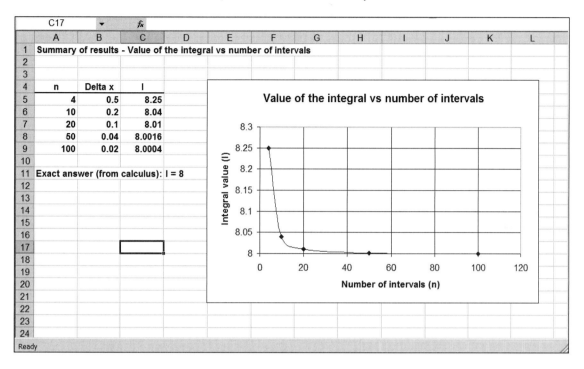

Figure 12.7 – Summary of results

Problems

12.1   Evaluate the integral

$$I = \int_1^5 x^3 dx$$

using each of the following methods:

(*a*)   Classical calculus.

(*b*)   The trapezoidal rule within an Excel worksheet, using eight equally spaced intervals.

(c)   The trapezoidal rule within an Excel worksheet, using 20 equally spaced intervals.

Compare the answers obtained in each case.

12.2   For the integral given in Prob. 12.1, how many equally spaced intervals would be required using the trapezoidal rule in order to obtain an answer that is within 2 percent of the exact answer obtained with classical calculus? Use Excel to carry out the integration using the trapezoidal rule.

12.3   Evaluate the integral

$$I = \int_0^{\pi} \sin x \, dx$$

using each of the following methods:

(a)   Classical calculus.

(b)   The trapezoidal rule within an Excel worksheet, using 10 equally spaced intervals.

(c)   The trapezoidal rule in an Excel worksheet, using 24 equally spaced intervals.

Compare the answers obtained in each case.

12.2 SIMPSON'S RULE

Simpson's rule is a widely used numerical integration technique that combines simplicity and accuracy. It is similar to the trapezoidal rule in the sense that it approximates an irregular area with a number of geometrically simple subintervals. Rather than construct single-interval rectangles, however, we now pass a second-order polynomial (i.e., a parabola) through successive groups of three adjacent, equally spaced data points, as illustrated in Fig. 12.8. The area under each polynomial can then be obtained by direct integration.

If the number of subintervals is even (i.e., if the number of data points is odd), then the repeated use of this parabolic approximation results in the following simple expression:

$$I = \int_a^b y \, dx = \frac{1}{3}(y_1 + 4y_2 + 2y_3 + 4y_4 + 2y_5 + \cdots + 2y_{n-2} + 4y_{n-1} + y_n)\Delta x \quad (12.15)$$

where n is the number of data points. Notice that the interior y-values are alternately multiplied by 4 and 2.

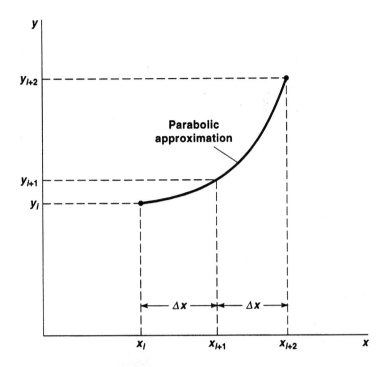

Figure 12.8 – Passing a parabola through three equally spaced points

Example 12.6 Comparing Simpson's Rule with the Trapezoidal Rule

Evaluate the integral

$$I = \int_0^1 e^{-x^2} dx$$

using both the trapezoidal rule and Simpson's rule, with 10 equally spaced intervals in each case. Compare the results obtained with each method with the tabulated answer, which is 0.7468. (This particular integral *cannot* be evaluated using the methods of classical calculus. The integral has, however, been evaluated numerically to a high degree of precision, and tabulated values are available in a number of reference books.)

Since there will be 10 subintervals, we will require 11 data points and Δx will have a value of 0.10. Hence we can construct the following table:

Point No.	x_i	y_i	Point No.	x_i	y_i
1	0	1.000	7	0.6	0.698
2	0.1	0.990	8	0.7	0.613
3	0.2	0.961	9	0.8	0.527
4	0.3	0.914	10	0.9	0.445
5	0.4	0.852	11	1.0	0.368
6	0.5	0.779			

Substituting these values into Equation (12.14) for the trapezoidal rule, we obtain

$$I = \left[\frac{1.000 + 0.368}{2} + (0.990 + 0.961 + \cdots + 0.527 + 0.445) \right] (0.1) = 0.7463$$

Similarly, if we substitute these same values into Equation (12.15) for Simpson's rule, we obtain

$$I = \frac{1}{3} \left[1.000 + 4(0.990) + 2(0.961) + 4(0.914) + \cdots + \right.$$

$$\left. 2(0.527) + 4(0.445) + 0.368 \right] (0.1) = 0.7469$$

Summarizing the results, we have

Tabulated answer:	0.7468
Trapezoidal rule:	0.7463
Simpson's rule:	0.7469

Thus, we see that Simpson's rule results in a more accurate answer with very little additional computational effort.

Simpson's rule is easy to implement in Excel. One way to do this is to enter the x-values in one column, the y-values in the next column, followed by an additional column containing the y-values multiplied by their appropriate constants (i.e., multiplied by 4 or 2, except for the end points). The value of the integral is then obtained by summing this last column and multiplying the sum by $\Delta x/3$. The details are illustrated in the following example.

Example 12.7 Simpson's Rule in Excel

In the last example we used Simpson's rule to evaluate the integral

$$I = \int_0^1 e^{-x^2} dx$$

based upon 10 equally spaced intervals. We now repeat the solution within an Excel worksheet.

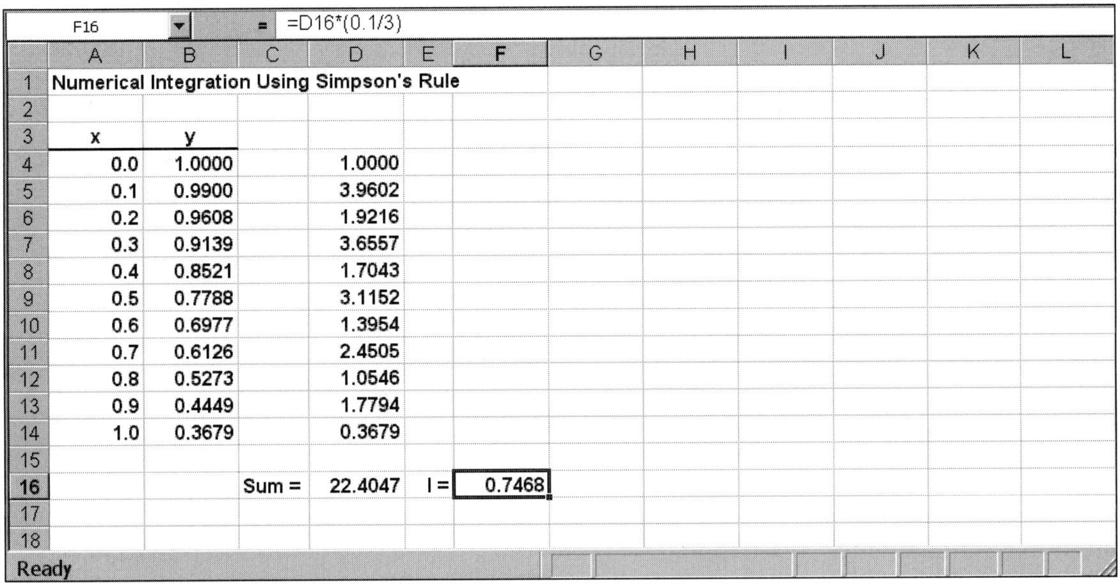

	F16		=	=D16*(0.1/3)							
	A	B	C	D	E	F	G	H	I	J	K	L
1	Numerical Integration Using Simpson's Rule										
2												
3	x	y										
4	0.0	1.0000		1.0000								
5	0.1	0.9900		3.9602								
6	0.2	0.9608		1.9216								
7	0.3	0.9139		3.6557								
8	0.4	0.8521		1.7043								
9	0.5	0.7788		3.1152								
10	0.6	0.6977		1.3954								
11	0.7	0.6126		2.4505								
12	0.8	0.5273		1.0546								
13	0.9	0.4449		1.7794								
14	1.0	0.3679		0.3679								
15												
16			Sum =	22.4047	I =	0.7468						
17												
18												

Ready

Figure 12.9 – Numerical integration using Simpson's rule

Figure 12.9 shows an Excel worksheet that incorporates the previously described layout. Column A contains the 11 equally spaced x-values, ranging from 0 to 1. All of these values, beginning with the second, are generated using cell formulas. The value in cell A5, for example, is generated using the cell formula =A4+0.1. This formula is copied to the remaining cells in column A, resulting in the values shown. Figure 12.10 shows the cell formulas used to generate the values shown in Fig. 12.9.

Column B in Fig. 12.9 contains the corresponding y-values. These values are all generated using cell formulas. For example, the value in cell B4 is generated using the cell formula =EXP(−($A4^2)). This formula is copied to the remaining cells in column B, as shown in Fig. 12.10.

Column D in Fig. 12.9 contains modifications of the formulas used to generate the values in column B. These formulas, like those in column B, refer to the x-values in column A. (Hence the use of absolute addressing with respect to column A.) Now, however, all of the cell formulas, other than the first and the last, are multiplied by either 4 or 2, as required by Simpson's rule. Thus, cell D5 contains the formula =4*EXP(−($A5^2)), cell D6 contains the formula =2*EXP(−($A6^2)), and so on, as shown in Fig. 12.10.

Returning to Fig. 12.9, the sum of the entries in column D is computed in cell D16. This value represents the sum of the terms shown in parentheses in Equation (12.15). This value (22.4047) is obtained using the cell formula =SUM(D4:D14).

	F16	▼	=	=D16*(0.1/3)			
	A	B	C	D	E	F	
1	Numerical Integratio						
2							
3	x	y					
4	0	=EXP(-($A4^2))		=EXP(-($A4^2))			
5	=A4+0.1	=EXP(-($A5^2))		=4*EXP(-($A5^2))			
6	=A5+0.1	=EXP(-($A6^2))		=2*EXP(-($A6^2))			
7	=A6+0.1	=EXP(-($A7^2))		=4*EXP(-($A7^2))			
8	=A7+0.1	=EXP(-($A8^2))		=2*EXP(-($A8^2))			
9	=A8+0.1	=EXP(-($A9^2))		=4*EXP(-($A9^2))			
10	=A9+0.1	=EXP(-($A10^2))		=2*EXP(-($A10^2))			
11	=A10+0.1	=EXP(-($A11^2))		=4*EXP(-($A11^2))			
12	=A11+0.1	=EXP(-($A12^2))		=2*EXP(-($A12^2))			
13	=A12+0.1	=EXP(-($A13^2))		=4*EXP(-($A13^2))			
14	=A13+0.1	=EXP(-($A14^2))		=EXP(-($A14^2))			
15							
16			Sum =	=SUM(D4:D14)		I =	=D16*(0.1/3)
17							
18							
Ready							

Figure 12.10 – The corresponding cell formulas

Cell F16 contains the value of the integral, $I = 0.7468$, as determined by Simpson's rule. This value was obtained with the cell formula =D16*(0.1/3), as indicated by Equation (12.15). (Note that the value 0.1 appearing within the parentheses represents the interval width, Δx.) The value obtained agrees with the exact answer to four significant figures. This value is slightly more accurate than the value obtained using Simpson's rule in Example 12.5 because of the greater precision in the spreadsheet solution.

The summation term in Equation (12.15) can be evaluated several other ways in Excel. For example, we might place the end points in one column, two times the interior points in a second column, and two times every other interior point in a third column. The desired summation can be obtained by adding the entries in each column and then adding the three column sums, as shown in Example 12.7.

Another approach is to make use of Excel's IF-THEN-ELSE feature (see Chap. 13) to provide the proper coefficients (alternating 4s and 2s) for the interior y-values. The selection of an appropriate coefficient (either 4 or 2) can be determined by numbering the interior points and then determining if each one is either odd or even. This method is illustrated in the next chapter.

Example 12.8 Simpson's Rule in Excel – Another Approach

Now let's consider another way to evaluate the integral

$$I = \int\limits_0^1 e^{-x^2}\,dx$$

within Excel, again using Simpson's rule with 10 equally spaced intervals. Now, however, we will use a different layout to determine the summation term in Equation (12.15).

An Excel worksheet similar to that presented in Example 12.6 is shown in Fig. 12.11. This worksheet is similar to that shown earlier, in Fig. 12.9. Again, columns A and B contain the x-values and the corresponding y-values, as explained in Example 12.6. Now, however, column D contains the y-values for the end points, column E contains two times the calculated y-values for *all* of the interior points, and column F contains two times the calculated y-values for *every other* interior point. Line 16 contains the sums of the values in columns D, E, and F. These three sums are added together in cell D18. Finally, the value of the integral is obtained in cell D20 by multiplying the value in cell D18 by (0.1/3). Cell D20 shows the value of the integral, $I = 0.7468$.

D20		=	=D18*(0.1/3)								
A	B	C	D	E	F	G	H	I	J	K	L
1 Numerical Integration Using Simpson's Rule											
2											
3 x	y		y	2y	2y						
4 0.0	1.0000		1.0000								
5 0.1	0.9900			1.9801	1.9801						
6 0.2	0.9608			1.9216							
7 0.3	0.9139			1.8279	1.8279						
8 0.4	0.8521			1.7043							
9 0.5	0.7788			1.5576	1.5576						
10 0.6	0.6977			1.3954							
11 0.7	0.6126			1.2253	1.2253						
12 0.8	0.5273			1.0546							
13 0.9	0.4449			0.8897	0.8897						
14 1.0	0.3679		0.3679								
15											
16	Column Sums =		1.3679	13.5563	7.4805						
17											
18	Overall Sum =		22.4047								
19											
20		I =	0.7468								
21											
22											
Ready											

Figure 12.11 – Another way to carry out Simpson's rule in Excel

Figure 12.12 shows the cell formulas used to generate the values shown in Fig. 12.11.

	D20	▾	=	=D18*(0.1/3)		
	A	B	C	D	E	F
1	Numerical Integratio					
2						
3	x	y		y	2y	2y
4	0	=EXP(-($A4^2))		=EXP(-($A4^2))		
5	=A4+0.1	=EXP(-($A5^2))			=2*EXP(-($A5^2))	=2*EXP(-($A5^2))
6	=A5+0.1	=EXP(-($A6^2))			=2*EXP(-($A6^2))	
7	=A6+0.1	=EXP(-($A7^2))			=2*EXP(-($A7^2))	=2*EXP(-($A7^2))
8	=A7+0.1	=EXP(-($A8^2))			=2*EXP(-($A8^2))	
9	=A8+0.1	=EXP(-($A9^2))			=2*EXP(-($A9^2))	=2*EXP(-($A9^2))
10	=A9+0.1	=EXP(-($A10^2))			=2*EXP(-($A10^2))	
11	=A10+0.1	=EXP(-($A11^2))			=2*EXP(-($A11^2))	=2*EXP(-($A11^2))
12	=A11+0.1	=EXP(-($A12^2))			=2*EXP(-($A12^2))	
13	=A12+0.1	=EXP(-($A13^2))			=2*EXP(-($A13^2))	=2*EXP(-($A13^2))
14	=A13+0.1	=EXP(-($A14^2))		=EXP(-($A14^2))		
15						
16		Column Sums =		=SUM(D4:D14)	=SUM(E4:E14)	=SUM(F4:F14)
17						
18		Overall Sum =		=SUM(D16:F16)		
19						
20			I =	=D18*(0.1/3)		
21						
22						
Ready						

Figure 12.12 – The corresponding cell formulas

Problems

12.4 Repeat Prob. 12.1 using Simpson's rule within an Excel worksheet. Compare the results obtained using Simpson's rule with those obtained earlier using the trapezoidal rule.

12.5 When solving Prob. 12.4, how many equally spaced intervals would be required using Simpson's rule in order to obtain an answer that is within 2 percent of the answer obtained with classical calculus? Use Excel to carry out the integration using Simpson's rule. Compare the results with those obtained in Prob. 12.2 using the trapezoidal rule.

12.6 Repeat Prob. 12.3 using Simpson's rule within an Excel worksheet. Compare the results obtained using Simpson's rule with those obtained earlier using the trapezoidal rule.

12.7 The charge within a capacitor is given by the expression

$$Q = \int_{0}^{t_{max}} i \, dt$$

where Q is the charge, in coulombs, i is the current, in amperes, and t is the time, in seconds. Suppose the variation of current with time is given by the equation

$$i = 0.2 \, e^{-0.1t}$$

Determine the charge stored within the capacitor during the first 200 seconds using each of the following methods:

(*a*) Classical calculus.

(*b*) The trapezoidal rule, within an Excel worksheet.

(*c*) Simpson's rule, within an Excel worksheet.

12.8 The total amount of water discharged from the bottom of a tank is given by the expression

$$V = \int_{0}^{t} q \, dt$$

where q, the volumetric flow rate in cubic feet per second, is given by the expression

$$q = 0.1(80 - t) \qquad 0 \le t \le 80 \text{ seconds}$$

and t is the time in seconds. Determine

(*a*) The total amount of water initially in the tank, assuming the entire tank is emptied during the first 80 seconds.

(*b*) The quantity of water discharged during the first 40 seconds.

(*c*) The quantity of water discharged between times $t = 15$ and $t = 60$ seconds.

Determine each value using each of the following methods:

(*a*) Classical calculus.

(*b*) The trapezoidal rule, within an Excel worksheet.

(*c*) Simpson's rule, within an Excel worksheet.

12.9   A group of students have designed a small rocket, which they plan to fire from the football field. The rocket contains enough fuel to burn for eight seconds. During this period, the rocket's vertical velocity will be determined by the expression

$$v = 6t^2 \qquad 0 \leq t \leq 8$$

where v represents the vertical velocity in ft/second and t represents the time in seconds.

After the 8-second burn, the rocket will fall to the ground under the force of gravity. Hence,

$$v = v_0 - 32.2\,(t - 8) \qquad t > 8$$

where v_0 is the velocity at the end of the eight-second burn.

Integrate these two expressions to determine the maximum height attained by the rocket and the time required for the rocket to return to the ground.

Calculate each value using each of the following methods:

(*a*)   Classical calculus.

(*b*)   The trapezoidal rule, within an Excel worksheet.

(*c*)   Simpson's rule, within an Excel worksheet.

12.10 The heat absorbed by a solid when its temperature is increased is given by the expression

$$Q = m \int_{T_1}^{T_2} C_p \, dT$$

In this expression, Q represents the total heat absorbed, in BTUs; m is the mass of the solid, in lbs; C_p is the specific heat of the solid, in BTU/(lb)(F°); and T is the temperature, in degrees Fahrenheit.

How much heat is absorbed when a 20-lb mass of copper is heated from 100°F to 500°F? The specific heat of copper is given by the expression

$$C_p = 0.0909 + 2 \times 10^{-5} T - 1 \times 10^{-9} T^2$$

Determine your answer using each of the following methods:

(*a*)   Classical calculus.

(*b*)   The trapezoidal rule, within an Excel worksheet.

(*c*)   Simpson's rule, within an Excel worksheet.

12.11   When water flows slowly within a circular pipe, its velocity increases with the square of the distance from the pipe wall. Thus, the velocity is zero at the wall and reaches a maximum at the center of the pipe.

Suppose the velocity distribution within a 12-inch-diameter pipe is given by the expression

$$v = 0.3\left(1 - \frac{r}{0.5}\right)^2$$

where v is the velocity, in ft/second, and r is the radial distance from the center of the pipe, in feet. (Hence, $0 \le r \le 0.5$.) Then the volumetric flow rate through the pipe, in cu ft/second, will be given by the expression

$$Q = \int v dA = 2\pi \int_0^{0.5} v r dr$$

Determine the volumetric flow within the pipe, using each of the following methods:

(*a*)   Classical calculus.

(*b*)   The trapezoidal rule, within an Excel worksheet.

(*c*)   Simpson's rule, within an Excel worksheet.

12.3 INTEGRATING MEASURED DATA

Numerical integration works very well when integrating an analytical function. It may also work reasonably well when integrating measured data with minimal scatter. (Note that there is usually *some* scatter associated with measured data.) When the scatter is significant, however, numerical integration based upon the measured data points can result in large errors. In such situations, a reasonably accurate value for the integral can often be obtained by passing a curve through the aggregate of the data, as discussed in Chap. 9, and then integrating the resulting curve. The integration can be carried out either analytically (which is preferable) or numerically.

Example 12.9 Integrating Measured Data

The force exerted by a spring as a function of displacement from its equilibrium position is given by the following data (reproduced from Example 9.3). Notice that the data are not evenly spaced.

Data Point No.	Distance (cm)	Force (N)
1	2	2.0
2	4	3.5
3	7	4.5
4	11	8.0
5	17	9.5

In Chap. 9 we used the method of least squares to pass the following trendline through the data:

$$y = 0.5147x + 1.2794$$

where y represents the force, in N, and x represents the distance from the equilibrium position, in cm. Figure 12.13 (reproduced from Example 9.5) shows a plot of the data and the trendline. Note that there is considerable scatter in the measured data.

Determine the work required to extend the spring within the measured interval; i.e., from $x = 2$ to $x = 17$.

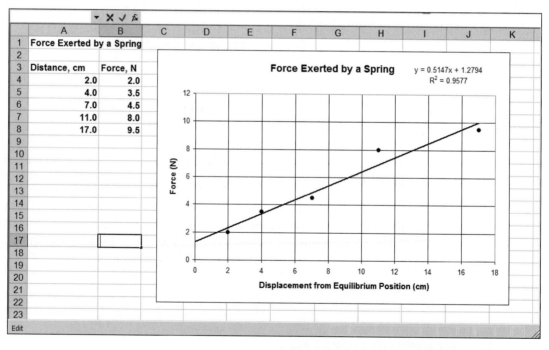

Figure 12.13 – A linear trendline passing through a set of measured data

The work required to extend the spring is determined by integrating the force over the appropriate interval. Thus,

$$W = \int_{2}^{17} y\,dx$$

Let us first evaluate the integral by applying the trapezoidal rule for unequally spaced data. We should not expect to obtain an accurate result this way because of the scatter in the data. It does, however, provide a basis for later comparison with the integral based upon the trendline.

Figure 12.14 shows an Excel worksheet for the trapezoidal rule. The approach is basically the same as the integral evaluated in Example 12.2 (see Fig. 12.3).

E13	▼	f_x =SUM(E7:E12)				
A	**B**	**C**	**D**	**E**	**F**	**G**
1 Work required to extend a spring						
2						
3 (Trapezoidal rule with unequally spaced intervals)						
4						
5						
6 x (cm)	y (N)	delta x (cm)	y bar (N)	area (N)(cm)		
7 2	2.0	2	2.75	5.5		
8 4	3.5	3	4.00	12.0		
9 7	4.5	4	6.25	25.0		
10 11	8.0	6	8.75	52.5		
11 17	9.5					
12						
13			Total =	95.0		
14						
15						
Ready						

Figure 12.14 – Applying the trapezoidal rule to the measured data

The value of the resulting integral is $W = 95.0$ (N)(cm).

Now let us evaluate this integral using classical calculus. If we use the trendline to represent the force, then the work is determined by evaluating the integral

$$W = \int_{2}^{17} y\,dx = \int_{2}^{17} (0.5147x + 1.2794)\,dx = 89.977 \text{ (N)(cm)}$$

This value is considerably less than the value obtained using the trapezoidal rule with the actual data. We cannot say with certainty which value is more accurate, but the value based upon the trendline seems more plausible because of the way it handles the scatter in the data. It is also easy to implement.

Problems

12.12 Using Excel and the methods described in Chapter 9, fit a polynomial to the voltage versus time data given in Example 12.1. Then integrate the polynomial over the interval $0 \leq t \leq 0.5$ seconds using classical calculus. Compare the ease of solution and the accuracy of the solution with the results obtained in Examples 12.1 and 12.2.

12.13 Using Excel, integrate the polynomial obtained in Problem 12.12 (not the actual data) using the trapezoidal rule. Carry out the integration with

(*a*)   ten equally spaced intervals.

(*b*)   thirty equally spaced intervals.

(*c*)   one hundred equally spaced intervals.

Compare your solution with the results obtained in Prob. 12.12. What can you conclude about the relationship between accuracy and the number of intervals?

12.14 Repeat Prob. 12.13 using Simpson's rule. Compare the results with those obtained in Probs. 12.12 and 12.13.

12.15 The following data, representing temperature as a function of vertical depth, are taken from Prob. 9.12.

Distance (cm)	Temperature (°C)
0.1	21.29
0.8	27.3
3.6	31.8
12	35.6
120	42.3
390	45.9
710	47.7
1200	49.2
1800	50.5
2400	51.4

Using Excel, determine the average temperature over the entire 2400-cm distance. Use whatever method you consider most appropriate. Compare this value with a simple arithmetic average of the temperatures.

12.16 The following data, representing reaction rate as a function of temperature, are taken from Example 9.13.

Temperature (°K)	Reaction Rate (moles/second)
253	0.12
258	0.17
263	0.24
268	0.34
273	0.48
278	0.66
283	0.91
288	1.22
293	1.64
298	2.17
303	2.84
308	3.70

Determine the average reaction rate over the entire temperature range. Carry out the integration within an Excel worksheet using the most appropriate method. Compare this value with a simple arithmetic average of the reaction-rate data.

12.17 The following data, taken from Prob. 9.24, represent the power produced by a wind-driven generator built by a group of students.

Wind Velocity (mph)	Power (watts)
0	0
5	0.26
10	2.8
15	7.0
20	15.8
25	28.2
30	46.7
35	64.5
40	80.2
45	86.8
50	88.0
55	89.2
60	90.3

Determine the average power generation over the given range of wind velocities. Evaluate the integral within an Excel worksheet using the most appropriate method.

12.18 The following stress-strain data, taken from Prob. 9.31, were obtained from a structural steel cylindrical sample with a diameter of 0.5 inch and a length of four inches.

Strain	Stress (psi)	Strain	Stress (psi)
0.02	29737	0.14	57046
0.04	37166	0.16	56593
0.06	44820	0.18	53448
0.08	44074	0.20	52103
0.10	49161	0.22	49185
0.12	53002	0.24	45386

Determine the average stress for the strain values ranging from 0 to 0.24. (Assume the data set begins at the origin; i.e., the stress is zero when the strain is zero.) Evaluate the integral within an Excel worksheet using the most appropriate method.

CHAPTER 13

MAKING LOGICAL DECISIONS (IF-THEN-ELSE)

Though we have already seen how to carry out many different kinds of calculations in earlier chapters of this book, the ability to carry out *branching operations* in Excel opens the door for numerous additional types of applications. If you've ever learned a programming language, you will be familiar with the concept of branching (often referred to as *if-then-else*). This feature allows computers to carry out logical decisions. In most programming languages, a branching operation is based upon a logical condition that is either true or false. A different course of action is then specified for each outcome.

In Excel, branching is carried out using the IF library function. In this chapter, we will see how this is done and why it can be useful.

13.1 LOGICAL (BOOLEAN) EXPRESSIONS

In Chapter 2 we learned about the *comparison operators* >, >=, <, <=, =, and <> (see Sec. 2.6). These operators are used to connect various operands, to form *logical expressions* (also known as *Boolean expressions*) that are either *true* or *false*. Consider, for example, the logical expression C1 > 100. If the numerical value in cell C1 is greater than 100, the expression will be *true*; otherwise (if the value in cell C1 is less than or equal to 100), the expression will be *false*. Similarly, the logical expression C1 = C2 will be true only if the values in cells C1 and C2 are the same; otherwise, the expression will be false.

Logical expressions can be combined using the *logical operators* AND, OR, and NOT. Consider, for example, the logical expression C1 > 0 AND C2 > C1. This expression will be true only if the value in cell C1 is greater than 0 *and* the value in cell C2 exceeds the value in cell C1 (both conditions must be true for the overall expression to be true). But the logical expression C1 > 0 OR C2 > C1 will be true if *either* the value in cell C1 exceeds 0 *or* the value in cell C2 exceeds the value in cell C1 (only one condition need be true for the overall expression to be true). Finally, the expression NOT (C1 > 0) will be true only if C1 is *not* greater than 0 (the NOT operator reverses the outcome of the true/false condition).

Another example of a logical expression is provided by the MOD function, which can be written as MOD(C1,C2). This function returns the remainder after dividing the value in cell C1 by the value in cell C2. The arguments can be constants or formulas as well as cell addresses. Thus, MOD(C1,2) returns the remainder after the numerical value in cell C1 is divided by 2. If the numerical value in cell C1 is a positive integer, then the only possible remainders are 0 (if the value is even), or 1 (if the value is odd).

Now suppose we incorporate this function into a logical expression by writing MOD(C1,2) = 0. This expression will be *true* if the remainder is 0 (i.e., if the numerical value in cell C1 is even), or *false* if the remainder is 1 (if the numerical value in cell C1 is odd). Hence, we have a convenient way to determine whether a positive integer is even or odd.

13.2 THE IF FUNCTION

The IF library function is used to carry out branching operations in Excel. The IF function requires three arguments: a logical expression, followed by two numerical expressions. If the logical function is *true*, then the first numerical expression is evaluated; if it is *false*, the second numerical expression is evaluated.

For example, suppose cell B5 contains the formula

=IF(C1 > 100, 50, C1/2)

If the value in cell C1 exceeds 100, the value 50 will be placed in cell B5. Otherwise (if the value in cell C1 does not exceed 100), the quantity C1/2 will be placed in cell B5.

The last two operands in the IF function may be labels (strings) as well as numerical expressions. Suppose, for example, that cell B5 contains the formula

=IF(C1 > 100, "Too Big", "Ok")

Then if the value in cell C1 exceeds 100, the label "Too Big" will be placed in cell B5. Otherwise (if the value in cell C1 is less than or equal to 100), the label "Ok" will be placed in cell B5.

13.3 NESTED IF FUNCTIONS

IF functions may be nested, one within another. The nesting may be up to eight deep (that is, as many as seven IF functions may appear as arguments in other IF functions). This feature allows us to carry out more complicated logical tests.

Suppose, for example, that cell A3 contains the temperature of a container of water, in degrees Celsius. We can identify the state of the water, based upon its temperature, as follows:

=IF(A3 < 0, "Ice", IF(A3 < 100, "Water", "Steam"))

Thus, a temperature less than 0°C is identified as "Ice," a temperature greater than or equal to zero but less than 100°C is identified as "Water," and a temperature greater than or equal to 100°C is identified as "Steam."

Example 13.1 Student Grades

Assign a final letter grade to each of the students listed in the worksheet in Example 2.4 (see Fig. 2.22). Base the letter grades on the following rules:

Overall Score	Grade
90 and above	A
80 to 89.9 . . .	B
70 to 79.9 . . .	C
60 to 69.9 . . .	D
Below 60	F

The letter grade conversion can easily be carried out with the following formula, which utilizes several nested IF functions:

=IF(E2 >= 90, "A", IF(E2 >= 80, "B", IF(E2 >= 70, "C", IF(E2 >= 60, "D", "F"))))

This formula assumes that cell E2 contains the numerical score being converted, as shown in Fig. 13.1. The formula itself appears in cell F2. The formula is copied into cells F3 through F8, to accommodate all of the remaining students.

Sometimes the restriction on nesting IF functions eight deep may be circumvented by the clever use of formulas. For example, in some problems it is possible to decompose a complicated set of *if-then-else* conditions into two or more subsets, and then combine the results after the logical outcome of each subset has been determined. Example 13.2 illustrates one such procedure.

	F2	▼		▪	=IF(E2>=90,"A",IF(E2>=80,"B",IF(E2>=70,"C",IF(E2>=60,"D","F"))))		
	A	B	C	D	E	F	G	H
1	Student	Exam 1	Exam 2	Final Exam	Overall Score	Final Grade		
2	Davis	82	77	94	84.3	B		
3	Graham	66	80	75	73.7	C		
4	Jones	95	100	97	97.3	A		
5	Meyers	47	62	78	62.3	D		
6	Richardson	80	58	73	70.3	C		
7	Thomas	74	81	85	80	B		
8	Williams	57	62	67	62	D		
9								
10	AVERAGE	71.6	74.3	81.3	75.7			
11								
12								
13								
14								
15								

Figure 13.1 – Using nested IF functions to assign letter grades

Example 13.2 Student Grades Revisited

Modify the previous example to include pluses and minuses (i.e., so that students may receive final grades of A–, C+, etc.). We will adopt the rule that, within any 10-point interval, a score of 3 or below will be assigned a minus, whereas a score of 7 or above will receive a plus. Thus, within the interval ranging from 80 to 89.9 . . . , a score of 83 or below will result in the grade B–, a score of 87 or above will result in the grade B+, and a score between 83 and 87 will correspond to an ordinary B grade.

The most direct way to proceed with this problem is to create a more complex set of nested IF functions, e.g.,

=IF(E2 >= 97, "A+", IF(E2 > 93, "A", IF(E2 >= 90, "A–", IF(E2 >= 87, "B+", . . . etc.))))

This approach is not valid, however, because the complete span of all possible grades (from A+ to D–) exceeds the restriction on nesting eight deep.

As an alternative, we can decompose the problem into the following two parts: First we can determine an ordinary letter grade, as we did in the last example. Then we can determine whether to append a + or – using the following nested IF functions:

IF(MOD(E2, 10) <= 3, "–", IF(MOD(E2, 10) >= 7, "+", " "))

We can then combine the results of these two nested IFs using the *string operator* (&) discussed in Sec. 2.6. Hence, the final formula can be written

$$=IF(E2 >= 90, \text{"A"}, IF(E2 >= 80, \text{"B"}, IF(E2 >= 70, \text{"C"}, IF(E2 >= 60, \text{"D"}, \text{"F"})))) \ \&$$

$$IF(MOD(E2, 10) <= 3, \text{"--"}, IF(MOD(E2, 10) >= 7, \text{"+"}, \text{" "}))$$

Neither of these nested IFs violates the restriction on maximum nesting.

Figure 13.2 shows the worksheet that is obtained when the above formula is used to determine the final grade for each of the students (compare with the worksheet shown in Fig. 13.1).

| F2 | | = | =IF(E2>=90,"A",IF(E2>=80,"B",IF(E2>=70,"C",IF(E2>=60,"D","F")))) |
| | | | &IF(MOD(E2,10)<=3,"-",IF(MOD(E2,10)>=7,"+"," ")) |

	A	B				
	Student	Exam 1	Exam 2	Final Exam	Overall Score	Final Grade
1						
2	Davis	82	77	94	84.3	B
3	Graham	66	80	75	73.7	C
4	Jones	95	100	97	97.3	A+
5	Meyers	47	62	78	62.3	D-
6	Richardson	80	58	73	70.3	C-
7	Thomas	74	81	85	80	B-
8	Williams	57	62	67	62	D-
9						
10	AVERAGE	71.6	74.3	81.3	75.7	
11						
12						
13						

Figure 13.2 – Using the string operator to combine two conditional strings

Example 13.3 Simpson's Rule Revisited

One of the methods for evaluating integrals that we studied in Chapter 12 was Simpson's Rule (see Sec. 12.2). Recall that Simpson's Rule involves a summation whose inner terms include alternating the coefficients 4 and 2.

We now consider another approach to Simpson's rule, using the IF function to determine the proper coefficients (either 4 or 2) for the interior summation terms. Figure 13.3 contains the Excel worksheet. This worksheet is similar to that shown in Fig. 12.9, except that the data points are now numbered in column C. These numbers will be tested within the IF function to determine whether each value is even or odd (i.e., to determine whether the appropriate coefficient is 4 or 2).

The results are identical to those obtained previously, in Chapter 12 (see Figs. 12.9 and 12.11). Now, however, the interior points are generated using a formula that includes the IF function. For example, the value in cell D5 is obtained using the formula

$$=IF(MOD(C5, 2)=0, 4, 2) * EXP(-(\$A5^2))$$

In this formula, the MOD function tests the integer quantity in cell C5 to determine if it is even or odd. The IF function then generates the appropriate coefficient, either 4 or 2, depending on the outcome of the even/odd test. The resulting coefficient then multiplies the exponential function EXP(−($A5^2)). A similar formula will be used to generate the values in cells D6 through D13.

	A	B	C	D	E	F	G	H	I	J	K
	F16	▼	=	=D16*(0.1/3)							
1	Numerical Integration Using Simpson's Rule										
2											
3	x	y	i								
4	0.0	1.0000	1	1.0000							
5	0.1	0.9900	2	3.9602							
6	0.2	0.9608	3	1.9216							
7	0.3	0.9139	4	3.6557							
8	0.4	0.8521	5	1.7043							
9	0.5	0.7788	6	3.1152							
10	0.6	0.6977	7	1.3954							
11	0.7	0.6126	8	2.4505							
12	0.8	0.5273	9	1.0546							
13	0.9	0.4449	10	1.7794							
14	1.0	0.3679	11	0.3679							
15											
16			Sum =	22.4047		I =	0.7468				
17											

Figure 13.3 – Integration with Simpson's rule utilizing the IF function

Figure 13.4 shows the cell formulas used to generate this worksheet.

	A	B	C	D	E	F
	F16	=	=D16*(0.1/3)			
1	Numerical Integration					
2						
3	x	y	i			
4	0	=EXP(-($A4^2))	1	=EXP(-($A4^2))		
5	=A4+0.1	=EXP(-($A5^2))	=C4+1	=IF(MOD(C5,2)=0,4,2)*EXP(-($A5^2))		
6	=A5+0.1	=EXP(-($A6^2))	=C5+1	=IF(MOD(C6,2)=0,4,2)*EXP(-($A6^2))		
7	=A6+0.1	=EXP(-($A7^2))	=C6+1	=IF(MOD(C7,2)=0,4,2)*EXP(-($A7^2))		
8	=A7+0.1	=EXP(-($A8^2))	=C7+1	=IF(MOD(C8,2)=0,4,2)*EXP(-($A8^2))		
9	=A8+0.1	=EXP(-($A9^2))	=C8+1	=IF(MOD(C9,2)=0,4,2)*EXP(-($A9^2))		
10	=A9+0.1	=EXP(-($A10^2))	=C9+1	=IF(MOD(C10,2)=0,4,2)*EXP(-($A10^2))		
11	=A10+0.1	=EXP(-($A11^2))	=C10+1	=IF(MOD(C11,2)=0,4,2)*EXP(-($A11^2))		
12	=A11+0.1	=EXP(-($A12^2))	=C11+1	=IF(MOD(C12,2)=0,4,2)*EXP(-($A12^2))		
13	=A12+0.1	=EXP(-($A13^2))	=C12+1	=IF(MOD(C13,2)=0,4,2)*EXP(-($A13^2))		
14	=A13+0.1	=EXP(-($A14^2))	=C13+1	=EXP(-($A14^2))		
15						
16			Sum =	=SUM(D4:D14)		I = =D16*(0.1/3)
17						
18						

Figure 13.4 – The corresponding cell formulas

Problems

13.1   Create an Excel worksheet containing a table of $f(x)$ versus x, where

$$f(x) = 2x, \qquad\qquad 0 \le x \le 3 \qquad\qquad (13.1)$$

and

$$f(x) = (x-3)^2 + 6, \qquad 3 < x \le 10 \qquad\qquad (13.2)$$

Then plot $f(x)$ versus x over the interval $0 \le x \le 10$. When constructing the table, choose the x values sufficiently close so that the functions can be plotted smoothly (*suggestion*: try $x = 0, 0.5, 1.0, \ldots, 9.5, 10.0$).

13.2   Repeat the curve-fitting exercise given in Prob. 9.9. Identify each data point that is 20 percent or more above or below the regression line with a single asterisk (*) placed in the first empty column. In addition, identify each data point that is 30 percent or more above or below the regression line with a double asterisk (**).

13.3   In Prob. 3.5 you were asked to create a worksheet containing a student grade report, based upon the information originally given in Prob. 2.3 and additional information given in Prob. 3.5. Modify this grade report so that each course receives a letter grade, in addition to the numerical grades included in the problem statements. Include + and − scores (e.g., B+, A−, etc.) where applicable, using the rules presented in Example 13.2. Use nested IF functions to determine the letter grades and to append the + and − signs, as described in Example 13.2.

13.4   Create an Excel worksheet that simulates throwing a pair of dice 20 times. To do so, enter the formula =RANDBETWEEN(1, 6) in the first 20 rows of column A, and again in column B. Each value will be a random integer between 1 and 6, representing the outcome of one random die. Within each row of column C, place the sum of the corresponding values in columns A and B. In column D, place one of the following labels within each row:

(*a*)   If the value in column C is 7 or 11, display "You Win."

(*b*)   If the value in column C is 2, 3, or 12, display "You Lose."

(*c*)   If the value is 4, 5, 6, 8, 9, or 10, display "Inconclusive."

Note: RANDBETWEEN(a, b) is an Excel library function that returns a random integer between the limits a and b (where $b > a$). To activate this function, run the Excel Setup program and install the Analysis ToolPak. Then enable the Analysis ToolPak by selecting Add-Ins from the Tools menu.

13.5 The *Fibonacci numbers* are an interesting sequence of positive integers where each value is the sum of the previous two values. We begin with

$$F_1 = F_2 = 1 \tag{13.3}$$

Each successive value is then obtained from the formula

$$F_i = F_{i-1} + F_{i-2}, \qquad i = 3, 4, 5, \ldots \tag{13.4}$$

Create an Excel worksheet containing the first 20 Fibonacci numbers. Test each Fibonacci number to determine if it is odd or even. Place the Fibonacci numbers in column A, and the corresponding odd/even labels in column B.

13.6 Create an Excel worksheet that will allow you to convert inches to one of the following metric units: millimeters, centimeters, meters, or kilometers, depending on the magnitude of the given value. Use the following rules to determine which metric unit will be shown:

(*i*) If the given value is less than 1 inch, convert to millimeters.

(*ii*) If the given value is greater than or equal to 1 inch but less than 100 inches, convert to centimeters.

(*iii*) If the given value is greater than or equal to 100 inches but less than 10,000 inches, convert to meters.

(*iv*) If the given value is greater than or equal to 10,000 inches, convert to kilometers.

Use the worksheet to carry out the following conversions:

(*a*) 0.04 inch

(*b*) 32 inches

(*c*) 787 inches

(*d*) 15,500 inches

13.7 Here is a common problem in consumer economics: Suppose you deposit *P* dollars in a bank account, and allow the money to remain in the bank for *n* years. If the bank pays interest at an annual interest rate *i* (expressed as a decimal), how much money will accumulate after *n* years?

The amount of money accumulated will vary, depending on the frequency of compounding. The following equations apply:

Annual compounding:

$$F = P(1+i)^n \tag{13.5}$$

Quarterly compounding:

$$F = P(1 + i/4)^{4n} \tag{13.6}$$

Monthly compounding:

$$F = P(1 + i/12)^{12n} \tag{13.7}$$

Daily compounding:

$$F = P(1 + i/365)^{365n} \tag{13.8}$$

In each of these equations, n is always the number of *years*, and i is the *annual* (*yearly*) interest rate, expressed as a decimal. (*Note*: compound interest calculations are discussed more extensively in Chapter 15.)

Construct an Excel worksheet that will calculate a value for F, based upon the following information entered directly into the worksheet by the user: P, i, n, and the single letter A, Q, M, or D (indicating annual, quarterly, monthly, or daily compounding). Use an Excel formula containing nested IF functions to obtain the correct equation, based upon the letter that indicates the frequency of compounding.

Build the worksheet two different ways:

(*a*) Assuming that the frequency of compounding will always be specified by a valid uppercase letter (A, Q, M, or D).

(*b*) Assuming that the frequency of compounding may be either an uppercase or a lowercase letter, which may or may not be valid (i.e., the frequency of compounding may be some letter other than A, Q, M, or D). Display the message "Data Error" if the letter indicating the frequency of compounding is invalid.

Label all of the information clearly.

As a test case, determine the value of F when $P = \$5000$, $i = 0.05$ (corresponding to an interest rate of 5 percent per year), and $n = 20$ years. Solve for F using annual, quarterly, monthly, and daily compounding. Does the frequency of compounding have a significant influence on the result?

13.8 The quadratic equation

$$ax^2 + bx + c = 0 \tag{13.9}$$

normally has two roots, though the roots depend on the values assigned to the coefficients a, b, and c.

If $b^2 > 4ac$, then the roots are given by the well-known quadratic formula

$$x_1 = \frac{-b + \sqrt{b^2 - 4ac}}{2a} \qquad x_2 = \frac{-b - \sqrt{b^2 - 4ac}}{2a} \qquad (13.10)$$

If $b^2 < 4ac$, the roots are complex, given by

$$x_1 = \frac{-b}{2a} + \frac{\sqrt{4ac - b^2}}{2a} i \qquad x_2 = \frac{-b}{2a} - \frac{\sqrt{4ac - b^2}}{2a} i \qquad (13.11)$$

where i represents the imaginary number $\sqrt{-1}$.

If $b^2 = 4ac$, there is one repeated root, given by

$$x = -b/2a \qquad (13.12)$$

Finally, if $a = 0$, there is only one root, given by

$$x = -c/b \qquad (13.13)$$

Create an Excel spreadsheet that will determine the roots of a quadratic equation in terms of the coefficients a, b, and c. In the case of complex roots, use the TEXT library function to convert the imaginary number to a string (i.e., a label), and then combine the resulting string with the letter i using the string operator (&).

Enter the following sets of coefficients in your worksheet, and determine the roots corresponding to each set:

(a) $a = 2$, $b = 4$, $c = 1$

(b) $a = 4$, $b = 2$, $c = 3$

(c) $a = 2$, $b = 4$, $c = 2$

(d) $a = 0$, $b = 3$, $c = 2$

CHAPTER **14**

RECORDING AND RUNNING MACROS

A *macro* is a sequence of consecutive keystrokes and/or mouse actions that is recorded for later playback. Thus, a macro allows you to automate a series of instructions within an Excel spreadsheet. Generally speaking, the more complicated the instructions, the more useful the macro. Every macro is stored under its own name and has a *shortcut key* (such as Ctrl-P, Ctrl-X, etc.) associated with it. Once a macro has been recorded, its instructions can be executed automatically simply by selecting the macro name or by pressing its shortcut key.

14.1 RECORDING A MACRO

To record a macro, select Macro from the Tools menu. Then select Record New Macro... from the resulting submenu, as shown in Fig. 14.1.

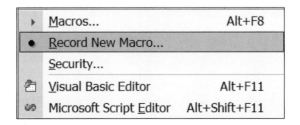

Figure 14.1 – The Macro submenu

The Record Macro dialog box will then appear with default entries similar to those shown in Fig. 14.2.

Record Macro

Macro name:
Macro1

Shortcut key: Store macro in:
Ctrl+   This Workbook

Description:
Macro recorded 4/13/2005 by Byron Gottfried

OK   Cancel

Figure 14.2 – The Record Macro dialog box

This dialog box allows you to provide a macro name. You may accept a default name (Macro1, Macro2, etc.), but you should probably provide a name that is more descriptive of the action taken by the macro. (Do not include blank spaces within the macro name).

You may also provide a *keyboard shortcut*, which will allow you to access (i.e., to execute) the macro by pressing the Ctrl key together with the shortcut key rather than accessing the macro by name. The shortcut key must be a letter, either upper- or lowercase. Note that the shortcut keys are case sensitive; i.e., an uppercase letter is not the same as a lowercase letter.

The Record Macro dialog box also allows you to designate where the macro will be stored (more about this later), and it allows you to provide a brief description of the action taken by the macro. Figure 14.2 shows where all of these features are entered.

After the required information has been entered into the Record Macro dialog box, click on OK. The Stop Recording Bar shown in Fig. 14.3 will then appear, and the Status Bar will display the message Recording. You may then begin entering instructions (i.e., keystrokes or mouse movements) anywhere within the worksheet. All of the completed instructions will be recorded within the macro.

Stop button

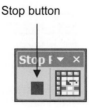

Figure 14.3 – The Stop Recording Bar

Note: If the Stop Recording Bar does not appear when you click on OK in the Record Macro dialog box, go to the View menu and choose Toolbars. Then activate the Stop Recording Bar by placing a check in the appropriate box.

Once the macro recording is finished, the recording is terminated by clicking on the Stop button within the Stop Recording Bar (or by selecting Macro/Stop Recording from the Tools Menu). The macro will then be saved in the designated place, with the designated name.

14.2 EXECUTING A MACRO

Once a macro has been saved, it can be executed simply by holding down the Ctrl key and pressing the shortcut key, or by selecting Macro/Macros... from the Tools menu, selecting a macro from the list shown in the resulting dialog box, and then selecting Run (see Fig. 14.4). The instructions within the macro will then be executed *within the cells specifically referenced when the macro was recorded*. It is also possible to record and execute the instructions in a more flexible manner by referencing a broader range of cells, as discussed in Sec. 14.3.

If you receive a message indicating that macros are disabled because the security level is too high, do the following: Select Options from the Tools menu, then select the Security tab. Click on the Macro Security button and select the Security Level tab. Then choose a lower security level.

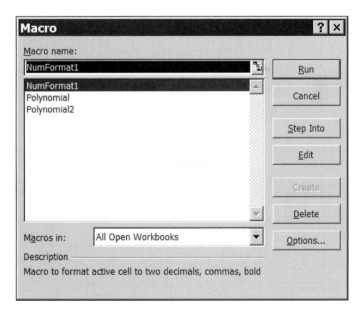

Figure 14.4 – The Macro dialog box

Example 14.1 Formatting a Number within a Cell

Let us record and execute a macro that will format any number in the following manner: display two decimal places, include commas to separate thousands, millions, etc., and display the number in boldface.

We begin by selecting any empty cell, say cell B3, and then select Macro/Record New Macro... from the Tools menu (see Figs. 14.1 and 14.2). This results in the Record Macro dialog box being superimposed over the worksheet, as shown in Fig. 14.5. We then supply the macro name (FormatNumber), the shortcut key (a), the macro storage location (Personal Macro Workbook – more about this later) and a description of the macro. The entries are shown in Fig. 14.5. Once this information is specified, we press OK to begin recording. The Stop Recording Bar will then appear in place of the Record Macro dialog box, and Recording will appear in the status bar, next to Ready. Figure 14.6 illustrates the recording mode.

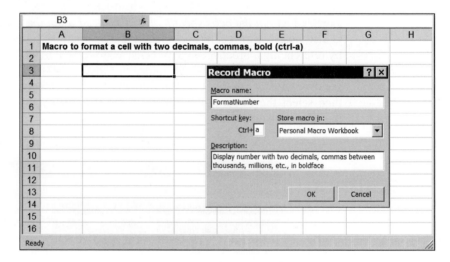

Figure 14.5 – Preparing to record the macro

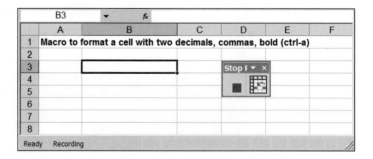

Figure 14.6 – Recording the macro

Once the recording begins, we will carry out the following steps:

1. Right-click on cell B3 and select Format Cells... (Or select Cells... from the Format Menu.) This causes the Format Cells dialog box to be displayed with the Number tab active, as shown in Fig. 14.7.

2. Within the Format Cells dialog box, specify two decimal places, and check the box labeled Use 1000 Separator (,). Then press OK. Figure 14.7 shows the correct entries.

3. Click on the Bold (**B**) button in the Formatting Toolbar (see Fig. 2.1). (Or select Cells from the Format menu, click on the Font tab, and select Bold from the drop-down list.)

4. End the recording by clicking on the Stop button within the Stop Recording Bar (see Fig. 14.6). The macro is now complete and ready for use.

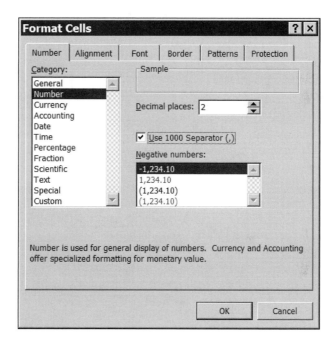

Figure 14.7 – The Format Cells dialog box

We are now ready to place numerical values in the worksheet and use the macro to format the cells that contain these values. Note that cell B3 is already formatted – it was formatted when we recorded the instructions that comprise the macro. Hence, if we enter the number 12345.6789 into cell B3, it will appear as **12,345.68** as shown in Fig. 14.8(*a*). However, if we enter this same value into any other cell, it will appear as we entered it; i.e., 12345.6789. For example, Fig. 14.8(*b*) shows this value entered in cell B5. But if we click on cell B5 to make it active and then press Ctrl-a, the number will be formatted as **12,345.68** as shown in Fig. 14.8(*c*).

B3	▼	fx	12345.6789		
A	B	C	D	E	F
1 Macro to format a cell with two decimals, commas, bold (ctrl-a)					
2					
3	12,345.68				
4					
5					
6					
7					
Ready					

Figure 14.8(*a*) – Entering a number into a previously formatted cell

B5	▼	fx	12345.6789		
A	B	C	D	E	F
1 Macro to format a cell with two decimals, commas, bold (ctrl-a)					
2					
3					
4					
5	12345.6789				
6					
7					
Ready					

Figure 14.8(*b*) – Entering a number into an unformatted cell

B5	▼	fx	12345.6789		
A	B	C	D	E	F
1 Macro to format a cell with two decimals, commas, bold (ctrl-a)					
2					
3					
4					
5	12,345.68				
6					
7					
Ready					

Figure 14.8(*c*) – Formatting the number by pressing Ctrl-a

14.3 CELL ADDRESSING WITHIN A MACRO

If a macro does not reference any cells other than the currently active cell, then it can be applied to any cell within the worksheet. The macro created in Example 14.1 falls into this category. But if a macro *does* include references to other cells, however, then some care must be taken in how the cell addressing is carried out.

A macro normally uses *absolute addressing* to reference other cells when it is recording. Therefore, the instructions will always be carried out using the same cells that were referenced when the macro was recorded. This may restrict the execution of the macro to certain cells within the worksheet.

This restriction may be removed, however, by using *relative addressing* rather than absolute addressing when recording the macro. The use of relative addressing is often desirable for certain types of applications. To record using relative addressing, you must activate the Relative Reference button within the Stop Recording Bar before the recording begins.

The Relative Reference button is a *toggle* that turns relative addressing on and off. (Click once to turn relative addressing on, click again to turn it off.) Figures 14.9(*a*) and 14.9(*b*) illustrate the two states of the Relative Reference button. Note that they look very similar. In Fig. 14.9(*b*), however, we see that a thin rectangle surrounds the Relative Reference button when it is in the relative addressing mode. (Presumably, this creates the illusion of the button being depressed, though the graphic is not very convincing.) Unfortunately, there is no other way to determine whether relative addressing is active or inactive.

Relative Reference button set for *absolute addressing*

Figure 14.9(*a*) – The Stop Recording Bar in the absolute addressing mode

Relative Reference button set for *relative addressing*

Figure 14.9(*b*) – The Stop Recording Bar in the relative addressing mode

Example 14.2 Evaluating a Polynomial

Create a macro to evaluate the polynomial

$$y = 3x^2 - 2x + 5 \tag{14.1}$$

for a given value of x. Create the macro two different ways:

1. Using the default recording method (absolute addressing), and

2. Using relative addressing.

Absolute Addressing

We first set up a worksheet that will accept the value of x in cell B3 and then place the corresponding value of y in cell B5. With cell B5 active, we then select Macro/Record New Macro... from the Tools menu. The Record Macro dialog box will then appear. We then supply the macro name (Polynomial), the shortcut key (p), the macro storage location (Personal Macro Workbook), and a description of the macro. Finally, we press OK to begin recording. Figure 14.10 shows the worksheet and the superimposed Record Macro dialog box.

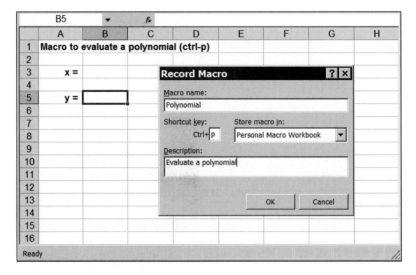

Figure 14.10 – Preparing to record the Polynomial macro

Once the recording begins, we simply enter the formula

=3*B3^2 – 2*B3 + 5

into cell B5 and press Enter. We then click on the Stop button to end the recording (see Fig. 14.3). This completes the macro using absolute addressing, which is the default recording mode.

Now suppose we enter a value for x in cell B3. If we then press Ctrl-p from any cell within the worksheet, the corresponding value of y will appear in cell B5. For example, if we enter the value 7 in cell B3 and press Ctrl-p from anywhere else in the worksheet, the value 138 will appear in cell B5, as shown in Fig. 14.11. From Equation (14.1), you can easily verify that $y = 138$ when $x = 7$.

However, if the value of x is entered anywhere other than cell B3, the macro will not return the correct result. This macro will always return a y-value in cell B5 corresponding to the x-value placed in cell B3. If cell B3 is empty, then pressing Ctrl-p will result in the value 5 appearing in cell B5. This is the value of y corresponding to $x = 0$.

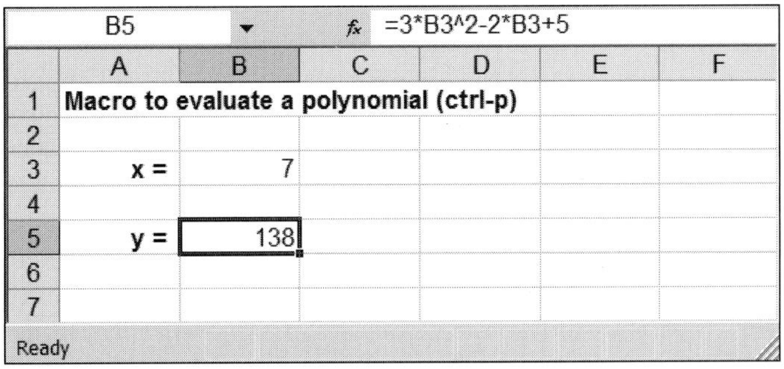

Figure 14.11 – Evaluating the polynomial by pressing Ctrl-p

Relative Addressing

Now let us create another macro to evaluate the same polynomial, this time using relative addressing. We will call this macro Polynomial2, and we will select the letter r as the shortcut key. Figure 14.12 shows the worksheet and the superimposed Record Macro dialog box.

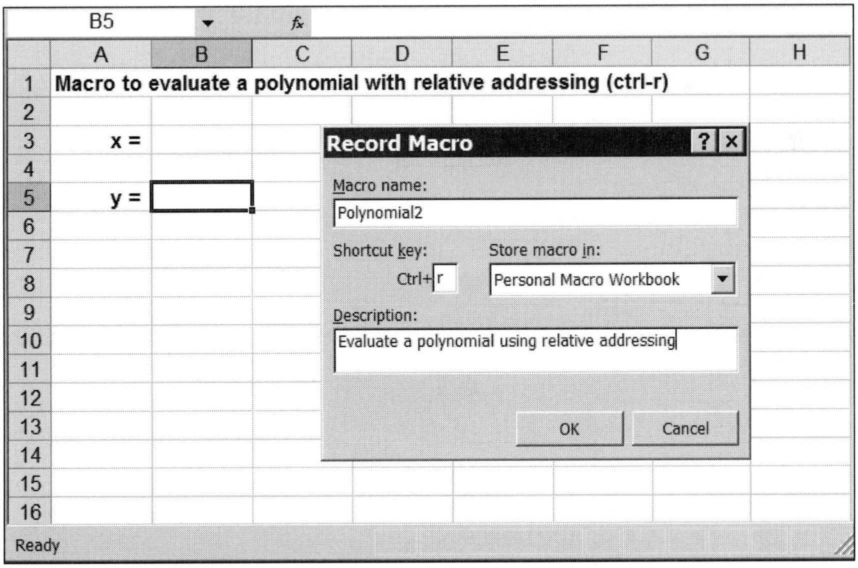

Figure 14.12 – Preparing to record the Polynomial2 macro

When the Stop Recording Bar first appears after closing the Record Macro dialog box, we will click on the Relative Reference button, as shown in Fig. 14.9(*b*). This activates relative addressing when recording the macro. We then enter the formula

=3*B3^2 – 2*B3 + 5

into cell B5 and press Enter, as before. We then click on the Stop button to end the recording. This completes the recording of the relative addressing macro. (*Note*: you can still enter an absolute cell address if you use a $ sign, as explained in Sec. 3.1.)

To see how this macro works, let's enter the value 7 in cell B3 as before. We then move down to cell B5 and press Ctrl-r. This results in the value 138 in cell B5, as shown in Fig. 14.13. So far this macro appears to behave in the same manner as the previous macro, which used absolute addressing. Since this macro uses relative addressing, however, *it can be used in any cell within the worksheet*. The only restriction is that *there must be a value of x entered two cells above the cell in which the macro is executed*. Otherwise, the macro will return a value of $y = 5$, corresponding to $x = 0$.

	B5	▼	f_x	=3*B3^2-2*B3+5		
	A	B	C	D	E	F	G
1	Macro to evaluate a polynomial with relative addressing (ctrl-r)						
2							
3	x =	7					
4							
5	y =	138					
6							
7							
Ready							

Figure 14.13 – Evaluating the polynomial by pressing Ctrl-r

Figure 14.14 shows a worksheet with several *x*-values entered in row 3. The corresponding *y*-values are shown in row 5. Each *y*-value was obtained by pressing Ctrl-r after entering the corresponding *x*-value directly above it in row 3.

	F5	▼	f_x	=3*F3^2-2*F3+5		
	A	B	C	D	E	F	G
1	Macro to evaluate a polynomial with relative addressing (ctrl-r)						
2							
3	x =	7	4	-6	12	3	
4							
5	y =	138	45	125	413	26	
6							
7							
Ready							

Figure 14.14 – Evaluating the polynomial for several different *x*-values

14.4 SAVING A MACRO

When you prepare to record a macro, the Record Macro dialog box asks you where you want to store (i.e., save) the macro. Three choices are available (see Fig. 14.15).

1. The currently active workbook.

2. A new workbook.

3. A special file called the Personal Macro Workbook.

If the macro is saved in the currently active workbook or a new workbook, it will not be available for use in other workbooks. But if you select the Personal Macro Workbook, the macro will be stored in a file called PERSONAL.XLS. Macros stored in this file are accessible to all Excel workbooks on the current computer.

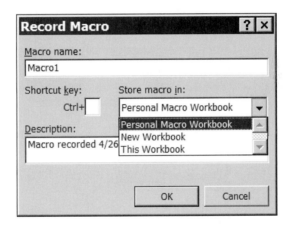

Figure 14.15 – The choices available for saving a macro

Example 14.3 Analyzing Student Exam Scores

Back in Example 2.4 we built a worksheet to record and average a set of student exam grades (see Fig. 2.22). Let us now create and execute a macro that will compare each student's overall score with the class average. In other words, let us determine the difference between each student's overall score (shown in column E) and the class average (cell E10), to determine the number of points by which the student is above or below the class average.

The recording is initiated by selecting Macro/Record New Macro... from the Tools menu, resulting in the dialog box shown in Fig. 14.16. Note that the macro will be named Class_Standing, its shortcut key will be s, and it will be stored within the current worksheet. A description of the macro is also provided at the bottom of the dialog box.

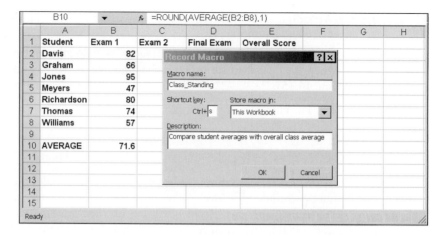

Figure 14.16 – Preparing to record the Class_Standing macro

	A	B	C	D	E	F	G	H
	B10		f_x	=ROUND(AVERAGE(B2:B8),1)				
1	Student	Exam 1	Exam 2	Final Exam	Overall Score			
2	Davis	82	77	94	84.3			
3	Graham	66	80	75	73.7			
4	Jones	95	100	97	97.3			
5	Meyers	47	62	78	62.3			
6	Richardson	80	58	73	70.3			
7	Thomas	74	81	85	80			
8	Williams	57	62	67	62			
9								
10	AVERAGE	71.6	74.3	81.3	75.7			
11								
12								
13								
14								
15								

Ready   Recording

Figure 14.17 – Recording the Class_Standing macro

Once we press OK, the recording process begins. Note that the Stop Recording bar appears on top of the worksheet and the message Recording appears within the Status Bar, as shown in Fig. 14.17. At this point, we will carry out the following actions:

1. Enter the title Comparison in cell F1.

2. Enter the formula =E2–E$10 in cell F2.

3. Copy this formula into cells F3 through F8.

4. Widen column F.

5. Select cells F1 through F8 and apply a bold font.

6. Adjust the size of the worksheet, eliminating some excess empty space.

When all of these actions have been completed, click on the Stop button within the Stop Recording Bar, or select Macro/Stop Recording from the Tools menu. The macro will then be saved with the current worksheet, under the name Class_Standing.

Once the macro has been recorded and stored, you can carry out the actions specified by the macro simply by pressing Ctrl-s. This produces the information shown in Fig. 14.18, column F.

	F1	▼	*fx* Comparison					
	A	B	C	D	E	F	G	H
1	Student	Exam 1	Exam 2	Final Exam	Overall Score	Comparison		
2	Davis	82	77	94	84.3	8.6		
3	Graham	66	80	75	73.7	-2		
4	Jones	95	100	97	97.3	21.6		
5	Meyers	47	62	78	62.3	-13.4		
6	Richardson	80	58	73	70.3	-5.4		
7	Thomas	74	81	85	80	4.3		
8	Williams	57	62	67	62	-13.7		
9								
10	AVERAGE	71.6	74.3	81.3	75.7			
11								
12								
13								
14								
15								
Ready					Sum=-2.84217E-14			

Figure 14.18 – The completed worksheet after applying the macro

Note that this macro is unique to this particular application. Hence, there would be no reason to save it within the Personal Macro Workbook, as in the previous examples.

Problems

14.1 Create a macro similar to FormatNumber, shown in Example 14.1. Now, however, use scientific notation with six decimal places and display the number in 12-point, bold italic type. Name the macro FormatNumber2 and select A (uppercase) as a shortcut key. Save the macro in your Personal Macro Workbook. Test it to be sure that it works correctly.

14.2 For the worksheet described in Example 2.2, create a macro that will transform the appearance of the worksheet from that shown in Fig. 2.18 to that shown in Fig. 3.11. Save the macro with the worksheet, using the name Fancy. Use F as a shortcut key. Run the macro, and compare the resulting output with that shown in Fig. 3.11.

14.3 Create a macro to carry out linear interpolation using the formula

$$y = y_1 + \frac{y_2 - y_1}{x_2 - x_1}(x - x_1)$$

Use relative addressing, assuming that the worksheet layout will be the same as that shown in Example 9.2. Name the macro Interpolate and select L (uppercase) as a shortcut key. Save the macro in your Personal Macro Workbook. Test the macro using the data given in Example 9.2 (see Fig. 9.2(*a*)).

14.4 Create a macro that will solve for the real roots of the quadratic equation

$$ax^2 + bx + c = 0$$

using the formulas

$$x_1 = \frac{-b + \sqrt{b^2 - 4ac}}{2a}, \qquad x_2 = \frac{-b - \sqrt{b^2 - 4ac}}{2a}$$

Base the macro on a worksheet layout that includes the known values of *a*, *b*, and *c*.

14.5 Create a macro that will convert a temperature measurement (*not* a temperature difference) from Fahrenheit to Celsius using the formula

$$°C = (5/9)\,(°F - 32)$$

Use relative addressing, so that the original Fahrenheit temperature may appear anywhere on the worksheet.

14.6 Create a macro that will convert pressure from psi to Pascals (see Example 7.7). Use relative addressing, so that the value of the pressure in psi may appear anywhere in the worksheet.

14.7 Create a macro that will determine the mean, median, mode, min, max, and standard deviation for a list of 20 numbers. Assume that the worksheet layout is the same as that shown in Example 8.1 (see Fig. 8.1).

14.8 Create a macro that will determine the inverse of a 3×3 matrix. Use relative addressing, so that the given 3×3 *A*-matrix may be placed anywhere on the worksheet.

14.9 Expand Prob. 14.8 to solve a system of three linear algebraic equations in three unknowns using matrix inversion. Assume the worksheet layout will be similar to that shown in Example 11.8 (see Fig. 11.2).

14.5 VIEWING A MACRO

The instructions that constitute a macro are members of a special programming language developed by Microsoft, called *Visual Basic for Applications* (commonly referred to as *VBA*). This language uses an instruction set that is built upon classical BASIC, though VBA includes a great many enhancements designed specifically for Microsoft Windows and the Microsoft Office environment. The operators listed in Sec. 2.6 are included within the VBA language, as are many of the functions listed in Sec. 2.7 and the Appendix. Mastery of VBA is complicated by the language's syntax and the many VBA identifiers that have special predefined meanings. We can, however, learn a great deal about VBA macros by viewing their contents within the VBA environment.

To view the instructions within a macro, select Macro/Macros... from the Tools menu. This will produce the Macro dialog box shown in Fig. 14.4. You may then view the macro by selecting it and then clicking the Edit button. This will bring up the VBA environment, as shown in Fig. 14.19. (Note that the VBA environment will appear *on top of* the familiar Excel environment. Hence Excel will still be active, though it will not be visible. To return to Excel, simply close or minimize the VBA window.)

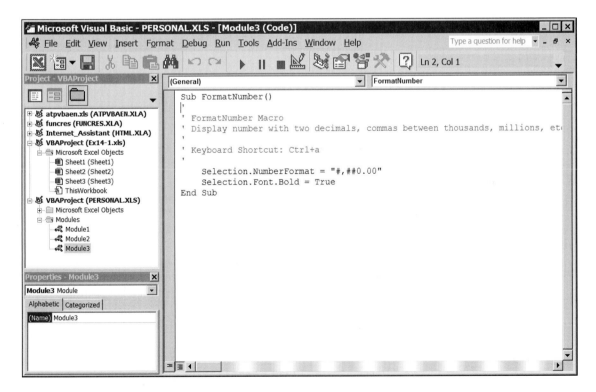

Figure 14.19 – The VBA environment

The large window on the right side of Fig. 14.19 contains the instructions (i.e., the *statements*) that constitute the macro. These instructions are stored within VBA as a *module*. Every macro begins with a Sub statement and ends with an End Sub statement, as seen in Fig. 14.19. Note that the Sub statement includes the macro name, followed by an empty pair of parentheses. Following the Sub statement is a block of *comments*. Comments are easy to identify because each line begins with an apostrophe.

Following the comments are the statements that carry out the actions taken by the module. A detailed discussion of these statements is beyond the scope of this text. But we can consider some general characteristics of the language that appear in this simple macro. For example, the instructions

    Sub FormatNumber()

and

    End Sub

include the two VBA *keywords* Sub and End. Some other keywords are For, Next, With, True, False, and so on. Keywords have certain predefined meanings and they cannot be used for any other purpose within a VBA program.

Many statements include one or more periods. The periods are separators that combine the components of an instruction. For example, in Fig. 14.19 we see the instruction

    Selection.NumberFormat = "#,##0.00"

This statement specifies the numerical format discussed in Example 14.1 (commas separating three-digit groups to the left of the decimal point, and the fractional portion shown with two decimal places). The first portion of the statement (Selection) is an *identifier* that refers to an *object* (in this case, the currently selected cell), and the second portion (NumberFormat) is a *property* (i.e., an *attribute*) of that object. The equal sign is used to *assign* whatever is on the right-hand side (in this case, a format specification) to the property (in this case, Selection.NumberFormat) on the left-hand side. Thus, this instruction assigns a numerical format to the object that represents the currently selected cell.

Sometimes an identifier is followed by a *method* rather than a property. A method specifies an action that will affect the associated object. For example, the instruction

    Range("F1").Select

causes cell F1 to become the currently active cell.

Note: If the macro you wish to view is stored within the PERSONAL.XLS file (where it is accessible to all Excel worksheets), you may need to "unhide" this file before its contents can be viewed. To do so, choose Unhide... from the Windows menu and select PERSONAL.XLS, as shown in Fig. 14.20.

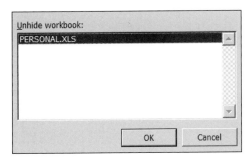

Figure 14.20 – The Windows/Unhide dialog box

Example 14.4 Viewing the Class_Standing Macro

Let us now examine the VBA instructions that make up the Class_Standing macro that we recorded and utilized in Example 14.3. To do so, we must first bring up the Excel file containing this macro. (Recall that this macro was *not* stored in PERSONAL.XLS; hence, we must bring up the worksheet containing the macro.) We then select Macro/Macros... from the Tools menu, resulting in the Macro dialog box shown in Fig. 14.21. If we then select the Class_Standing macro and click on the Edit button, we obtain the *VBA editing window* (see Fig. 14.22) showing a listing of the Class_Standing macro.

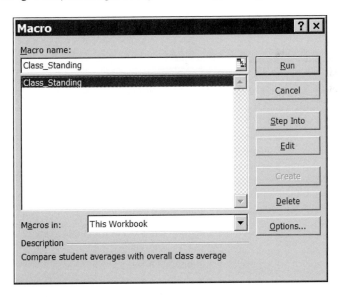

Figure 14.21 – The Macro dialog box showing the Class_Standing macro

```
(General)                                    ▼    Class_Standing                              ▼

Sub Class_Standing()
'
' Class_Standing Macro
' Compare student averages with overall class average
'
' Keyboard Shortcut: Ctrl+s
'
    Range("F1").Select
    ActiveCell.FormulaR1C1 = "Comparison"
    With ActiveCell.Characters(Start:=1, Length:=10).Font
        .Name = "Arial"
        .FontStyle = "Bold"
        .Size = 10
        .Strikethrough = False
        .Superscript = False
        .Subscript = False
        .OutlineFont = False
        .Shadow = False
        .Underline = xlUnderlineStyleNone
        .ColorIndex = xlAutomatic
    End With
    Range("F2").Select
    ActiveCell.FormulaR1C1 = "=RC[-1]-R10C[-1]"
    Selection.AutoFill Destination:=Range("F2:F8"), Type:=xlFillDefault
    Range("F2:F8").Select
    Columns("F:F").ColumnWidth = 11.56
    Columns("F:F").ColumnWidth = 12.22
    Selection.Font.Bold = True
End Sub
```

Figure 14.22 – The VBA listing of the Class_Standing macro

The module begins with the Sub Class_Standing() statement and ends with the End Sub statement. Beneath the Sub Class_Standing() statement we see the following block of comments:

```
'
' Class_Standing Macro
' Compare student averages with overall class average
'
' Keyboard Shortcut: Ctrl+s
'
```

The actual macro instructions appear beneath the comments. We now explain the meaning of these instructions.

The first two statements

```
Range("F1").Select
ActiveCell.FormulaR1C1 = "Comparison"
```

cause the heading Comparison to appear in cell F1.

These statements are followed by the following block of statements that assigns a typeface (Arial) and a style (10-point, bold) to cell F1.

```
With ActiveCell.Characters(Start:=1, Length:=10).Font
    .Name = "Arial"
    .FontStyle = "Bold"
    .Size = 10
    .Strikethrough = False
    .Superscript = False
    .Subscript = False
    .OutlineFont = False
    .Shadow = False
    .Underline = xlUnderlineStyleNone
    .ColorIndex = xlAutomatic
End With
```

Several of these statements (.Strikethrough = False through .ColorIndex = xlAutomatic) are not required for this particular macro, but are generated automatically by the macro recorder when assigning properties to cell F1.

The next two statements

```
Range("F2").Select
ActiveCell.FormulaR1C1 = "=RC[-1]-R10C[-1]"
```

place the formula =E2–E$10 into cell F2. The term =RC[-1]-R10C[-1] is the way the macro recorder expresses mixed addressing: R10 refers to row 10 as an absolute address, whereas C[-1] refers to column E relative to column F, which is currently active.

The following two statements

```
Selection.AutoFill Destination:=Range("F2:F8"), Type:=xlFillDefault
Range("F2:F8").Select
```

cause the formula to be copied into cells F2 through F8. Finally, the remaining three statements

```
Columns("F:F").ColumnWidth = 11.56
Columns("F:F").ColumnWidth = 12.22
Selection.Font.Bold = True
```

cause column F to be widened and displayed in boldface.

From this discussion it should be obvious that the macro recorder greatly simplifies the creation of an Excel macro. Without the recorder, a thorough working knowledge of VBA would be required in order to accomplish the same thing.

14.6 EDITING A MACRO

Editing a macro involves adding or changing some VBA statements within an existing macro. In general this is not a simple process because of the complexity of the VBA language. However, you can often make minor changes to an existing macro by viewing the macro and making an educated guess regarding the changes that are required. Or, if you wish to add new features, you may be able to copy the required statements from one macro to another.

To edit a macro, you must first open it within the VBA editing environment, as described in Sec. 14.5. The changes can then be made in the same way you would use a text editor or a word processor, by altering existing statements or by adding statements from other macros. When this process is complete, you may save the edited macro by clicking on the Save button within the VBA Standard Toolbar (see Fig. 14.19) or by selecting Save from the VBA File menu. The macro will be saved using whatever name appears in the Sub statement. (Hence, if you wish to save the edited macro as a new macro rather than a replacement for an existing macro, be sure to change the macro name in the Sub statement.)

Saving a macro with a new name does not assign a shortcut key. To do so, leave the VBA editor (after saving the new macro) and return to Excel. Then choose Macro/Macros... from the Tools menu. When the Macro dialog box appears (see Fig. 14.21), click on the Options... button. You may then specify a shortcut key. The entire procedure is illustrated in the following example.

Example 14.5 Editing a Macro

In Example 14.2 we created two macros to evaluate the polynomial

$$y = 3x^2 - 2x + 5 \qquad (14.1)$$

for a given value of x. Recall that the second version, which used relative addressing, was named Polynomial2 and its shortcut key was r.

Let us now utilize the VBA editor to alter this macro, replacing the original equation with

$$y = 3(\sin x + \cos x)$$

We will retain the worksheet layout used in Example 14.2, in which the value of y will be displayed two rows below the corresponding value of x, within the same column. In addition, we will display the calculated value of y in red, using a bold italic, 12-point font. We will again use relative addressing, as in Polynomial2. Let us call this macro TrigFunction, and let us choose t as a shortcut key. We can use the VBA statements in the macro FormatNumber as a guide for the numerical display (see Example 14.1).

Figure 14.23 shows the VBA listing of the original macro (Polynomial2). Note that the designation R1C1 within the statement

ActiveCell.FormulaR1C1 = "=3*R[-2]C^2-2*R[-2]C+5"

refers to relative addressing. Similarly, the designation R[–2]C refers to the second row above the current row (i.e., current row number minus 2), and the current column.

```
(General)                              ▼   Polynomial2                    ▼

  Sub Polynomial2()                                                        ▲
  '
  ' Polynomial2 Macro
  ' Evaluate a polynomial using relative addressing
  '
  ' Keyboard Shortcut: Ctrl+r
  '
      ActiveCell.Select
      ActiveCell.FormulaR1C1 = "=3*R[-2]C^2-2*R[-2]C+5"
      ActiveCell.Offset(1, 0).Range("A1").Select
  End Sub
                                                                           ▼
```

Figure 14.23 – The VBA listing of the Polynomial2 macro

Figure 14.24 shows the VBA listing of the macro FormatNumber, developed in Example 14.1.

```
(General)                              ▼   FormatNumber                   ▼

  Sub FormatNumber()                                                       ▲
  '
  ' FormatNumber Macro
  ' Display number with two decimals, commas between thousands,
  ' millions, etc., in boldface
  '
  ' Keyboard Shortcut: Ctrl+a
  '
      Selection.NumberFormat = "#,##0.00"
      Selection.Font.Bold = True
  End Sub
                                                                           ▼
```

Figure 14.24 – The VBA listing of the FormatNumber macro

We now modify the statements taken from Polynomial2 in the following manner. For emphasis, those statements that have been modified are shown in italics. Thus, we have replaced the original polynomial equation with the given trigonometric equation, and we have altered the comments to reflect these changes. (Note that the ActiveCell.Offset statement has been removed, as it is not needed.)

```
Sub TrigFunction()
'
' TrigFunction Macro
' Evaluate a trigonometric formula using relative addressing
'
' Keyboard Shortcut: Ctrl+t
'
    ActiveCell.Select
    ActiveCell.FormulaR1C1 = "=3*(sin(R[-2]C)+cos(R[-2]C))"
End Sub
```

We must still add the formatting statements. From Figure 14.24, we can add the following statements:

```
    Selection.Font.Bold = True
    Selection.Font.Italic = True
    Selection.Font.Name = "Arial"
    Selection.Font.Size = 12
    Selection.Font.ColorIndex = 3
```

The meaning of these statements should be readily apparent with the exception of the last statement. In VBA, color designations are referred to by a numerical color index. Red is represented by the value 3. Hence, this statement specifies the color red.

Here is the entire macro (without the italics).

```
Sub TrigFunction()
'
' TrigFunction Macro
' Evaluate a trigonometric formula using relative addressing
'
' Keyboard Shortcut: Ctrl+t
'
    ActiveCell.Select
    ActiveCell.FormulaR1C1 = "=3*(sin(R[-2]C)+cos(R[-2]C))"

    Selection.Font.Bold = True
    Selection.Font.Italic = True
    Selection.Font.Name = "Arial"
    Selection.Font.Size = 12
    Selection.Font.ColorIndex = 3
End Sub
```

Once the alterations are finished, we can save the new macro by selecting Save PERSONAL.XLS from the File menu, as shown in Fig. 14.25. (If the original macro had been stored within a different worksheet, then Save would refer to that worksheet rather than the PERSONAL.XLS file.) The new macro will automatically be saved using the name TrigFunction, because TrigFunction appears within the Sub statement. However, a new shortcut key is not automatically specified.

To specify a shortcut key for the new macro, we close the VBA environment. We then select Macro/Macros... from the Tools menu, resulting in the Macro dialog box shown in Fig. 14.26. We then select the TrigFunction macro and click on Options.... We can then select t as the new shortcut key within the resulting Macro Options dialog box. Figure 14.26 shows the sequence of events.

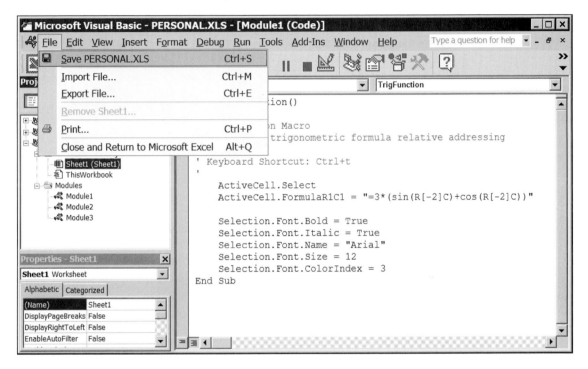

Figure 14.25 – Saving the edited macro in the PERSONAL.XLS file

Problems

14.10   Record a macro similar to the FormatNumber macro created in Example 14.1. Now, however, format the number so that it is displayed in red using a 12-point, bold italic Arial font. View the macro within the VBA editor. Compare the resulting VBA statements with those shown in Example 14.5. Note, in particular, how the red color is designated.

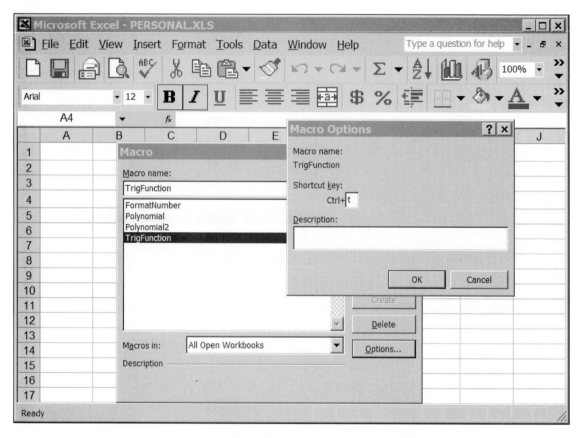

Figure 14.26 – Specifying a shortcut key for the edited macro

14.11 Repeat Prob. 14.10, but this time display the number in blue using a 14-point, italic Times Roman font. View the macro within the VBA editor. Compare the resulting VBA statements with those in Prob. 14.10 and Example 14.5. Note, in particular, how the blue color is designated.

14.12 View the first macro created in Example 14.2 (Polynomial) within the VBA editor. Recall that this macro uses absolute addressing, in contrast with the second macro (Polynomial2), which uses relative addressing and is listed in Fig. 14.23. Compare the VBA statements within the respective macros. Note in particular how absolute addressing and relative addressing differ from each other within VBA.

14.13 Using the VBA editor, alter the Class_Standing macro created in Example 14.3 (and listed in Fig. 14.22) so that the Comparison column appears in column G rather than column F.

14.14 View the linear interpolation macro created in Prob. 14.3 within the VBA editor. Be sure that you understand how relative addressing is carried out within VBA.

14.15 Using the VBA editor, alter the temperature conversion macro created in Prob. 14.5 so that it converts a temperature in degrees Celsius to degrees Fahrenheit. Use the formula

$$°F = (1.8 °C) + 32$$

to carry out the calculation. Test the macro using a few different Celsius temperatures.

14.16 Using the VBA editor, alter the macro created in Prob. 14.7 in the following manner:

(*a*) Remove the calculations of the min and max

(*b*) Add a calculation of the variance

Test the macro using the data given in Example 8.1. Verify that the variance is correct by squaring the standard deviation.

CHAPTER **15**

COMPARING ECONOMIC ALTERNATIVES

A new engineering design frequently results in a potential investment opportunity for the sponsoring company. Thus, the desirability of the proposed design is often measured in economic rather than technical terms. Furthermore, if several different designs have been proposed, the desirability of one investment opportunity over another is generally based upon certain economic criteria. Engineers must therefore have some knowledge of engineering economic analysis as well as a mastery of technical fundamentals.

Spreadsheets provide an excellent means for carrying out engineering economic analysis since they can accommodate complex investment patterns projected over future time periods. They also include library functions for carrying out the required calculations. Excel is particularly effective in this area.

In this chapter, we will present some fundamental concepts in engineering economic analysis. In particular, this chapter discusses compound interest and the time value of money, illustrates how these concepts are used to analyze complicated cash flows, and discusses how Excel can conveniently be used to carry out these studies.

15.1 COMPOUND INTEREST

Money is a commodity that is used to purchase goods and services. It therefore has *value*. If a person or an organization loans, deposits, or invests a sum of money, the lender is entitled to a payment for the use of that money. Similarly, if a person or an organization borrows a sum of money, the borrower must pay for the use of the money. A sum of money that is loaned or borrowed is called *principal*; a payment that is made for the use of someone else's money is known as *interest*.

Interest calculations are always based upon an *interest rate*. The interest rate is ordinarily expressed as a percentage, but it must be converted to decimal form in order to carry out an economic analysis calculation. To do so, we first express any fractional percentage points in decimal form and then divide the entire interest rate by 100. For example, an annual interest rate of 5½ percent would be expressed in fractional form as 0.055. (Notice that we have not stated how often the interest is paid; for example, annually, quarterly, monthly, etc.)

The determination of an appropriate interest rate is normally based upon two factors: the degree of risk associated with the intended investment and general economic conditions. Risky investments are associated with higher interest rates than conservative, safe investments.

Economic calculations are generally based upon the use of *compound interest*. When carrying out compound interest calculations, it is assumed that the overall time span is broken up into several consecutive interest periods (e.g., several consecutive years) and that interest accumulates from one interest period to the next.

Suppose we invest a given amount of principal (P), which earns interest at a constant rate (i) during each of several consecutive interest periods. Within a given interest period, the current interest (I) is determined as a fraction of the total accumulation (i.e., the principal plus the previously accumulated interest). Thus, for the first interest period, the interest earned is determined as

$$I_1 = iP \tag{15.1}$$

and the total amount of money that has accumulated at the end of the first interest period is

$$F_1 = P + I_1 = P + iP = P(1 + i) \tag{15.2}$$

For the second interest period, the interest is determined as

$$I_2 = iF_1 = iP(1 + i) \tag{15.3}$$

and the total amount accumulated at the end of the second interest period is

$$F_2 = P + I_1 + I_2 = P + iP + iP(1 + i)$$

$$= P[(1 + i) + i(1 + i)] = P(1 + i)^2 \tag{15.4}$$

For the third interest period, we obtain

$$I_3 = iP(1 + i)^2 \tag{15.5}$$

$$F_3 = P(1 + i)^3 \tag{15.6}$$

and so on.

In general, if there are n interest periods, the total amount of money accumulated at the end of the last interest period is given by

$$F_n = F = P(1 + i)^n \tag{15.7}$$

This is the so-called "law" of compound interest. Equation (15.7) can be used to determine either the amount of money accumulated, in the case of an investment, or the amount of money owed, in the case of a loan. It is easily implemented within a spreadsheet.

Excel includes the FV function, which provides a simple way to evaluate Equation (15.7). (See Table 15.1 later in this section.)

Example 15.1 Accumulating Compound Interest

A student invests $2000 in a bank account earning 5 percent interest, compounded annually. If the student does not make any subsequent deposits or withdrawals, how much money will accumulate after 20 years? Calculate and plot the amount of money accumulated on a year-by-year basis.

In this problem, we know that $P = \$2000$, $i = 0.05$, and $n = 20$. Therefore, we can determine the amount of money accumulated after 20 years using Equation (9.7):

$$F = \$2000 \, (1 + 0.05)^{20} = \$2000 \, (1.05)^{20} = \$5306.60$$

Note that the interest accumulated over the 20-year period is

$$F - P = \$5306.60 - \$2000.00 = \$3306.60$$

which is substantially more than the original deposit.

The easiest way to determine the year-by-year accumulation is to construct an Excel worksheet based upon the repeated use of Equation (15.7), as shown in Fig. 15.1. The amount of money accumulated at the end of each year is shown in column E, beginning with the end of year 0 (the time of the initial deposit) and ending with the end of year 20. Each value in column E is obtained by applying Equation (15.7) to the principal entered in cell B3, the interest rate in cell B5, and the number of years in the adjoining cell in column D. (Notice, for example, the formula used to obtain the final value, 5306.60, which is shown in the highlighted cell E25.) Thus, changing a value in cell B3 or cell B5 will result in different values for all of the accumulations shown in column E.

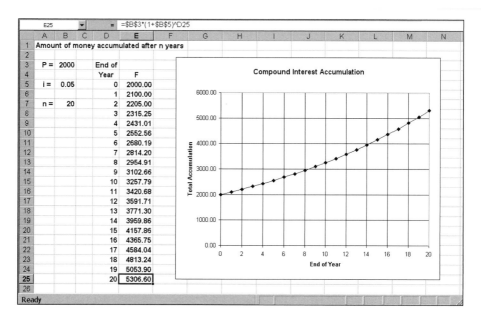

	E25		▼		=$B$3*(1+$B$5)^D25									
	A	B	C	D	E	F	G	H	I	J	K	L	M	N

1 Amount of money accumulated after n years

3	P =	2000		End of	
4				Year	F
5	i =	0.05		0	2000.00
6				1	2100.00
7	n =	20		2	2205.00
8				3	2315.25
9				4	2431.01
10				5	2552.56
11				6	2680.19
12				7	2814.20
13				8	2954.91
14				9	3102.66
15				10	3257.79
16				11	3420.68
17				12	3591.71
18				13	3771.30
19				14	3959.86
20				15	4157.86
21				16	4365.75
22				17	4584.04
23				18	4813.24
24				19	5053.90
25				20	5306.60
26					

Ready

Figure 15.1 – The accumulation of money due to compounding interest

Equation (15.7) can be used to determine the cost of a single-payment loan as well as the value of an accumulation. When dealing with a loan of this type, the principal (P) represents the amount of money originally borrowed, and the future accumulation (F) represents the amount of money that must be repaid (principal plus accumulated interest) after n interest periods.

Example 15.2 A Single-Payment Loan

A small manufacturing firm plans to borrow $500,000 in order to expand its loading dock. Interest will be charged at the rate of 7 percent per year, compounded annually. The company plans to allow the interest to accumulate for four years, at the end of which the company will repay the principal plus all of the accumulated interest. How much must the company repay at the end of the loan period? How much money will this loan cost the company, in terms of accumulated interest?

This problem can be solved using Equation (15.7), with $P = \$500,000$, $i = 0.07$, and $n = 4$. Substituting these values into Equation (15.7), we obtain

$$F = \$500,000 \, (1 + 0.07)^4 = \$500,000 \, (1.07)^4 = \$655,398.$$

Hence, the company must repay $655,398 at the end of four years. The cost of the loan will be $655,398 – $500,000 = $155,398 in accumulated interest.

Although interest rates are typically specified on an annual basis, the interest is often compounded more frequently. Thus, if i is the interest rate, expressed on an annual basis, m is the number of compounding periods per year (e.g., $m = 12$ for monthly compounding), and n is the total number of compounding periods (i.e., $n = m \times$ number of years), Equation (15.7) can be modified to read

$$F = P(1+i/m)^{n} \tag{15.8}$$

Thus, if the interest is compounded quarterly, Equation (15.8) would be written as

$$F = P(1+i/4)^{n} \tag{15.9}$$

where n is the total number of quarters. Similarly, if the interest is compounded monthly, Equation (15.8) becomes

$$F = P(1+i/12)^{n} \tag{15.10}$$

where n now represents the total number of months.

The interest can even be compounded daily. In such situations, Equation (15.8) becomes

$$F = P(1+i/365)^{n} \tag{15.11}$$

where n represents the total number of days. (The banking industry sometimes uses a factor of 360 rather than 365 for daily compounding since 360 is a multiple of both 4 and 12. This results in a simple relationship between the annual interest rate, the quarterly interest rate, the monthly interest rate, and the daily interest rate. Obviously, this practice precedes the use of computers in the banking industry.)

Example 15.3 Compound Interest: Frequency of Compounding

In Example 15.1, we considered a $2000 deposit earning interest at 5 percent per year, compounded annually. After 20 years, we found that $5306.60 had accumulated. Let us now recalculate the amount of money accumulated, based upon an annual interest rate of 5 percent and

(*a*)   Quarterly compounding

(*b*)   Monthly compounding

(*c*)   Daily compounding

The accumulation resulting from quarterly compounding is determined using Equation (15.9); thus,

$$F = \$2000 \, (1 + 0.05 / 4)^{4 \times 20} = \$5402.97$$

If the compounding is monthly, we use Equation (15.10) to obtain

$$F = \$2000 \, (1 + 0.05 / 12)^{12 \times 20} = \$5425.28$$

Similarly, if the compounding is daily, we use Equation (15.11) to obtain

$$F = \$2000 \, (1 + 0.05 / 365)^{365 \times 20} = \$5436.19$$

The results are summarized below:

Frequency of Compounding	Accumulation
Annual	$5306.60
Quarterly	$5402.97
Monthly	$5425.28
Daily	$5436.19

Comparing annual compounding with daily compounding, we see that an additional $130 (approximately) is obtained as a result of the more frequent compounding.

G3		=	=$B$3*(1+$B$5)^$B$7							
	A	B	C	D	E	F	G	H	I	J	K
1	Compound Interest: Varying Frequency of Compounding										
2											
3	P =	2000		Annual Compounding:			5306.60				
4											
5	i =	0.05		Quarterly Compounding:			5402.97				
6											
7	n =	20		Monthly Compounding:			5425.28				
8											
9				Daily Compounding:			5436.19				
10											
11											
Ready											

Figure 15.2 – The effect of different compounding frequencies on a 20-year accumulation

The calculations can be carried out within an Excel worksheet, as shown in Fig. 15.2. The accumulated results, which appear in column G, were obtained using the appropriate compound interest equation for the given compounding period. Notice, for example, the formula used to obtain the value within the highlighted cell G3. This formula corresponds to Equation (15.7), which is the appropriate equation for annual compounding.

Figure 15.3 contains the cell formulas used to obtain the values shown in Fig. 15.2. (The cell widths have been modified to display their entire contents.) Notice the cell formulas used to generate the values appearing in column G. These formulas correspond to Equations (15.7), (15.9), (15.10), and (15.11).

Note that we could have used Excel's FV function (see Table 15.1 below) to solve this problem rather than Equations (15.7), (15.9), (15.10), and (15.11).

	G3		=	=$B$3*(1+$B$5)^$B$7					
	A	B	C	D	E	F	G		H
1	Compound Interest:								
2									
3		P = 2000		Annual Compounding:			=$B$3*(1+$B$5)^$B$7		
4									
5		i = 0.05		Quarterly Compounding:			=$B$3*(1+$B$5/4)^(4*$B$7)		
6									
7		n = 20		Monthly Compounding:			=$B$3*(1+$B$5/12)^(12*$B$7)		
8									
9				Daily Compounding:			=$B$3*(1+$B$5/365)^(365*$B$7)		
10									
11									
Ready									

Figure 15.3 – The corresponding cell formulas

Sometimes F and P are both specified, and it is necessary to solve for either an interest rate (i) or the number of interest periods (n). If n is given and i is unknown, we can rearrange Equation (15.7) to read

$$i = (F/P)^{1/n} - 1 \qquad (15.12)$$

On the other hand, if i is given and n is unknown, we can solve Equation (15.7) for n, resulting in

$$n = \frac{\log(F/P)}{\log(1+i)} \qquad (15.13)$$

Equations (15.9), (15.10), and (15.11) can also be solved for i and n, with similar results.

Note that Excel's RATE and NPER functions offer a simple way to solve for i and n. (See Table 15.1 in Sec. 15.3.)

Example 15.4 A Single-Payment Loan with Monthly Compounding

In Example 15.2, we determined the cost of a four-year $500,000 loan at an interest rate of 7 percent per year, compounded annually. At the end of four years, we found that the total amount of money that must be repaid is $655,398.

(a) What annual interest rate would the company be charged if the total amount due is $700,000 at the end of four years, assuming monthly compounding?

(b) Suppose the company will be charged 7 percent per year, compounded monthly, but the company cannot afford to repay more than $625,000. For how many months should the loan be taken out?

The first part of this question can be answered by solving Equation (15.10) for the annual interest rate, resulting in

$$i = 12 \left[(F/P)^{1/n} - 1 \right]$$

Substituting numerical values into this expression,

$$i = 12 \left[(700{,}000/500{,}000)^{1/(12 \times 4)} - 1 \right] = 12 \left[(1.4)^{1/48} - 1 \right] = 0.0844$$

Hence, the interest rate is 8.44 percent per year, based upon monthly compounding.

To answer the second part of the question, we solve Equation (15.10) for n, resulting in

$$n = \frac{\log(F/P)}{\log(1 + i/12)}$$

Substituting numerical values into this expression and taking logarithms to the base 10, we obtain

$$n = \frac{\log(620{,}000/500{,}000)}{\log(1 + 0.07/12)} = \frac{\log(1.250)}{\log(1.0058)} = \frac{0.0969100}{0.00252602} = 38.36$$

Therefore, the loan should be taken out for 38 months. (Note that we could have obtained this same result using natural logarithms rather than base-10 logarithms.)

As an alternative, we can use Excel's Goal Seek feature (see Chapter 10) to solve for either i or n when the other three parameters are known. The details are provided in the following example.

Example 15.5 A Single-Payment Loan with Monthly Compounding in Excel

Repeat the calculations in Example 15.4 within an Excel worksheet. Use Excel's Goal Seek feature to solve for the interest rate and the number of months, as required.

Figure 15.4 shows an Excel worksheet containing the necessary information. Note that the values $P = 500{,}000$, $i = 0.07$, and $n = 48$ are entered directly into cells B3, B5, and B7, respectively. The corresponding value $F = 661{,}026.94$ shown in cell B9 is obtained by a cell formula, which is an expression of Equation (15.9).

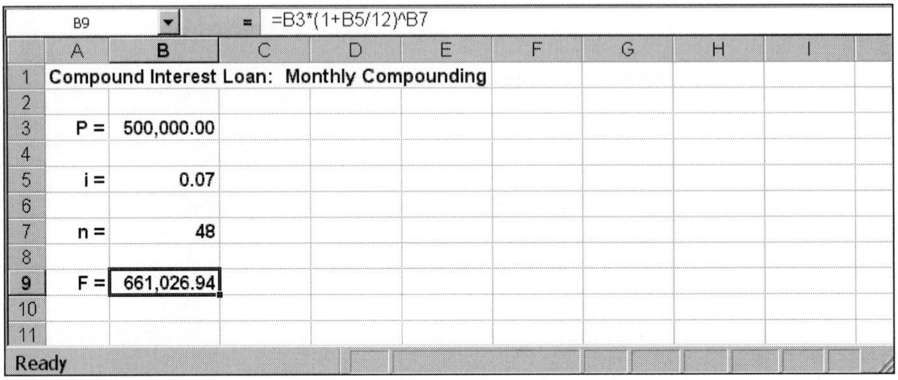

Figure 15.4 – Future value with monthly compounding

Figure 15.5 shows the Goal Seek dialog box, obtained by selecting Goal Seek from the Tools menu. The dialog box specifies that the formula in cell B9 must result in a value of 700,000. This value will be obtained by altering the value of the interest rate in cell B5.

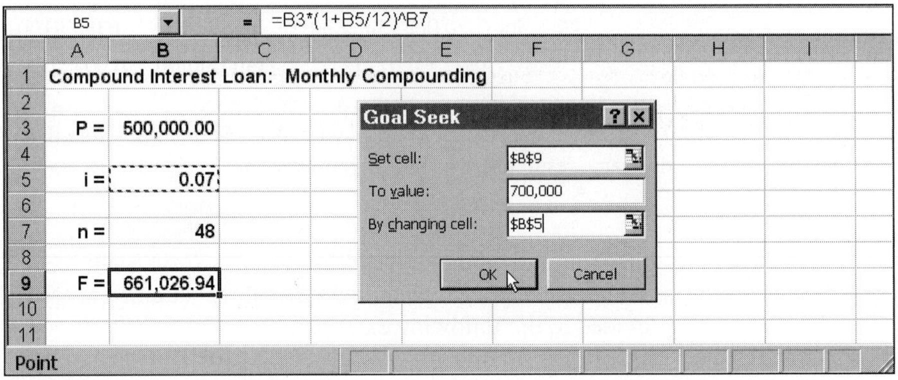

Figure 15.5 – Preparing to solve for the interest rate using Goal Seek

Figure 15.6 shows the resulting solution. We see that the calculated accumulation is $700,000, as required. To obtain this result, Goal Seek selected a value of 0.0844 (i.e., an annual interest rate of 8.44 percent) in cell B5. This corresponds to the value obtained in Example 15.4.

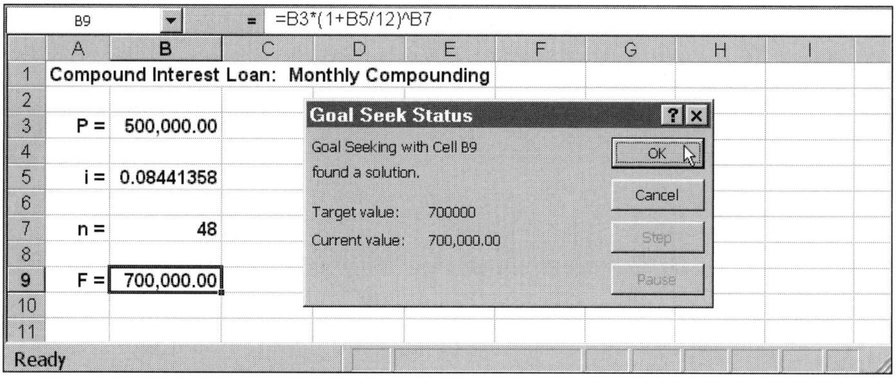

| B9 | ▼ | = | =B3*(1+B5/12)^B7 |

	A	B	C	D	E	F	G	H	I
1	Compound Interest Loan: Monthly Compounding								
2									
3	P =	500,000.00							
4									
5	i =	0.08441358							
6									
7	n =	48							
8									
9	F =	700,000.00							
10									
11									

Goal Seek Status   ? ×

Goal Seeking with Cell B9
found a solution.

     OK   Cancel

Target value: 700000
Current value: 700,000.00

Step

Pause

Ready

Figure 15.6 – Annual interest rate that will yield a desired future value

To solve the second part of the problem, we reset the annual interest rate in cell B5 to 0.07 and again select Goal Seek from the Tools menu. Now, however, we require that the formula in cell B9 result in a value of 625,000. To obtain this value, we must alter the value in cell B7 (the number of months). Figure 15.7 shows the Goal Seek dialog box for this problem.

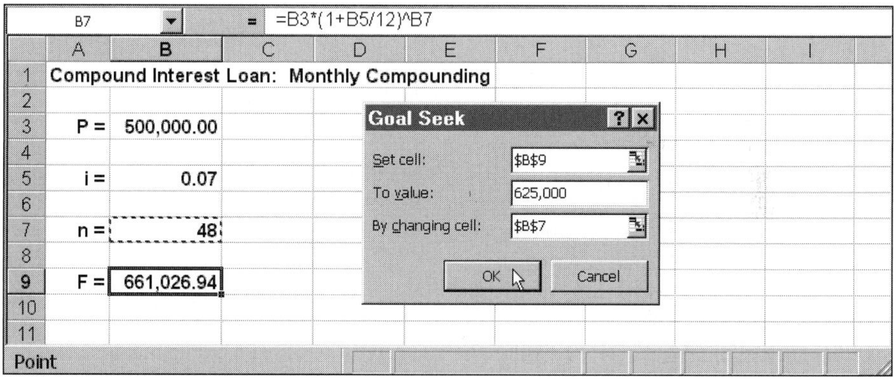

| B7 | ▼ | = | =B3*(1+B5/12)^B7 |

	A	B	C	D	E	F	G	H	I
1	Compound Interest Loan: Monthly Compounding								
2									
3	P =	500,000.00							
4									
5	i =	0.07							
6									
7	n =	48							
8									
9	F =	661,026.94							
10									
11									

Goal Seek   ? ×

Set cell: $B$9
To value: 625,000
By changing cell: $B$7

    OK   Cancel

Point

Figure 15.7 – Preparing to solve for the number of compounding periods

The solution is shown in Fig. 15.8. Notice that the calculated accumulation is $625,000, as required. This value was obtained by changing the length of the loan from 48 months (the value originally entered into cell B7) to 38.36 months. This result is consistent with the value obtained in Example 15.4. Rounding to the nearest lower integer, we conclude that the length of the loan should not exceed 38 months.

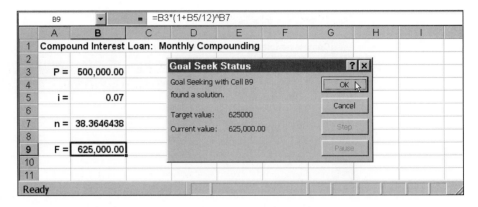

Figure 15.8 – Number of months that will yield a desired future value

Excel's Financial Functions

Excel includes a large number of financial functions, some of which are very useful in engineering economic analysis. These functions are summarized in Table 15.1. When using these functions, note that the interest rate must be expressed as a decimal, as in the equations developed earlier. However, the interest rate refers to the *specific compounding period*; it is not an annual interest rate (unless, of course, the interest is compounded annually).

Table 15.1   Commonly Used Financial Functions

Function	*Purpose*
FV(i, n, A)	Returns the future value of a series of n payments of x A dollars each at interest rate i.
IRR($A_1, A_2, \ldots$)	Returns the internal rate of return for a series of cash flows.
NPER(i, A, P)	Returns the number of payment periods for a loan of P dollars with constant payments of A dollars each at interest rate i.
NPV($i, A_1, A_2, \ldots$)	Returns the net present value of a series of cash flows at interest rate i.
PMT(i, n, P)	Returns the periodic (e.g., monthly) payment for an n-payment loan of P dollars at interest rate i.
PV(i, n, A)	Returns the present value of a series of n payments of A dollars each at interest rate i.
RATE(n, A, P)	Returns the interest rate for series of n equal payments of A dollars each with present value P.

Problems

15.1   In Example 15.1, suppose the student seeks a bank that will pay interest more often than once a year. Using Excel, determine the additional amount of money that will accumulate if the interest is compounded

(*a*)   Quarterly

(*b*)   Monthly

(*c*)   Daily

Assume an annual interest rate of 5 percent in each case. Compare the total accumulation with that obtained in Example 15.1 for annual compounding.

15.2   A student received some cash gifts when she graduated from college. She hopes to accumulate $50,000 by the time she retires, 40 years later. If the local bank pays interest at 4½ percent per year, compounded monthly, how much money must the student deposit when she graduates in order to meet her goal? Solve using Excel.

15.3   Suppose a student deposits $5000 in a bank account when she graduates from college. Using Excel, determine the required annual interest rate for the student to accumulate $50,000 after 40 years. Assume the interest is compounded

(*a*)   Annually

(*b*)   Quarterly

(*c*)   Monthly

(*d*)   Daily

15.4   Suppose a student deposits $5000 in a bank account when she graduates from college. The bank pays interest at 5 percent per year. Using Excel, determine how long it will take for her money to double if the interest is compounded

(*a*)   Annually

(*b*)   Quarterly

(*c*)   Monthly

(*d*)   Daily

15.5   Suppose a student deposits $5000 in a bank account when she graduates from college. She plans to keep the money in the bank, accumulating interest, for 40 years. Using Excel, determine how much money will accumulate after 40 years if the annual interest rate is (*a*) 4 percent

(*b*) 5 percent

(*c*) 8 percent

Assume quarterly compounding in each case.

15.6   Suppose you borrow $3000 at the beginning of your senior year to meet college expenses. If you make no payments for 10 years and then repay the entire amount of the loan, including accumulated interest, how much money will you owe? Use Excel to determine your solution. Assume interest is 6 percent per year, compounded

(*a*) Annually

(*b*) Quarterly

(*c*) Monthly

(*d*) Daily

How significant is the frequency of compounding?

15.7   A company plans to borrow $150,000 to equip a product-testing lab. If the company must repay $275,000 after seven years, what interest rate is being charged? Assume annual compounding. Use Excel to determine a solution.

15.8   A company plans to borrow $250,000 to promote a new product. The current interest rate is 8 percent per year, compounded monthly. Suppose the company does not plan to repay any of the loan until the end of the loan period, at which time the entire loan will be repaid. If the company cannot repay more than $325,000, what is the maximum permissible loan period?

15.9   A company has recently developed an inexpensive, battery-powered lawn mower. At present, the company's annual production rate is only 40 percent of total production capacity. However, the company expects its annual sales (and hence its annual production rate) to increase by 9 percent each year, as the demand for battery-powered lawn mowers increases. If so, how long will it take for the company to be producing lawn mowers at its full capacity?

15.10   A local beach is monitored each year for chemical pollutants. Currently, the concentration of a certain hazardous chemical is 6 ppm (parts per million). The maximum safe level of this chemical is 20 ppm. If the concentration increases at 5 percent per year, how long will it take before the beach must be closed?

15.2 THE TIME VALUE OF MONEY

Because money has the ability to earn interest, its value increases with time. Thus, $100 today is equivalent to $105 one year from now if the interest rate is 5 percent per year, compounded annually. We therefore say that the *future value* of $100 is $105 if $i = 5$ percent, compounded annually, and $n = 1$.

If money *increases* in value as we move from the present to the future, it must *decrease* in value as we move from the future to the present. Thus, $100 one year from now is equivalent to $95.24 today if the interest rate is 5 percent per year, compounded annually. (Note that $95.24 \times 1.05 = $100.) Thus, we say that the *present value* of $100 is $95.24 if $i = 5$ percent, compounded annually, and $n = 1$. The present value is also referred to as the *net present value*.

Example 15.6 Present Value of a Future Sum of Money

A student will inherit $10,000 in three years. The student has a savings account that pays 5½ percent per year, compounded annually. What is the present value of the student's inheritance?

This problem can be solved by rearranging Equation (15.7) into the following form:

$$P = F / (1 + i)^n$$

Substituting numerical values into this expression,

$$P = \$10,000 / (1 + 0.055)^3 = \$8516.14$$

Thus, the present value of the student's inheritance is $8516.14. (If the student were to deposit $8516.14 in a savings account that pays 5½ percent per year, compounded annually, $10,000 would accumulate at the end of three years.)

Engineers frequently compare the present value of one investment strategy with the present value of another. Thus, the present value provides a basis of comparison for various economic alternatives. This concept is particularly useful when the time frame of one investment strategy differs from that of another.

Example 15.7 Comparing Two Economic Alternatives

The services of an engineering consulting firm are being sought by two different clients. The first client wants the firm to carry out a study on the design of a new water purification system. The study will result in a payment of $700,000 in two years. The second client wants the firm to conduct a series of tests on an existing water purification system and then make recommendations for redesigning the system, based upon the results of the tests. This study will result in a payment of \$800,000 in five years. If the consulting firm can accept only one additional client, which one should it be? Solve using Excel, assuming the firm normally earns 8 percent per year on its assets.

We wish to determine the present value for each of the proposed payments, using the 8 percent interest rate. This allows us to compare one possible payment received two years in the future with another possible payment received five years in the future.

Figure 15.9 contains an Excel worksheet showing the present value of each of the proposed payments. The numerical values associated with the two proposals are given in columns B and E, respectively. Thus, the present value associated with the first client is shown in cell B9 as P_1 = \$600,137.17, and the present value associated with the second client is shown in cell E9 as P_2 = \$544,466.56. Since P_1 exceeds P_2, the consulting firm should choose the first client.

The values in cells B9 and E9 were obtained by solving Equation (15.7) for P, using the values for F, i, and n given in each column. The cell formulas are shown in Fig. 15.10.

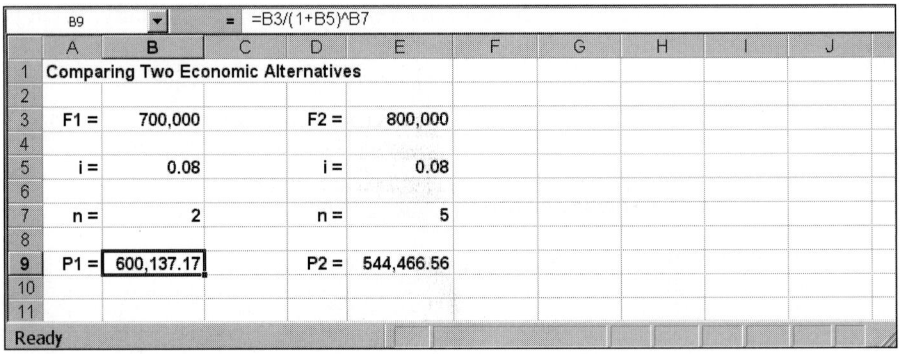
B9 = =B3/(1+B5)^B7

	A	B	C	D	E
1	Comparing Two Economic Alternatives				
2					
3	F1 =	700,000		F2 =	800,000
4					
5	i =	0.08		i =	0.08
6					
7	n =	2		n =	5
8					
9	P1 =	600,137.17		P2 =	544,466.56
10					
11					

Ready

Figure 15.9 – The present value of two proposed payments

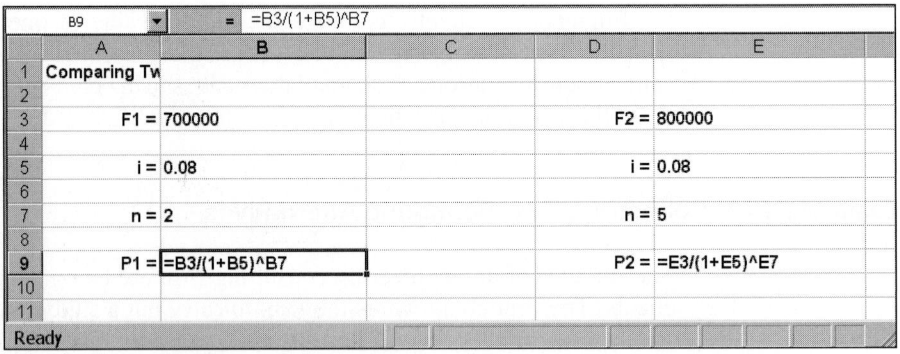
B9 = =B3/(1+B5)^B7

	A	B	C	D	E
1	Comparing Tw				
2					
3	F1 =	700000		F2 =	800000
4					
5	i =	0.08		i =	0.08
6					
7	n =	2		n =	5
8					
9	P1 =	=B3/(1+B5)^B7		P2 =	=E3/(1+E5)^E7
10					
11					

Ready

Figure 15.10 – The corresponding cell formulas

Problems

15.11   Which would have a greater present value?

(*a*) $10,000 five years from now.

(*b*) $12,500 eight years from now.

Assume an interest rate of 7 percent, compounded annually.

15.12   Repeat Prob. 15.11 using an interest rate of 8 percent, compounded annually. Compare your answer with that obtained for Prob. 15.11.

15.13   Repeat Prob. 15.11 using an annual interest rate of 7 percent, compounded daily. Compare your answer with that obtained for Prob. 15.11.

15.14   Which would have a greater present value?

(*a*) $10,000 five years from now.

(*b*) $12,500 nine years from now.

Assume an interest rate of 7 percent, compounded annually. Compare your answer with that obtained for Prob. 15.11.

15.3 UNIFORM, MULTIPAYMENT CASH FLOWS

Most realistic economic alternatives involve a number of cash inflows (*receipts*) and/or cash outflows (*disbursements*) over a period of several years. These items may all be identical, they may all be different, or they may involve some repeated items within an overall irregular pattern. A particularly common investment pattern involves an initial cash outflow (i.e., an initial investment) followed by a series of cash inflows at later times.

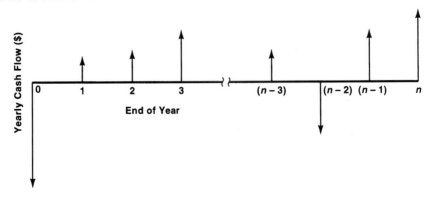

Figure 15.11 – A cash flow diagram

Collectively, the cash inflows and cash outflows associated with a proposed investment are called *cash flows*. It is often convenient to represent a series of cash flows graphically, by means of a *cash flow diagram*, as illustrated in Fig. 15.11. In a cash flow diagram, we plot the individual cash flow items as vertical arrows along the time axis. Cash *inflows* are represented as *upward*-pointing arrows, and cash *outflows* are represented as *downward*-pointing arrows.

Of particular interest is a cash flow pattern consisting of a single initial investment followed by *n* uniform payments, where each payment is made at the end of a compounding period. This cash flow pattern might represent a loan, where the initial investment is the amount of the loan (an investment, from the lender's perspective), and the uniform periodic payment is the amount repaid at the end of each compounding period (e.g., at the end of each month). Figure 15.12 shows a diagram of this cash flow from the lender's perspective.

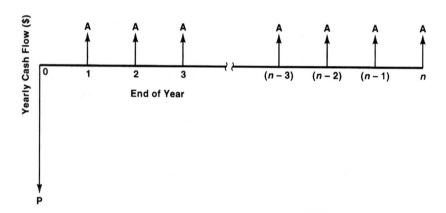

Figure 15.12 – Cash flow diagram of a loan (lender's perspective)

This cash flow pattern can be represented mathematically by the expression

$$A = P \left[\frac{(i/m)(1+i/m)^n}{(1+i/m)^n - 1} \right] \qquad (15.14)$$

where P represents the initial investment, A is the amount of each payment, i is the annual interest rate, m is the number of compounding periods per year, and n is the total number of payments (i.e., the total number of compounding periods).

Example 15.8 Determining the Cost of a Loan

Suppose you want to borrow $10,000 to buy a car, which you plan to repay on a monthly basis over three years (i.e., 36 equal monthly payments). If the current interest rate is 8 percent per year, compounded monthly, how much will you have to repay each month?

This question can be answered through the use of Equation (15.14). Thus,

$$A = \$10,000 \left[\frac{(0.08/12)(1+0.08/12)^{12\times3}}{(1+0.08/12)^{12\times3} - 1} \right] = \$313.36$$

We therefore conclude that the amount of each monthly payment is $313.36. (Note that the total amount of money repaid, including interest, is $313.36 × 36 = $11,280.96. Hence, the total interest paid is $11,280.96 − $10,000 = $1280.96.)

Equation (15.14) can easily be evaluated in Excel. However, Excel includes two library functions that simplify uniform payment loan calculations even further. In particular, the PMT function, written as PMT(i, n, P), returns the periodic payment (A) for an n-payment loan of P dollars at interest rate i. (Note that i represents the interest rate for each *payment period*, not the *annual* interest rate.) Similarly, the PV function, written as PV(i, n, A), returns the present value (P) of a series of n payments of A dollars each, at interest rate i. The direct use of Equation (15.14) and the use of the PV function are illustrated in the next example.

Example 15.9 Present Value of a Proposed Investment in Excel

A company is considering investing $1 million in a new product. This investment will result in a return of $140,000 at the end of each year for the next 12 years. The company normally expects an 8 percent return per year, compounded annually, on its investments. Should the company invest in this product? Solve using Excel.

We can solve this problem by calculating the present value of the 12 yearly payments of $140,000 each and then comparing this value with the required initial cost of $1 million. If the present value of the 12 payments is higher than the actual initial cost, the investment would be desirable.

Figure 15.13 contains an Excel worksheet showing the solution, $P = \$1,055,051$ (rounded to the nearest dollar). For illustrative purposes, the solution has been obtained two different ways: using Equation (15.14) directly (shown in cell B9) and using the PV function (cell B11). The value obtained using the PV function is negative, to indicate a cash outflow. (When using Excel financial functions, the signs of the cash flow components are consistent with the sign convention in cash flow diagrams, as in Fig. 15.11.) Since the present value of the proposed investment ($1,055,051) exceeds its cost ($1 million), we conclude that the company *should* invest in this product.

Figure 15.14 shows the cell formulas used to obtain the results in Fig. 15.13.

B11 = =PV(B5,B7,B3)

	A	B	C	D	E	F	G	H	I
1	Present Value of an Investment								
2									
3	A =	140,000							
4									
5	i =	0.08							
6									
7	n =	12							
8									
9	P =	1,055,050.92							
10									
11	P =	-1,055,050.92							
12									

Ready

Figure 15.13 – Present value of a proposed investment

B11 = =PV(B5,B7,B3)

	A	B	C	D
1	Present Value of an			
2				
3	A =	140000		
4				
5	i =	0.08		
6				
7	n =	12		
8				
9	P =	=B3*((1+B5)^B7-1)/(B5*(1+B5)^B7)		
10				
11	P =	=PV(B5,B7,B3)		
12				

Ready

Figure 15.14 – The corresponding cell formulas

Another cash flow pattern that has been studied extensively is the future value of n uniform payments. The payments can be viewed as cash outflows, and the accumulation as a cash inflow. This cash flow pattern represents a bank savings plan, where a fixed amount of money, A, is deposited at the end of each month, accumulating interest at an annual rate, i, compounded m times per year. The accumulation, F, is then withdrawn at the end of the year. Figure 15.15 shows the cash flow diagram.

Equation (15.15) provides the relationship between F and A. This equation can be implemented directly in Excel, though Excel includes an equivalent library function, FV (see Table 15.1). The use of Equation (15.15) and the FV function are illustrated in the next example.

$$F = A\left[\frac{(1+i/m)^n - 1}{(i/m)}\right] \tag{15.15}$$

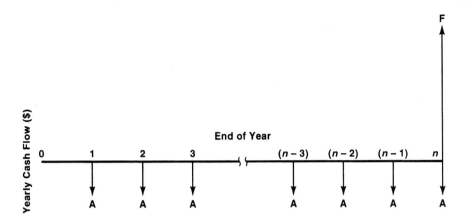

Figure 15.15 – Cash flow diagram of a savings plan

Example 15.10 Future Value of a Series of Uniform Payments

A student has opened a savings plan at a local bank. The bank requires the student to deposit $100 into the savings plan at the end of each month (January through November) for 11 months. The student then receives the total amount of the accumulation at the end of the calendar year (end of December). How much will the student accumulate, if the annual interest rate is 5 percent, compounded monthly?

Equation (15.15) can be used to answer this question. Note, however, that Equation (15.15) applied to this problem would provide the amount of money accumulated from 12 uniform monthly payments. The last payment would occur at the end of December, at the same time as the withdrawal (a mathematical artifice). In reality, only 11 payments will be made. Hence, we must subtract the amount of one monthly payment ($100) from the calculated value of F obtained from Equation (15.15).

Substituting into Equation (15.15), we obtain

$$F = \$100 \left[\frac{(1 + 0.05/12)^{12} - 1}{(0.05/12)} \right] = \$1227.89$$

Subtracting $100 for the nonexistent December payment, we conclude that the accumulation will be $1127.89.

Figure 15.16 shows an Excel worksheet containing the solution. The value shown in cell B11 was obtained directly from Equation (15.15), whereas the value in cell B13 was obtained using the FV function. Both methods result in a value of $F = \$1127.89$, which is consistent with our earlier result.

B13	▼	*fx* =FV(B5/B7,B9,-B3)-B3							
A	B	C	D	E	F	G	H	I	J

	A	B	C	D	E	F	G	H	I	J
1	Future Value of a Monthly Savings Plan									
2										
3	A =	100								
4										
5	i =	0.05								
6										
7	m =	12								
8										
9	n =	12								
10										
11	F =	1,127.89								
12										
13	F =	1,127.89								
14										

Ready

Figure 15.16 – The future value of a series of uniform payments

A1	▼	*fx* Future Value of a Monthly Savings Plan	

	A	B	C
1	Future Value of a Monthly Savings Plan		
2			
3	A =	100	
4			
5	i =	0.05	
6			
7	m =	12	
8			
9	n =	12	
10			
11	F =	=B3*((1+B5/B7)^B9-1)/(B5/B7)-B3	
12			
13	F =	=FV(B5/B7,B9,-B3)-B3	
14			

Ready

Figure 15.17 – The corresponding cell formulas

Figure 15.17 shows the corresponding cell formulas. The formula shown in cell B11 corresponds to Equation (15.15). Note that one monthly payment has been subtracted to account for the nonexistent December payment.

Cell B13 shows the formula using the FV function. The function arguments represent the monthly interest rate, the number of compounding periods, and the monthly payment, respectively. Notice that the monthly payment is entered as a negative quantity since it represents a cash outflow. Also, note that one monthly payment has been subtracted from the FV function to correct for the nonexistent December payment.

Problems

15.15 Redraw the cash flow shown in Fig. 15.12 from a borrower's perspective.

15.16 Redraw the cash flow shown in Fig. 15.15 from the bank's perspective.

15.17 For the investment described in Example 15.9, how much would the company have to recover each year in order to break even after 12 years? Solve using Excel.

15.18 An engineer has accumulated $300,000 in a retirement plan and is now considering an early retirement. To do so, the engineer plans to withdraw a fixed amount of money at the end of each year for 20 years. If the accumulated money earns interest at 6 percent per year, compounded annually, what is the maximum amount that can be withdrawn each year? Solve using Excel.

15.19 An electronics firm is planning to invest $1,500,000 to develop a new product. It is assumed that the company will receive a return of $250,000 a year for 10 years from the sale of this product, beginning at the end of the first year. To what interest rate does this correspond, assuming annual compounding? Is this a good investment? Solve using Excel.

15.20 A company plans to borrow $150,000 at 9 percent per year, compounded annually, in order to finance the installation of pollution reduction equipment. The loan is to be repaid in 8 equal annual payments. Using Excel, determine the amount of each payment.

15.21 Suppose the company described in Problem 15.20 cannot repay more than $15,000 each year. For how long should the loan be taken, assuming an interest rate of 9 percent per year, compounded annually?

15.22 An engineer plans to borrow $80,000 in order to buy a house. Suppose the interest rate is 9 percent per year, compounded monthly.
  (*a*) What will be the engineer's monthly payment if the payback period is 30 years?
  (*b*) How much interest will be paid on the loan over the entire 30-year period?
  Solve using Excel.

15.23 A recent engineering graduate is planning to buy a sports car. To do so, the engineer must borrow $12,000, which will be repaid monthly over a four-year period. Using Excel, determine (*a*) The monthly payment if the bank charges interest at the rate of 10 percent per year, compounded monthly.

(*b*) The total amount of interest paid over the entire life of the loan.

15.24 An engineer deposits $3000 into a retirement account at the end of each year. The account pays interest at the rate of 6 percent per year, compounded annually. Using Excel, determine

(*a*) How much money will have accumulated over 30 years.

(*b*) How long it will take for the engineer to accumulate $300,000.

15.25 An engineer deposits a predetermined amount of money into a savings account at the end of each year. The bank pays interest at the rate of 6 percent per year, compounded annually. How much money must the engineer save each year in order to accumulate $300,000 after 30 years? Solve using Excel.

15.26 Repeat Prob. 15.25 assuming that the money is deposited at the *beginning* of each year. Solve using Excel. Compare your answer with the result obtained in Prob. 15.25.

15.27 A country has sufficient reserves of an essential mineral to last for 400 years at the current rate of consumption. It is expected, however, that the rate of consumption will increase by 5 percent each year. If this is so, how long will it take for the country's mineral reserves to be depleted? Solve using Excel.

15.28 A chemical company has developed a new lightweight, high-strength plastic. The company is planning to spend $20 million for a manufacturing plant in order to begin full-scale production. As a result, the company expects to receive $2.4 million a year for 10 years, beginning at the end of the first year. Use Excel to determine the corresponding interest rate, assuming annual compounding.

15.29 A student has just won $1 million in the state lottery. She is given the choice of receiving $50,000 a year for 20 years or accepting a lump-sum payment of $650,000. Suppose the student is able to deposit the money into a money market fund that pays 5½ percent per year, compounded monthly. Using Excel, determine the best investment policy.

15.30 A company is considering two investment alternatives. Each involves an initial investment of $10,000,000. It is estimated that the first alternative will return $1,800,000 at the end of each year for 10 years. The second alternative should return $1,500,000 at the end of each year for 12 years.

If the company normally receives 10 percent per year on its investments, which alternative should the company select? Solve using Excel.

15.31 Suppose the company described in Prob. 15.30 anticipates that it can receive 13 percent per year on its investments during the next 12 years. If so, should the company invest in either of the proposed alternatives? If so, which one? If not, why not? Solve using Excel.

15.4 IRREGULAR CASH FLOWS

The cash flows associated with engineering investments usually do not follow the simple cash flow patterns described in the last section. Rather, they tend to be irregular, with a different payment corresponding to each payment period. Thus, it is much more difficult to evaluate the present value of an irregular cash flow since the simple mathematical relationships presented in the last section do not apply. The use of a spreadsheet is particularly helpful in situations of this type.

One way to evaluate the present value of an irregular cash flow within an Excel worksheet is to enter the cash flow components in one column and the present value of each component in an adjoining column. Equation (15.7), or a related equation [i.e., Equations (15.9) through (15.11)] can be used to determine the present value of each component. The present value of the entire cash flow can then be obtained by summing the present value of the individual components. The following example illustrates the procedure.

Example 15.11 Present Value of an Irregular Cash Flow in Excel

A manufacturing company is considering bringing out a new product. To do so, the company must invest $1 million at the beginning of year 1 (i.e., at the end of year 0) and another $8 million one year later. Marketing studies suggest that this investment will generate an irregular series of revenues from years 3 through 12. The anticipated cash flow components are summarized below:

End of Year	Revenue	End of Year	Revenue
0	−$10,000,000	7	$5,000,000
1	−8,000,000	8	6,000,000
2	0	9	5,000,000
3	1,000,000	10	4,000,000
4	2,000,000	11	3,000,000
5	3,000,000	12	2,000,000
6	4,000,000	13	1,000,000

Enter this cash flow into an Excel worksheet and determine the present value of this proposed investment, assuming the company normally receives a return of 8 percent per year, compounded annually. (Note that the initial investments are shown as negative cash flow components at the end of years 0 and 1.) Based upon the present value of the cash flow, determine whether or not this new product would represent an attractive investment strategy.

Figure 15.18 shows an Excel worksheet containing the cash flow components in column B and the present value of the cash flow in column D. A bar graph of the cash flow components (column D) is also shown.

Cell D21 contains the sum of the present values of the cash flow components. This value ($2,380,571) is positive, indicating that the present value of the future cash inflows exceeds the present value of the initial investments. Hence, this new product represents an attractive investment opportunity.

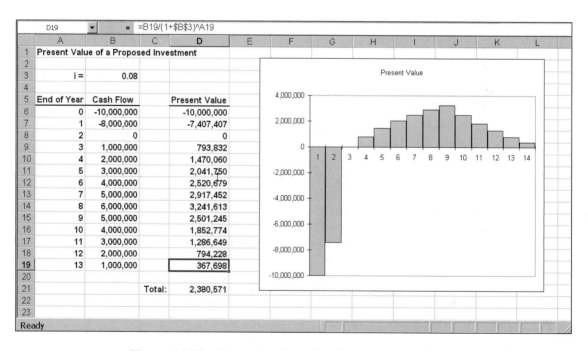

Figure 15.18 – Present value of an irregular cash flow

Figure 15.19 shows the cell formulas that were used to generate the present values in Fig. 15.18. The formulas shown in cells D7 through D19 each represent Equation (15.7), solving for P as a function of F (in column B), i (cell B3), and n (column A). The formula in cell D21 makes use of the SUM function to determine the sum of the individual present values.

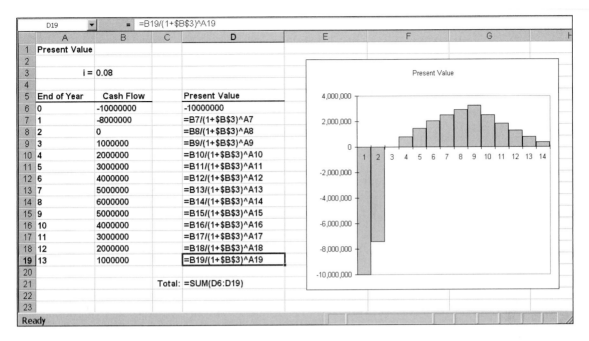

	D19	▼	=	=B19/(1+$B$3)^A19					
	A	B	C	D	E	F	G	H
1	Present Value							
2								
3	i =	0.08						
4								
5	End of Year	Cash Flow		Present Value				
6	0	-10000000		-10000000				
7	1	-8000000		=B7/(1+$B$3)^A7				
8	2	0		=B8/(1+$B$3)^A8				
9	3	1000000		=B9/(1+$B$3)^A9				
10	4	2000000		=B10/(1+$B$3)^A10				
11	5	3000000		=B11/(1+$B$3)^A11				
12	6	4000000		=B12/(1+$B$3)^A12				
13	7	5000000		=B13/(1+$B$3)^A13				
14	8	6000000		=B14/(1+$B$3)^A14				
15	9	5000000		=B15/(1+$B$3)^A15				
16	10	4000000		=B16/(1+$B$3)^A16				
17	11	3000000		=B17/(1+$B$3)^A17				
18	12	2000000		=B18/(1+$B$3)^A18				
19	13	1000000		=B19/(1+$B$3)^A19				
20								
21			Total:	=SUM(D6:D19)				
22								
23								
Ready								

Figure 15.19 – The corresponding cell formulas

The NPV function can also be used to determine the present value of an irregular cash flow in Excel. This function is written as NPV(i, $A1$, $A2$, . . .), where i represents the interest rate, $A1$ represents the cash flow component at the end of the first period, $A2$ represents the cash flow component at the end of the second period, and so on (see Table 15.1). The total number of cash flow components (n) need not be specified.

Note that the arguments for the NPV function do not include a cash flow component at the beginning of the first year (i.e., at the end of year 0). Hence, if a cash flow includes an initial investment at the beginning of the first year (as many do), this value must be subtracted from the value that is returned by the NPV function. The next example illustrates this procedure.

Example 15.12   Use of the NPV Function in Excel

Use Excel's NPV function to determine the present value of the cash flow given in Example 15.11.

Figure 15.20 shows an Excel worksheet containing the desired solution. The present value of the cash flow appears in cell B21 as $2,380,571. This value agrees with that obtained in Example 15.11, as expected.

The formula used to obtain this value is shown in the formula bar as =NPV(B3,B7:B19)+B6. The first argument refers to the annual interest rate in cell B3 (0.08). The second argument indicates the range of cash flow components in cells B7 through B19. Note that these components range from the end of year 1 to the end of year 13. The initial investment in cell B6, which occurs at the end of year 0, is not included as an argument to the NPV function. Hence, this value must be subtracted from the value returned by the NPV function. (The cell address B6 is *added* to the cell formula because it contains a negative quantity, resulting in a decrease in the present value of the cash flow.)

	B21	▼		f_x =NPV(B3,B7:B19)+B6								
	A	B	C	D	E	F	G	H	I	J	K	
1	Present Value of a Proposed Investment											
2												
3	i =	0.08										
4												
5	End of Year	Cash Flow										
6	0	-10,000,000										
7	1	-8,000,000										
8	2	0										
9	3	1,000,000										
10	4	2,000,000										
11	5	3,000,000										
12	6	4,000,000										
13	7	5,000,000										
14	8	6,000,000										
15	9	5,000,000										
16	10	4,000,000										
17	11	3,000,000										
18	12	2,000,000										
19	13	1,000,000										
20												
21	NPV =	2,380,571										
22												
23												
Ready												

Figure 15.20 – Present value of an irregular cash flow using the NPV function

Use of the NPV function is particularly convenient when comparing one investment opportunity with another. The cash flow associated with each investment can be entered into a separate column in an Excel worksheet, with the present value of the cash flow at the bottom of the column. The investment with the largest positive present value is selected as the most desirable. If none of the proposed investments result in a positive present value, none of the proposals are considered desirable. (The company should then invest in whatever it normally does to earn the interest rate used in the calculations.)

Example 15.13 Comparing Two Investment Opportunities

A company has developed two promising new products but can afford to manufacture and market only one of them. Each product requires an initial investment of $3,500,000, and each is expected to provide a return of $7,200,000 over a six-year period. The expected revenues are distributed differently with respect to time, however, as indicated by the following cash flows. If the company can normally earn the equivalent of 10 percent per year, compounded annually, in which product should the company invest? Solve using Excel.

End of Year	Revenue, Proposal A	Revenue, Proposal B
0	−$3,500,000	−$3,500,000
1	1,200,000	600,000
2	1,200,000	900,000
3	1,200,000	1,100,000
4	1,200,000	1,300,000
5	1,200,000	1,500,000
6	1,200,000	1,800,000

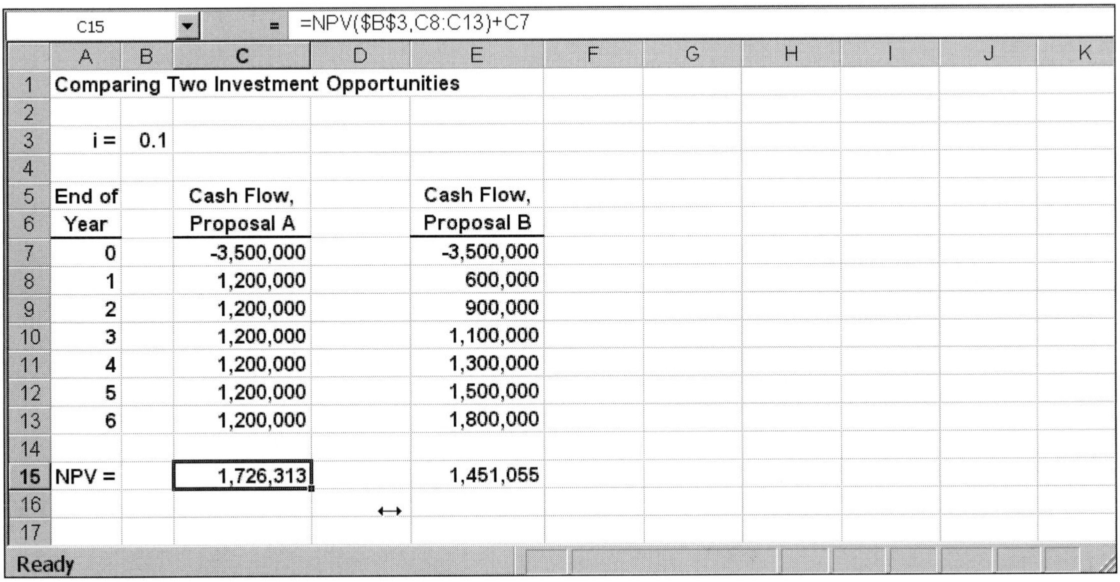

C15   =NPV($B$3,C8:C13)+C7

	A	B	C	D	E	F	G	H	I	J	K
1	Comparing Two Investment Opportunities										
2											
3	i =	0.1									
4											
5	End of		Cash Flow,		Cash Flow,						
6	Year		Proposal A		Proposal B						
7	0		-3,500,000		-3,500,000						
8	1		1,200,000		600,000						
9	2		1,200,000		900,000						
10	3		1,200,000		1,100,000						
11	4		1,200,000		1,300,000						
12	5		1,200,000		1,500,000						
13	6		1,200,000		1,800,000						
14											
15	NPV =		1,726,313		1,451,055						
16											
17											

Ready

Figure 15.21 – Comparing two different investment opportunities using NPV

Figure 15.21 contains the solution. The cash flow associated with Proposal A is entered in Column C, and the cash flow associated with Proposal B is entered in Column E. Beneath each cash flow is the present worth, as determined by the NPV function. Thus, cell C15 shows that the present value for Proposal A is $1,726,313. This value was obtained using the formula =NPV($B$3,C8:C13)+C7. Similarly, cell E15 shows a value of $1,451,055 for Proposal B, obtained from the formula =NPV($B$3,E8:E13)+E7. In each formula, the first argument refers to the annual interest rate in cell B3, and the second argument indicates the range of cells containing the cash flow components. Note that the initial investment (at end of year 0) is not included in the range of cash flow components and must therefore be added separately at the end of each formula.

The present value of each proposal is seen to be positive; hence, either proposal represents a viable investment opportunity. Since the present value of Proposal A exceeds the present value of Proposal B, the correct decision is to select proposal A.

Problems

15.32 Determine the present value of the following cash flow, using an interest rate of 6 percent per year, compounded annually. Solve with Excel, using

(*a*)   Equation (15.7).

(*b*)   The NPV function.

End of Year	Revenue
0	−$1,000,000
1	0
2	200,000
3	250,000
4	300,000
5	350,000
6	400,000

15.33 Repeat Prob. 15.32 using an interest rate of 12 percent per year, compounded annually. Compare your answer with that obtained in Prob. 15.32. Explain why the two solutions are fundamentally different.

15.34 Using Excel, determine which of the following two investment opportunities is more desirable, based upon their present values. Assume an interest rate of 12 percent per year, compounded annually.

End of Year	Revenue, Proposal A	Revenue, Proposal B
0	−$500,000	0
1	200,000	−$800,000
2	200,000	0
3	200,000	400,000
4	200,000	400,000
5	0	400,000

Compare with the conclusion that would be reached if the time value of money were not taken into account.

15.35 Use Excel to determine which of the following two investment opportunities is more desirable, based upon their present values. Assume an interest rate of 10 percent per year, compounded annually. (Note that both cash flows sum to the same value, though the order of the components is different for each proposal.)

End of Year	Revenue, Proposal A	Revenue, Proposal B
0	−$25,000	−$25,000
1	20,000	10,000
2	12,000	12,000
3	10,000	15,000
4	15,000	20,000
5	20,000	20,000

15.36 Use Excel to determine which of the following two investment opportunities is the most desirable.

End of Year	Revenue, Proposal A	Revenue, Proposal B
0	−$500,000	−$500,000
1	150,000	50,000
2	150,000	70,000
3	150,000	100,000
4	150,000	200,000
5	150,000	400,000

Assume annual compounding, using an annual interest rate of

(*a*)   0 percent

(*b*)   6 percent

(*c*)   12 percent

Is the choice influenced by the interest rate?

15.37 Use Excel to determine which of the following two investment opportunities is the most desirable, based upon an interest rate of 8 percent per year, compounded annually.

End of Year	Revenue, Proposal A	Revenue, Proposal B
0	−$750,000	−$450,000
1	0	−400,000
2	150,000	100,000
3	150,000	150,000
4	200,000	200,000
5	200,000	250,000
6	200,000	250,000
7	200,000	250,000

How would the two investments compare if the time value of money were not considered?

15.38 A car manufacturer plans to invest $27 million in order to expand its production facilities. Two different proposals are being considered. The first proposal is to expand a current Midwestern assembly plant. This will allow the manufacturer to produce an additional 50,000 cars a year, at an average net profit of $200 per car. The second proposal is to build a new assembly plant on the West Coast. This facility will have an annual output of only 40,000 cars per year, but the profit is expected to be $300 per car.

    The Midwestern plant can begin production during the first year (i.e., the same year that the initial investment is made). However, the West Coast plant requires a one-year startup, so that production will not begin until the second year. For planning purposes, each assembly plant is assumed to have a lifetime of five years of actual production.

    If the company uses an annual interest rate of 12 percent, compounded annually, which is the better proposal? Solve using Excel.

15.39 In the preceding problem, suppose the expected profit is reduced to $265 per car at the West Coast plant because of higher labor and utility costs. Which proposal will now be more desirable?

15.40 A large regional office of a computer company requires an additional 400,000 square feet of storage space. One proposal is to erect a 400,000-square-foot building now, at a cost of $25 per square foot. Another proposal is to erect a 250,000-square-foot building now at a cost of $26 per square foot, and to add another 150,000 square feet three years from now at a cost of $30 per square foot. Which is the better proposal, based upon an annual interest rate of 8 percent? Solve using Excel.

15.41 For the situation described in the previous problem, suppose the annual interest rate is 12 percent rather than 8 percent. Which would be the better proposal?

15.5 INTERNAL RATE OF RETURN

The *internal rate of return* (IRR) is another criterion that is often used to compare one investment opportunity with another. Like net present value (NPV), it is derived from the cash flow associated with an investment. Unlike net present value, however, it is not necessary to specify an interest rate when determining the internal rate of return. The internal rate of return is a popular criterion for this reason.

To understand the internal rate of return, let us begin with a cash flow consisting of an initial cash outflow (i.e., an initial investment), followed by a series of cash inflows, not necessarily uniform, whose sum exceeds the initial cash outflow. Hence, if we neglect the time value of money (i.e., if we select an interest rate of zero), the cash flow will have a positive net value.

Now suppose we plot the net present value of the cash flow as a function of interest rate, as shown in Fig. 15.22. We see that the net present value decreases with increasing interest rate, eventually crossing the abscissa and becoming negative. The internal rate of return is the *crossover point*; that is, *the IRR is the value of the interest rate at which the net present value is zero*. Figure 15.22, for example, shows an internal rate of return slightly in excess of 15 percent.

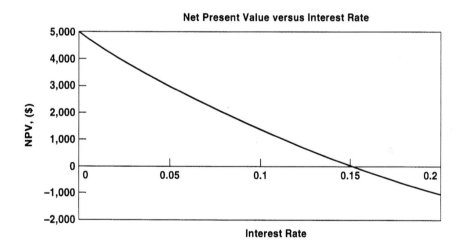

Figure 15.22 – Graphical representation of the internal rate of return

When the interest rate is zero, the net present value is equal to the algebraic sum of the cash flow components. This value will be positive when the sum of the cash inflows exceeds the initial cash outflow. When the interest rate becomes very large, however, the net present value will asymptotically approach the (negative) value of the initial cash outflow.

Example 15.14 Calculating the Internal Rate of Return

Determine the internal rate of return for the following cash flow within an Excel worksheet. To do so, proceed as follows:

(a)   Determine the present value of the cash flow for each of several different interest rates.

(b)   Plot the net present value of the cash flow as a function of interest rate.

(c)   From the resulting graph, determine the internal rate of return.

End of Year	Revenue
0	−$100,000
1	15,000
2	20,000
3	25,000
4	30,000
5	35,000
6	40,000

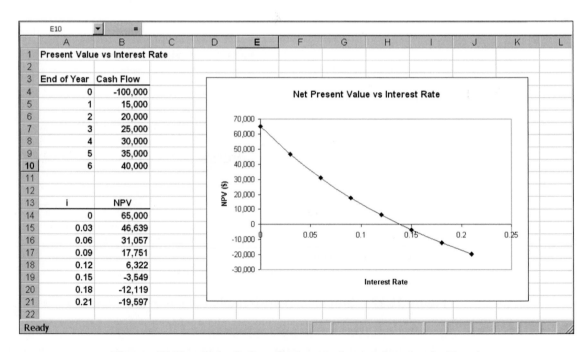

Figure 15.23 – Calculating the internal rate of return in Excel

Figure 15.23 contains the desired information. The cash flow components are entered into cells B4 through B10. Cells A14 through A21 contain various interest rates. The corresponding NPV values, obtained using the NPV function, are shown in Cells B14 through B21. Notice that the value in cell B14, which corresponds to an interest rate of zero, is simply the sum of the individual cash flow components. As the interest rate increases, the NPV value decreases, becoming negative for interest rate values above 0.12.

The figure also contains a plot of NPV versus interest rate. The NPV value decreases with increasing interest rate, as expected, crossing the abscissa at about $i = 0.14$. Hence, we conclude that the internal rate of return is about 0.14 (14 percent).

Figure 15.24 shows the cell formulas that were used to obtain the NPV values shown in Fig. 15.23.

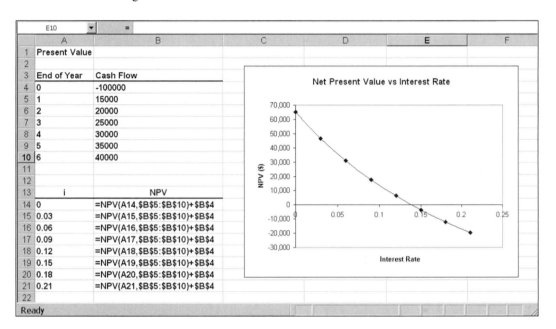

Figure 15.24 – The corresponding cell formulas

The internal rate of return, like the present value, is particularly useful for selecting among multiple investment opportunities. The cash flow having the largest internal rate of return is considered the more desirable alternative. In most situations, this is equivalent to selecting the cash flow that has the greatest net present value.

Excel includes a library function (IRR) that determines the internal rate of return directly, thus avoiding the need to calculate the present value of the cash flow as a function of interest rate. This function is written as IRR($P, A1, A2, . . .$), where P represents the initial cash outflow and $A1, A2, . . .$ represent the various cash inflows (see Table 15.1). Note that this function, in contrast to the NPV function, *does* include the initial cash outflow in its list of arguments. All cash outflows, including the initial cash outflow, must be expressed as negative quantities.

Example 15.15 Comparing Two Investment Opportunities Using IRR

Example 15.13 presented the cash flows associated with two different investment opportunities and compared them on the basis of their present value. Let us now compare these alternatives on the basis of their internal rate of return. To do so, we will modify the Excel worksheet shown in Fig. 15.21 to include the use of the IRR function as well as the NPV function. The cash flows are reproduced below for your convenience.

End of Year	Revenue, Proposal A	Revenue, Proposal B
0	−$3,500,000	−$3,500,000
1	1,200,000	600,000
2	1,200,000	900,000
3	1,200,000	1,100,000
4	1,200,000	1,300,000
5	1,200,000	1,500,000
6	1,200,000	1,800,000

	G21	▼	=								
	A	B	C	D	E	F	G	H	I	J	K
1	Comparing Two Investment Opportunities										
2											
3	i =	0.1									
4											
5	End of		Cash Flow,		Cash Flow,						
6	Year		Proposal A		Proposal B						
7	0		-3,500,000		-3,500,000						
8	1		1,200,000		600,000						
9	2		1,200,000		900,000						
10	3		1,200,000		1,100,000						
11	4		1,200,000		1,300,000						
12	5		1,200,000		1,500,000						
13	6		1,200,000		1,800,000						
14											
15	NPV =		1,726,313		1,451,055						
16											
17	IRR =		26%		21%						
18											
Ready											

Figure 15.25 – Comparing two investment opportunities using IRR

Figure 15.25 shows an Excel worksheet containing the two cash flows. Both the present value and the internal rate of return are shown for each of the cash flows. The present values were determined using the NPV function, and the internal rates of return were determined with the IRR function. The conclusions based upon each criterion are consistent, with Proposal A appearing more attractive in each case. That is, the present value of Proposal A ($1,726,313) is higher than that of of Proposal B ($1,451,055), and the internal rate of return of proposal A (26 percent) exceeds that of Proposal B (21 percent).

Figure 15.26 reveals the cell formulas used to generate the NPV and IRR values shown in Fig. 15.25. Notice that the cash flow components are entered into the two formulas differently. In particular, the initial cash outflow (end of year 0) is not included in the list of arguments in the NPV function. Instead, this value is added to the value of the future cash inflows that is returned by the NPV function. (Since the initial cash outflow is negative, we are really subtracting this dollar amount rather than adding it.) The IRR function, on the other hand, accepts *all* of the cash flow components, including the initial cash outflow, as arguments. Thus, there is no need to treat this cash flow component separately.

	G21		▼	=				
	A	B		C		D	E	
1	Comparing Tw							
2								
3	i =	0.1						
4								
5	End of			Cash Flow,			Cash Flow,	
6	Year			Proposal A			Proposal B	
7	0			-3500000			-3500000	
8	1			1200000			600000	
9	2			1200000			900000	
10	3			1200000			1100000	
11	4			1200000			1300000	
12	5			1200000			1500000	
13	6			1200000			1800000	
14								
15	NPV =			=NPV($B$3,C8:C13)+C7			=NPV($B$3,E8:E13)+E7	
16								
17	IRR =			=IRR(C7:C13)			=IRR(E7:E13)	
18								
Ready								

Figure 15.26 – The corresponding cell formulas

The internal rate of return is actually a root of a polynomial equation in terms of the interest rate (*i*). Thus, in order to determine the internal rate of return, we must solve a polynomial equation for a positive, real root. This is carried out automatically when using the IRR function. Remember, however, that some polynomials have no real roots, while others have multiple real roots. Therefore, there are some cash flows for which the internal rate of return cannot be determined. Moreover, other cash flows have multiple values for the internal rate of return. (In this situation, the usual procedure is to select the smallest positive value.) This places a limitation on the use of the internal rate of return as a basis for comparing cash flows. The present value, on the other hand, can always be calculated for any cash flow.

Convergence can also be a problem when solving for the internal rate of return. Since the governing polynomial equation is generally solved iteratively (see Sec. 10.6), there is no assurance that the solution will converge to the smallest positive real root or to any real root at all. This limitation applies to the use of the IRR function as well as a solution that might be attempted using Excel's Goal Seek feature. On the other hand, a graphical solution, though tedious, can always be obtained if a real root exists.

Example 15.16 Internal Rate of Return as the Root of a Polynomial

In Example 15.14, we calculated the internal rate of return for the following cash flow:

End of Year	Revenue
0	−$100,000
1	15,000
2	20,000
3	25,000
4	30,000
5	35,000
6	40,000

Show that the equation for the present value of this cash flow is a polynomial in i and that the internal rate of return is a positive real root of this polynomial.

Let us make use of Equation (15.7) to determine the present value of this cash flow. Thus, we can write the present value as

$$NPV = -100,000 + 15,000\frac{1}{(1+i)} + 20,000\frac{1}{(1+i)^2} + 25,000\frac{1}{(1+i)^3} +$$

$$30,000\frac{1}{(1+i)^4} + 35,000\frac{1}{(1+i)^5} + 40,000\frac{1}{(1+i)^6}$$

The internal rate of return is the value of i that causes NPV to equal zero. Hence,

$$-100,000 + \frac{15,000}{(1+i)} + \frac{20,000}{(1+i)^2} + \frac{25,000}{(1+i)^3} + \frac{30,000}{(1+i)^4} + \frac{35,000}{(1+i)^5} + \frac{40,000}{(1+i)^6} = 0$$

If we multiply this equation by $(1 + i)^6$, we obtain

$$-100,\!000(1+i)^6 + 15,\!000(1+i)^5 + 20,\!000(1+i)^4 + 25,\!000(1+i)^3 +$$

$$30,\!000(1+i)^2 + 35,\!000(1+i) + 40,\!000 = 0$$

This equation is a sixth-degree polynomial in terms of the interest rate, i. It can be solved several different ways, including the techniques described in Chapter 10 (see Prob. 15.44 below).

Problems

15.42 Enter the cash flow given in Example 15.14 into an Excel worksheet. Determine the internal rate of return using the IRR function, as described in Example 15.15. Compare your answer with that obtained in Example 15.14.

15.43 The following cash flow was originally given in Example 15.11. It is reproduced below for your convenience.

End of Year	Revenue	End of Year	Revenue
0	−$10,000,000	7	$5,000,000
1	−8,000,000	8	6,000,000
2	0	9	5,000,000
3	1,000,000	10	4,000,000
4	2,000,000	11	3,000,000
5	3,000,000	12	2,000,000
6	4,000,000	13	1,000,000

(a) Enter this cash flow into an Excel worksheet. Using the NPV function, determine the present value of the cash flow for several different interest rates, as in Example 15.14 (see Fig. 15.23). Choose a large enough range of interest rates (beginning with $i = 0$) so that the present value becomes negative for large interest rates.

(b) Plot the present value against interest rate, as in Fig. 15.23.

(c) Determine the internal rate of return by observing where the curve crosses the abscissa (i.e., at what interest rate the present value of the cash flow is zero).

(d) Determine the internal rate of return using the IRR function. Compare your answer with that obtained in part (c).

15.44 Using Excel's Goal Seek feature, determine the smallest positive real root of the polynomial given in Example 15.16. Compare this value with the value obtained for internal rate of return in Example 15.14 and in Prob.

15.42. *Hint*: Rewrite the polynomial in terms of the variable u, where $u = (1 + i)$. Solve for u and then determine the internal rate of return as $i = (u - 1)$. Is the root of the polynomial equivalent to the internal rate of return?

15.45 Each of the following previous problems contains cash flows for two different investment opportunities. For each problem, construct an Excel worksheet to determine the more desirable investment opportunity, based upon the internal rate of return. Compare with the results obtained previously, based upon present value.

(*a*) Problem 15.34

(*b*) Problem 15.35

(*c*) Problem 15.37

15.46 The following cash flows describe two competing investment opportunities. (These cash flows were presented previously, in Prob. 15.36.)

End of Year	Revenue, Proposal A	Revenue, Proposal B
0	−$500,000	−$500,000
1	150,000	50,000
2	150,000	70,000
3	150,000	100,000
4	150,000	200,000
5	150,000	400,000

(*a*) Enter both cash flows into an Excel worksheet.

(*b*) Determine which investment is more attractive, based upon present value using an interest rate of 5 percent, compounded annually.

(*c*) Determine which investment is more attractive, based upon the internal rate of return.

(*d*) Explain any apparent anomalies by plotting present worth against interest rate for each of the cash flows.

15.47 Using Excel, solve Prob. 15.38 on the basis of internal rate of return. Compare with the results obtained earlier, using the present value and an annual interest rate of 12 percent.

CHAPTER 16

FINDING OPTIMUM SOLUTIONS

Many problems in engineering have multiple solutions, and selecting among them can be a major task. Normally, there is some criterion, such as cost, profit, weight, or yield, that distinguishes one solution from another. When expressed mathematically, this is known as an *objective function*. (It may also be called the *cost function* or the *performance criterion*.) In addition, there are certain conditions, such as conservation laws, capacity restrictions, or other technical relationships, that must always be satisfied. In mathematical form, these conditions are known as *constraints*. (They are also referred to as *auxiliary conditions* or *design requirements*.)

When selecting an optimum solution, the goal is to determine the particular solution that causes the objective function to be maximized or minimized while satisfying all of the constraints. Suppose, for example, we want to determine the lightest-weight bridge that will satisfy a given set of design requirements. The objective function will be the weight of the bridge, expressed in terms of the dimensions of the major structural members. The constraints will be the various force and moment relationships that must be satisfied. The problem is to determine the dimensions of the major structural members that will minimize the weight, subject to the applicable constraints. Problems of this type are known as *optimization problems*.

Excel includes a provision for solving a wide variety of optimization problems. Both maximization and minimization problems, linear or nonlinear, with or without constraints, can be solved very easily. However, the *formulation* of an optimization problem generally requires considerable care. In this chapter, we will see how optimization problems are formulated and then solved using Excel's Solver feature.

16.1 OPTIMIZATION PROBLEM CHARACTERISTICS

In general terms, an optimization problem can be written in the following manner: Determine the values of the independent variables $x_1, x_2, \cdots, x_n$ that will maximize or minimize the *objective function*

$$y = f(x_1, x_2, \ldots, x_n) \tag{16.1}$$

The independent variables are often referred to as *decision variables* or *policy variables*. Also, the objective function is sometimes called the *performance criterion*, *performance index*, *profit function* (for a maximization problem), or *cost function* (for a minimization problem).

In many problems, the optimum solution must satisfy one or more auxiliary conditions, called *constraints*. Each constraint may be expressed as an equation or as an inequality; that is,

$$g_j(x_1, x_2, \ldots, x_n) = 0 \tag{16.2}$$

or as

$$g_j(x_1, x_2, \ldots, x_n) \leq 0 \tag{16.3}$$

or

$$g_j(x_1, x_2, \ldots, x_n) \geq 0 \tag{16.4}$$

for $j = 1, 2, \ldots, m$, where m is the total number of constraints.

In addition, the permissible values of the independent variables are usually restricted to being nonnegative. Thus,

$$x_i \geq 0 \qquad i = 1, 2, \ldots, n. \tag{16.5}$$

For now, we will confine ourselves to problems that involve only two independent variables; that is, for which $n = 2$. Problems of higher dimensionality will be considered in the next section, where we discuss solution procedures based upon the use of Excel.

These concepts will become clearer in the following two examples, which present two different types of optimization problems and their respective solution characteristics.

Example 16.1 Scheduling Production to Maximize Profit

Suppose a company manufactures two products, A and B, which can be sold for $120 per unit and $80 per unit, respectively. Management requires that at least 1000 units be manufactured each month. Product A requires five hours of labor per unit, and product B requires three hours. The cost of labor is $12 per hour, and a total of 8000 hours are available per month. Determine a monthly production schedule that will maximize the company's profit.

To express this problem mathematically, we begin by defining the following variables:

$$x_1 = \text{units of product A manufactured per month}$$
$$x_2 = \text{units of product B manufactured per month}$$

These variables are referred to as decision variables or policy variables.

Now consider the objective function, which will represent the profit. Let us write the objective function as

$$y = [120 - (5 \times 12)]\, x_1 + [80 - (3 \times 12)]\, x_2$$

or

$$y = 60\, x_1 + 44\, x_2 \tag{16.6}$$

Next, we develop the constraints. The minimum production requirement can be expressed as

$$x_1 + x_2 \geq 1000 \tag{16.7}$$

Notice that this condition is expressed as an *inequality* rather than an equation (i.e., a strict equality) since the problem states that *at least* 1000 units must be manufactured per month. If the problem required a production schedule of *exactly* 1000 units per month, Equation (16.7) would be written as an equality rather than an inequality.

Similarly, the restriction on labor availability can be expressed as

$$5x_1 + 3x_2 \leq 8000 \tag{16.8}$$

Finally, we must restrict the decision variables to being nonnegative; that is,

$$x_1 \geq 0,\, x_2 \geq 0 \tag{16.9}$$

We can now state the optimization problem in the following concise manner: Determine the values of x_1 and x_2 that will maximize Equation (16.6), subject to the auxiliary conditions (constraints) expressed by Equations (16.7) through (16.9).

It is instructive to view this problem graphically. Thus, Fig. 16.1 shows a plot of x_2 versus x_1. Only the upper right quadrant is shown since this is the region that satisfies the nonnegativity conditions expressed by Equation (16.9). The lower diagonal line represents Equation (16.7). Any point on or above this line will satisfy the minimum production requirement. Similarly, the upper diagonal line represents Equation (16.8). Any point on or below this line will represent the labor availability restriction.

Notice that the three constraints form a polyhedron, which is shown as a shaded area in Fig. 16.1 This is called the *feasible region*. Any point within the feasible region will satisfy all of the constraints.

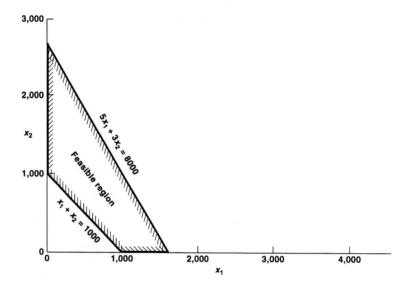

Figure 16.1 – A feasible region defined by three constraints

Now let us superimpose several lines representing different values of the objective function over the feasible region, as shown in Fig. 16.2. The different values of the objective function are shown as a family of parallel lines. Note that the value of the objective function increases as we move from lower left to upper right. We seek the line representing the largest possible value of the objective function that is within or on the feasible region.

The solution is the line representing the value $y = 11.7 \times 10^4$, which touches the upper left corner of the feasible region at $x_1 = 0$ and $x_2 = 2667$. Any point beneath this line would represent a lower value of y. Similarly, any point above this line would not lie within the feasible region, and the constraints would not be satisfied.

Notice that the solution falls on the line represented by Equation (16.8). Hence, the constraint represented by Equation (16.8) is a strict equality at the optimum. Therefore, we say that this constraint is binding (or is active). On the other hand, the constraint represented by Equation (16.7) is not binding (i.e., is inactive) since the optimum does not fall on this line.

We conclude, then, that the most profitable strategy is to manufacture 0 units of product A and 2667 units of product B per month. The resulting profit will be $117,000.

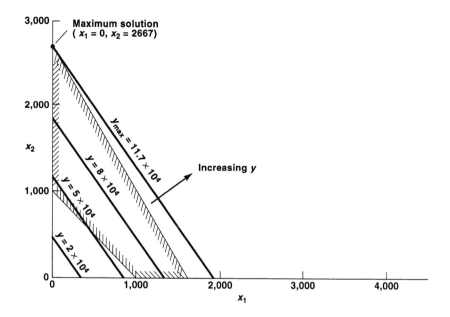

Figure 16.2 – The feasible region and a family of objective functions

Notice that the objective function and the constraints were represented by linear (i.e., straight-line) relationships in the preceding example. Optimization problems of this type are called *linear programming* problems. Linear programming has been studied extensively, and a great deal is known about the nature of the feasible region and the optimal solution. In particular:

1. The feasible region will always be a polyhedron, if it exists (i.e., if the constraints are not inconsistent).

2. The feasible region may be closed, or it may be open (unbounded). If the feasible region is closed, the optimal solution will always occur at one of the vertices.

3. If the feasible region is open, the optimal solution may be unbounded (infinite). Note, however, that an open feasible region does *not* necessarily *require* that the objective function be unbounded.

4. The optimal solution, if it exists, may not be unique. That is, the line (or plane) representing the objective function may be parallel to one of the lines (planes) that bound the feasible region. Hence, the same solution will be obtained at two or more vertices. In such situations, the *optimal value of the objective function* will be the same at each vertex, but the optimal *policy*, that is, the values of the independent variables, will be different.

Linear programming problems can be solved by a precise, though complex, solution procedure known as the *simplex algorithm* (there are several variations). This procedure is included in Excel's Solver feature. We will see how the Solver feature is used in the next section.

In the next example, we will examine another optimization problem, which involves nonlinearities. Nonlinear problems have solution characteristics that may differ markedly from linear programming problems. Hence the procedures used to obtain solutions are more complicated than the procedures used to solve linear programming problems.

Example 16.2 A Minimum-Weight Structure

Consider a structure that consists primarily of two cylindrical load-bearing columns whose radii, in feet, are x_1 and x_2, respectively. The weight of the structure, in thousands of pounds, is given by the expression

$$y = 10 + (x_1 - 0.5)^2 + (x_2 - 0.5)^2 \tag{16.10}$$

Determine the values of x_1 and x_2 that will minimize the weight of the structure, subject to the following requirements:

1. The combined cross-sectional area of the two columns must be at least 10 square feet; that is,

$$\pi(x_1^2 + x_2^2) \geq 10 \tag{16.11}$$

2. Structural considerations require that the radius of the first column not exceed 1.25 times the radius of the second column; that is,

$$x_1 \leq 1.25 \, x_2 \tag{16.12}$$

In addition, the radii must be nonnegative; that is,

$$x_1 \geq 0, x_2 \geq 0 \tag{16.13}$$

Figure 16.3 shows the feasible region defined by the constraints. Any point within the shaded region will satisfy all of the constraints, as expressed by Equations (16.11) through (16.13). Notice that the feasible region now has a curved boundary because of the nonlinear constraint expressed by Equation (16.11). In addition, note that the feasible region is not bounded (some linear problems also exhibit this characteristic).

The objective function represents a family of circles whose center is ($x_1 = 0.5$, $x_2 = 0.5$). Let us superimpose several circular arcs representing different values of the objective function over the feasible region, as shown in Fig. 16.4. Notice that the value of the objective function increases as we move radially outward from the center. The smallest value of the objective function that satisfies all of the constraints is the circle that is tangential to the feasible region at $x_1 = x_2 = 1.26$. This circle has a value of 11.16. Hence, the lightest structure that satisfies all of the imposed conditions weighs 11,160 lb. Each of the supporting columns will have a radius of 1.26 ft.

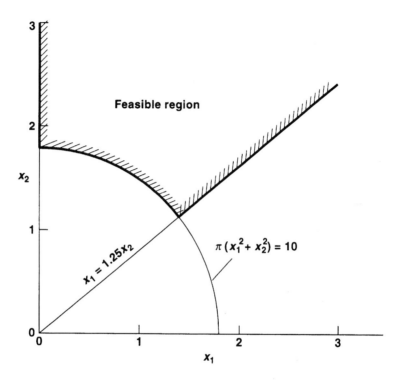

Figure 16.3 – A feasible region defined by three constraints

Note that the constraint represented by Equation (16.11) is active at the optimum, but the other constraint, represented by Equation (16.12), is inactive. If the family of objective functions were oriented differently, however (i.e., if the center of the circles were located elsewhere), the optimum solution might be at one of the corners, where Equation (16.11) intersects the x_2 axis or where Equation (16.11) and Equation (16.12) intersect. Moreover, the optimum solution might occur *within* the feasible region, where none of the constraints would be active.

This last example illustrates some fundamental differences between linear and nonlinear optimization problems. Namely, in nonlinear problems, the feasible region is not necessarily a polyhedron, and the optimum solution does not necessarily occur at a corner. In fact, the optimum solution may occur along one of the bounding surfaces or within the feasible region.

Moreover, some nonlinear problems exhibit multiple optimum points (called *local optima*). Thus, a maximum point obtained for this type of problem may represent the height of the nearest hill rather than the height of the highest hill around. These characteristics place greater demands on the mathematical procedures used to find the optimum of a nonlinear problem.

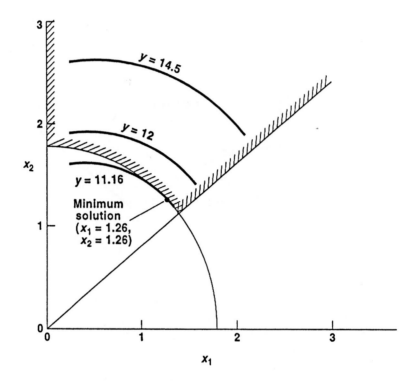

Figure 16.4 – The feasible region and a family of objective functions

Before leaving this section, we note that engineering optimization problems generally have two distinct characteristics. First, most realistic optimization problems involve many independent variables. Such multidimensional problems cannot be represented graphically; hence, we must depend upon complex mathematical procedures to obtain an optimum solution. Our use of simple two-dimensional problems was intended only to provide insight and understanding into the nature of optimization problems.

Secondly, we should understand that engineering optimization problems tend to fall into one of the following two categories:

1. Large linear optimization problems, which may have hundreds, perhaps thousands, of independent variables. Problems involving production scheduling, inventory control, product blending, etc., often fall into this category.

2. Problems involving one or more nonlinearities. As a rule, these problems have fewer independent variables than purely linear problems, though they may be much more difficult to solve because of the nonlinearities. Engineering design problems usually fall into this category.

Excel includes mathematical solution procedures for both problem categories. We will see how they are used in the next section.

Problems

16.1 Each of the following problems is a variation of the linear optimization problem given in Example 16.1. For each problem, draw the feasible region. Then determine the optimum solution by superimposing several lines or curves that represent different values of the objective function over the feasible region. Determine both the optimum value of the objective function and the corresponding optimum policy.

(a) Minimize

$$y = 60x_1 + 44x_2$$

subject to the following constraints:

$$x_1 + x_2 \geq 1000$$

$$5x_1 + 3x_2 \leq 8000$$

$$x_1 \geq 0, x_2 \geq 0$$

(b) Maximize

$$y = 60x_1 + 30x_2$$

subject to the following constraints:

$$x_1 + x_2 \geq 1000$$

$$5x_1 + 3x_2 \leq 8000$$

$$x_1 \geq 0, x_2 \geq 0$$

(c) Maximize

$$y = 50x_1 + 30x_2$$

subject to the following constraints:

$$x_1 + x_2 \geq 1000$$

$$5x_1 + 3x_2 \leq 8000$$

$$x_1 \geq 0, x_2 \geq 0$$

(d) Maximize

$$y = 60x_1 + 44x_2$$

subject to the following constraints:

$$x_1 + x_2 \geq 1000$$
$$5x_1 + 3x_2 \geq 8000$$
$$x_1 \geq 0, x_2 \geq 0$$

(e) Minimize

$$y = 60x_1 + 44x_2$$

subject to the following constraints:

$$x_1 + x_2 \geq 1000$$
$$5x_1 + 3x_2 \geq 8000$$
$$x_1 \geq 0, x_2 \geq 0$$

(f) Maximize

$$y = 60x_1 + 44x_2$$

subject to the following constraints:

$$x_1 + x_2 \leq 1000$$
$$5x_1 + 3x_2 \geq 8000$$
$$x_1 \geq 0, x_2 \geq 0$$

16.2   The following linear optimization problems are variations of one another. For each problem, draw the feasible region. Then determine the optimum solution by superimposing several lines or curves that represent different values of the objective function over the feasible region. Determine both the optimum value of the objective function and the corresponding optimum policy.

(a) Maximize

$$y = x_1 + x_2$$

subject to the following constraints:

$$x_1 + 2x_2 \leq 6$$
$$2x_1 + x_2 \leq 8$$
$$x_1 \geq 0, x_2 \geq 0$$

(b) Minimize

$$y = x_1 + x_2$$

subject to the following constraints:

$$x_1 + 2x_2 \leq 6$$
$$2x_1 + x_2 \leq 8$$
$$x_1 \geq 0, x_2 \geq 0$$

(c) Minimize

$$y = x_1 + x_2$$

subject to the following constraints:

$$x_1 + 2x_2 \geq 6$$
$$2x_1 + x_2 \geq 8$$
$$x_1 \geq 0, x_2 \geq 0$$

(d) Maximize

$$y = x_1 + x_2$$

subject to the following constraints:

$$x_1 + 2x_2 \geq 6$$
$$2x_1 + x_2 \geq 8$$
$$x_1 \geq 0, x_2 \geq 0$$

(e) Minimize

$$y = 2x_1 + 4x_2$$

subject to the following constraints:

$$x_1 + 2x_2 \geq 6$$
$$2x_1 + x_2 \geq 8$$
$$x_1 \geq 0, x_2 \geq 0$$

16.3 Each of the following problems is a variation of the nonlinear optimization problem given in Example 16.2. For each problem, draw the feasible region. Then determine the optimum solution by superimposing several lines or curves that represent different values of the objective function over the feasible region. Determine both the optimum value of the objective function and the corresponding optimum policy.

(a) Minimize
$$y = 10 + (x_1 - 0.5)^2 + (x_2 + 2)^2$$
subject to the following constraints:
$$\pi(x_1^2 + x_2^2) \geq 10$$
$$x_1 - 1.25x_2 \leq 0$$
$$x_1 \geq 0, x_2 \geq 0$$

(b) Minimize
$$y = 10 + (x_1 + 2)^2 + (x_2 - 0.5)^2$$
subject to the following constraints:
$$\pi(x_1^2 + x_2^2) \geq 10$$
$$x_1 - 1.25x_2 \leq 0$$
$$x_1 \geq 0, x_2 \geq 0$$

(c) Minimize
$$y = 10 + (x_1 - 0.5)^2 + (x_2 - 0.5)^2$$
subject to the following constraints:
$$\pi(x_1^2 + x_2^2) \leq 10$$
$$x_1 - 1.25x_2 \leq 0$$
$$x_1 \geq 0, x_2 \geq 0$$

(d) Maximize
$$y = 10 + (x_1 - 0.5)^2 + (x_2 - 0.5)^2$$
subject to the following constraints:
$$\pi(x_1^2 + x_2^2) \geq 10$$
$$x_1 - 1.25x_2 \leq 0$$
$$x_1 \geq 0, x_2 \geq 0$$

(e) Minimize
$$y = 10 + (x_1 - 0.5)^2 + (x_2 - 0.5)^2$$
subject to the following constraints:
$$\pi(x_1^2 + x_2^2) \leq 10$$
$$x_1 - 1.25x_2 \leq 0$$
$$x_1 \geq 0, x_2 \geq 0$$

(f)   Maximize

$$y = 10 + (x_1 - 0.5)^2 + (x_2 - 0.5)^2$$

subject to the following constraints:

$$\pi(x_1^2 + x_2^2) \leq 10$$

$$x_1 - 1.25x_2 \leq 0$$

$$x_1 \geq 0, x_2 \geq 0$$

16.4   Draw the feasible region for each of the following nonlinear optimization problems. Then determine the optimum solution by superimposing several lines or curves that represent different values of the objective function over the feasible region. Determine both the optimum value of the objective function and the corresponding optimum policy.

(a)   Minimize

$$y = x_1^2 + (x_2 - 2)^2$$

subject to the following constraints:

$$2x_1 - x_2 \geq 3$$

$$x_1 \geq 0, x_2 \geq 0$$

(b)   Minimize

$$y = x_1 - 2x_2$$

subject to the following constraints:

$$(2x_1 - x_2)^2 \geq 3$$

$$x_1 \geq 0, x_2 \geq 0$$

(c)   Maximize

$$y = (x_1 - 2x_2)^2$$

subject to the following constraints:

$$2x_1 - x_2 \leq 3$$

$$x_1 \geq 0, x_2 \geq 0$$

16.2 SOLVING OPTIMIZATION PROBLEMS IN EXCEL

Excel includes mathematical procedures for solving optimization problems within its Solver feature. However, Solver is an Excel add-in, like the Analysis Toolpak used to generate histograms (see Chapter 8). To install the Solver, choose Add-Ins from the Tools menu. Then select Solver Add-In from the resulting Add-Ins dialog box. (Once the Solver feature has been installed, it will remain installed unless it is removed by reversing the above procedure.)

To solve an optimization problem in Excel, proceed as follows:

1. Enter a value for each independent variable within a block of adjacent cells. These values will be used as an initial guess. (Initial values are not required when solving linear problems.)

2. Enter an equation for the objective function, expressed as an Excel formula. Within this formula, express the independent variables in terms of their cell addresses.

3. Enter an equation for each constraint, expressed as an Excel formula. The independent variables should again be expressed in terms of their cell addresses.

4. Select Solver from the Tools menu.

5. When the Solver Parameters dialog box appears, enter the following information:

   (*a*)   Enter the address of the cell containing the objective function in the Set Target Cell location.

   (*b*)   Select either Max or Min in the area labeled Equal to, beneath the Target Cell location.

   (*c*)   Enter the range of cell addresses containing the independent variables in the area labeled By Changing Cells.

   (*d*)   Enter the cell address containing each constraint, the type of constraint, and the value of the right-hand side (details given below). To add a constraint, click on the Add... button and provide the following information:

       (*i*)   Enter the cell address of the constraint in the Cell Reference location.

       (*ii*)   Specify the type of constraint (i.e., $\leq$, $\geq$, or =) from the pull-down menu.

       (*iii*)   Enter the value of the right-hand side in the Constraint location.

       Note that you can change or delete any constraint after it has been added if you wish.

   (*e*)   If the objective function and the constraints are linear, click on the Options... button and select Assume Linear Model from the Solver Options dialog box. Then select OK.

(*f*)   When all of the required information has been added correctly, select Solve. This will initiate the actual solution procedure.

6.   Once a solution has been obtained, the optimal values of the independent variables, the corresponding value of the objective function, and the value of each constraint will appear within their respective cells. A dialog box labeled Solver Results will also appear, with various options for displaying and saving the optimum solution. Respond by selecting Keep Solver Solution and by requesting an Answer report. The Answer report will then be generated and placed in a separate worksheet.

You may wish to select the feature labeled Show Iteration Results from the Solver Options dialog box. This feature will cause the solution procedure to pause after each iteration (i.e., each repeated step), thus providing a step-by-step history of the computation. This feature is particularly useful if you are taking a course or have had a course in mathematical optimization procedures.

You may also wish to select the Automatic Scaling feature in the Solver Options dialog box. This feature is very useful if the magnitude of the independent variables differs substantially from the magnitude of the objective function or the right-hand side of the constraints.

Example 16.3 Solving the Production Scheduling Problem in Excel

Solve the production scheduling problem given in Example 16.1 using Excel's Solver feature.

To recap briefly, we wish to solve the following problem:
Maximize

$$y = 60x_1 + 44x_2 \tag{16.6}$$

subject to the following constraints:

$$x_1 + x_2 \geq 1000 \tag{16.7}$$

$$5x_1 + 3x_2 \leq 8000 \tag{16.8}$$

$$x_1 \geq 0, x_2 \geq 0 \tag{16.9}$$

This problem is entirely linear. Hence, we will solve the problem using Solver's optional Assume Linear Model feature.

We begin by entering the model into an Excel worksheet, as shown in Fig. 16.5. Within this worksheet, column A contains labels for the items provided in column B. Within column B, notice the assumed values for the independent variables in cells B3 and B4, the corresponding value of the objective function in cell B6, and the values of the constraints represented by Equations (16.7) and (16.8) in cells B8 and B9, respectively.

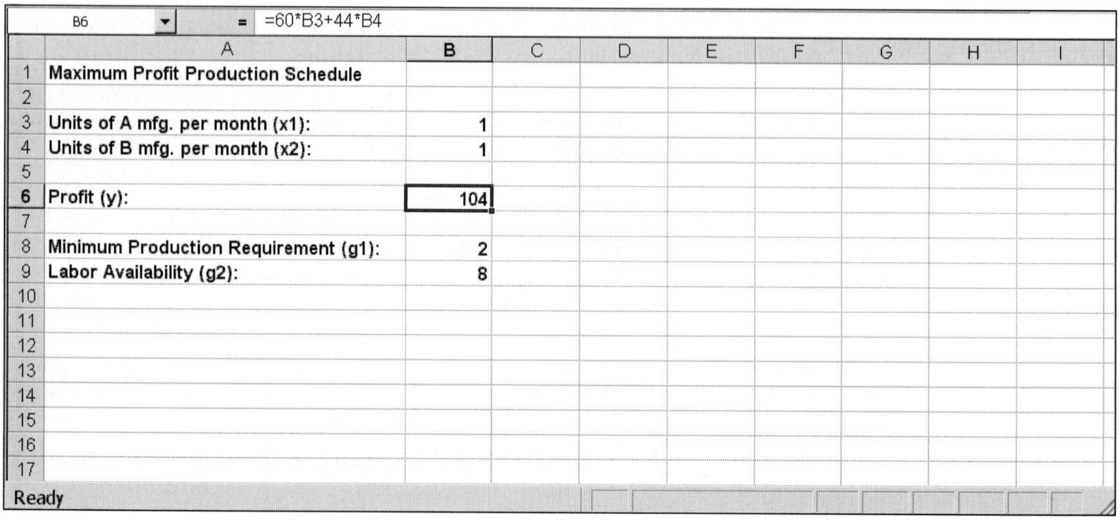

Figure 16.5 – Defining a linear optimization problem in Excel

The values shown in cells B6, B8, and B9 result from cell formulas corresponding to Equations (16.6) through (16.8). These values result from the initial values of the independent variables provided in cells B3 and B4. Figure 16.6 shows the cell formulas.

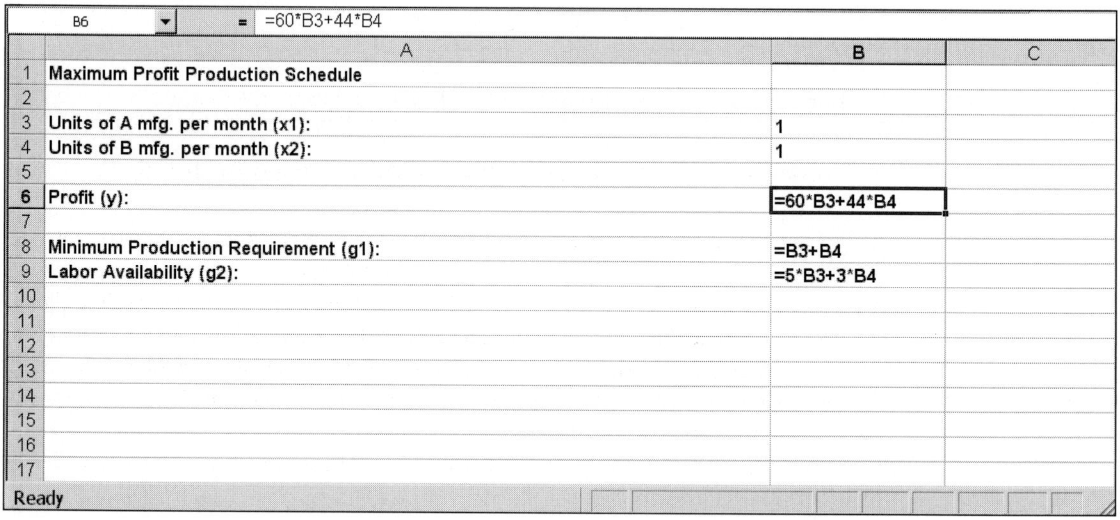

Figure 16.6 – The corresponding cell formulas

Once the problem specification has been entered into the worksheet, the Solver is invoked from the Tools menu. Figure 16.7 shows the resulting dialog box. Note that the address of the objective function (B6) is entered at the top, in the area labeled Set Target

Cell. The type of problem is entered beneath the cell address. For this problem, we select Max. The range of cell addresses containing the independent variables (B3:B4) is then entered in the area labeled By Changing Cells.

The constraints are then added, one at at time, by pressing the Add button and providing the requested information. The cell address, the type of constraint ($\leq$, $\geq$, or =), and the value of the right-hand side are added for each constraint. (The cell addresses will automatically be rearranged in ascending order once all of the constraints have been entered.) Four individual constraints, corresponding to Equations (16.7) through (16.9), are specified for this problem. Note that Equation (16.9) involves two separate constraints.

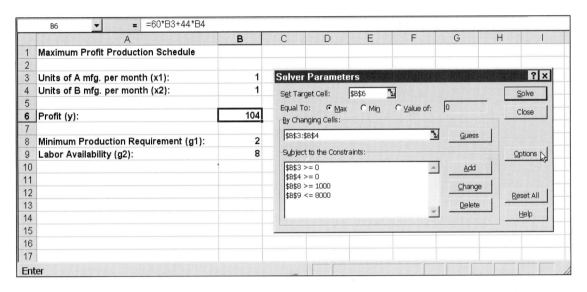

Figure 16.7 – Preparing to solve the linear optimization problem with Solver

Since this is a linear optimization problem, we select the Options button after entering all of the information requested by the Solver Parameters dialog box. This results in the Solver Options dialog box shown in Fig. 16.8. Most of the items in this dialog box should not be changed since they refer to options that affect the mathematical solution techniques. For this problem, however, we select Assume Linear Model, thus invoking the more efficient solution procedure that is designed for purely linear problems. We then select OK to return to the Solver Parameters dialog box.

From the Solver Parameters dialog box, we select Solve to initiate the actual computation. The optimum values then appear in the various cells within the worksheet, as shown in Fig. 16.9. Thus, we see that the maximum profit is $117,333, obtained by manufacturing 0 units of A and 2667 units of B per month (the production rate of B is rounded up to the nearest integer). Moreover, we see that the left-hand side of Equation (16.7) is 2667 (rounded up), and the left-hand side of Equation (16.8) is 1000. We therefore conclude that the first constraint is inactive (not binding), but the second constraint is active (binding).

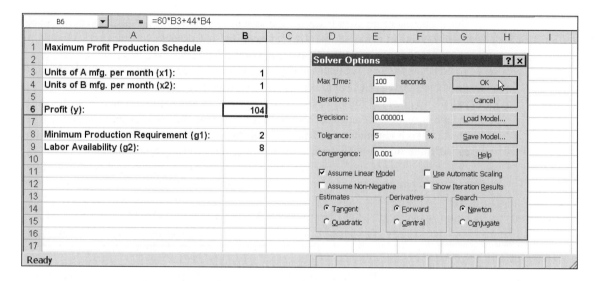

Figure 16.8 – Selecting Solver options

Superimposed over the worksheet containing the optimum solution is the Solver Results dialog box, as shown in Fig. 16.9. This dialog box allows us to either keep the optimum solution within the worksheet or restore the original values. It also allows us to generate additional reports in separate worksheets if we wish.

B6		=	=60*B3+44*B4						
	A		B	C	D	E	F	G	H	I
1	Maximum Profit Production Schedule								
2									
3	Units of A mfg. per month (x1):		0						
4	Units of B mfg. per month (x2):		2666.667						
5									
6	Profit (y):		117333.3						
7									
8	Minimum Production Requirement (g1):		2666.667						
9	Labor Availability (g2):		8000						

Solver Results dialog box:
Solver found a solution. All constraints and optimality conditions are satisfied.

Reports: Answer, Sensitivity, Limits

● Keep Solver Solution
○ Restore Original Values

OK   Cancel   Save Scenario...   Help

Figure 16.9 – The optimum solution

In this problem, we will keep the optimum solution and generate a report containing more detailed information about the final solution. The appropriate responses are shown in Fig. 16.9. We then select OK at the bottom of the dialog box, resulting in the report shown in Fig. 16.10. Notice that this report appears in a separate worksheet labeled Answer Report 1. (The original worksheet appeared in the worksheet labeled Sheet 1.)

The report shown in Fig. 16.10 contains both the original values and the final values for the independent variables, the objective function, and each of the constraints. The status of each constraint (binding or nonbinding) is also indicated. In addition, a value called *slack* is shown for each constraint. For an inequality-type constraint, this is the difference in magnitude between the limiting value expressed by the right-hand side of the constraint equation and the actual constraint value. Slack can be thought of as excess capacity, for a ≤ type constraint, or as excess production, for a ≥ type constraint. (Equality constraints, on the other hand, are always binding. Hence, the slack associated with an equality constraint is always zero.) The slack values reported in Fig. 16.10 indicate a weekly production rate of 1667 (rounded) units in excess of the minimum required production rate, with labor working to full capacity (zero slack).

Note that the report shown in Fig. 16.10 is nicely formatted, making it suitable for printing and inclusion within a more comprehensive written report. The attractive formatting is provided automatically when the report is generated.

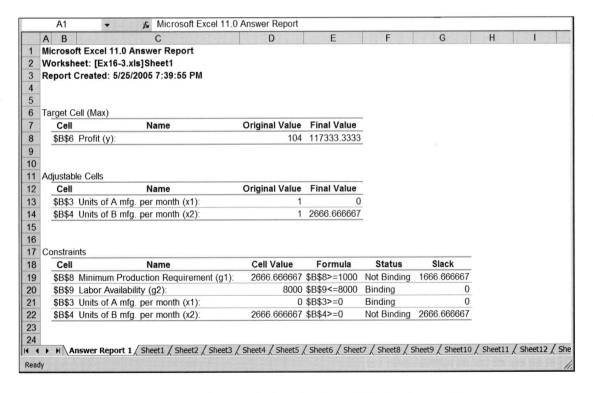

Figure 16.10 – Viewing solution details within the Answer Report

Remember that Solver includes different procedures for finding the optimum of linear and of nonlinear problems. Since the solution procedure used to solve linear problems is more efficient than the solution procedure used for nonlinear problems, *it should always be used when solving linear optimization problems.* However, this method *cannot* be used to solve *nonlinear* problems.

Moreover, if a nonlinear optimization problem has multiple optima, the solution that is obtained may be dependent upon the initial guess. Thus, an initial guess should be specified, and it should be as close as possible to the desired optimum.

Example 16.4 Solving a Nonlinear Minimization Problem in Excel

Solve the following nonlinear optimization problem using Excel:

Minimize

$$y = x_1 + x_2$$

subject to

$$2x_1^2 + 3x_2^2 \geq 12$$

$$x_1 \geq 0, x_2 \geq 0$$

The solution to this problem is shown graphically in Fig. 16.11. The feasible region lies above the ellipse represented by the first constraint. This particular problem has two local minima—one at each corner. The values associated with each local minimum are shown below.

First Local Minimum	*Second Local Minimum*
$y = 2$	$y = 2.45$
$x_1 = 0$	$x_1 = 2.45$
$x_2 = 2$	$x_2 = 0$

The first point, $y = 2$ at $x_1 = 0$, $x_2 = 2$, represents the global (absolute) minimum.

Now let us see how this problem can be solved in Excel. The procedure is very similar to that used in Example 16.3, except that we do *not* select the Assume Linear Model option. We will see that the optimum solution will depend upon the initial values selected for x_1 and x_2.

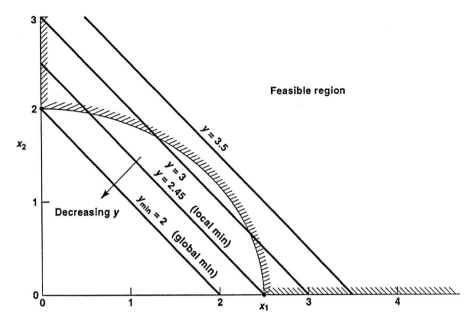

Figure 16.11 – Graphical representation of a nonlinear optimization problem

Figure 16.12 shows the worksheet containing the problem specification, with initial values of $x_1 = x_2 = 2$. Note that the objective function shown in cell B6 and the nonlinear constraint shown in cell B8 are generated from formulas. The cell formulas are shown in Fig. 16.13.

	B6	▼	=	=B3+B4					
	A	B	C	D	E	F	G	H	I
1	Nonlinear Optimization Problem								
2									
3	x1 =	2							
4	x2 =	2							
5									
6	y =	4							
7									
8	g1 =	20							
9									
10									
Ready									

Figure 16.12 – Defining the nonlinear optimization problem in Excel

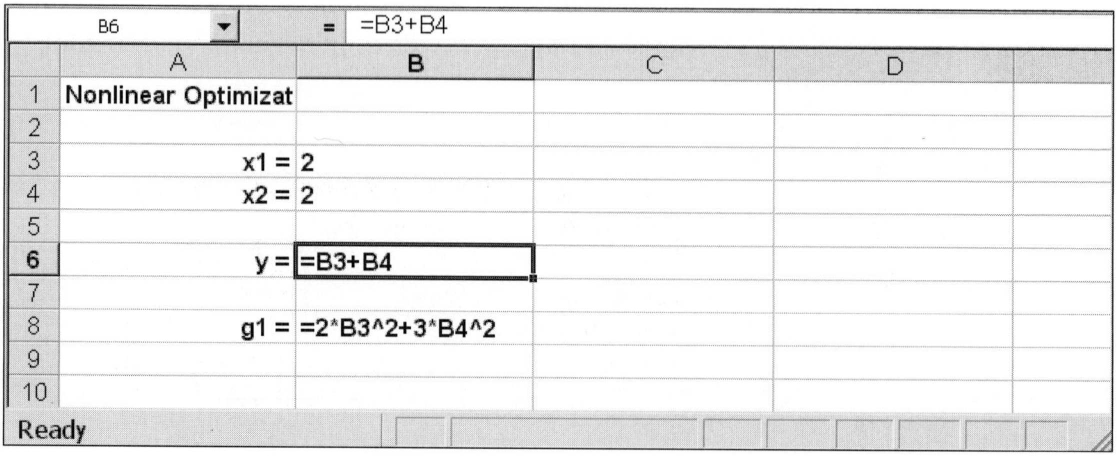

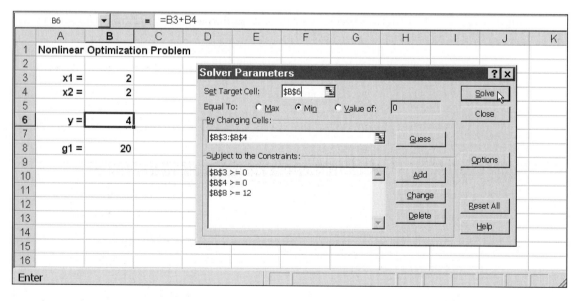

Figure 16.13 – The corresponding cell formulas

Figure 16.14 shows the Solver dialog box (obtained by selecting Tools/Solver from the menu bar), with the appropriate entries for this problem.

Figure 16.14 – Preparing to solve the nonlinear optimization problem with Solver

Once all of the information has been entered correctly, we press the Solve button, resulting in the solution shown in Fig. 16.15. Within this figure, we see that the desired minimum solution is $y = 2$ at $x_1 = 0$, $x_2 = 2$. This is one of the two local minima (in fact, the better of the two solutions) that we found at the beginning of this example when we solved the problem graphically.

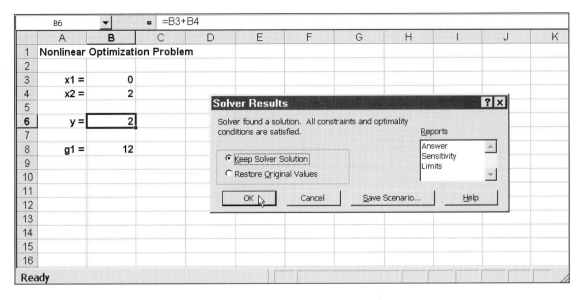

Figure 16.15 – The optimum solution

Now let us see what happens if we repeat the entire solution procedure from a starting point of $x_1 = 2$ and $x_2 = 0$. From our earlier graphical analysis, we know that this starting point is close to the other local optimum, at $x_1 = 2.45$, $x_2 = 0$. (In more realistic, multidimensional problems, however, we might not have this same insight as to where the minimum solution is likely to be found.)

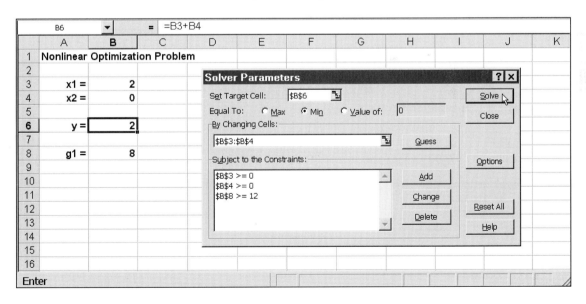

Figure 16.16 – Initiating the optimization from a different starting point

Figure 16.16 shows the worksheet containing the problem setup. The resulting solution is shown in Fig. 16.17 and Fig. 16.18 (the latter contains results rounded to two decimals). The solution (rounded) is $y = 2.45$ at $x_1 = 2.45$, $x_2 = 0$, as expected.

Unfortunately, there is no indication that this solution represents a local minimum, and is not the global minimum. This can be a serious deficiency when working with higher dimensionality problems. It is not a shortcoming of Excel but is a characteristic of the complex nature of nonlinear optimization problems.

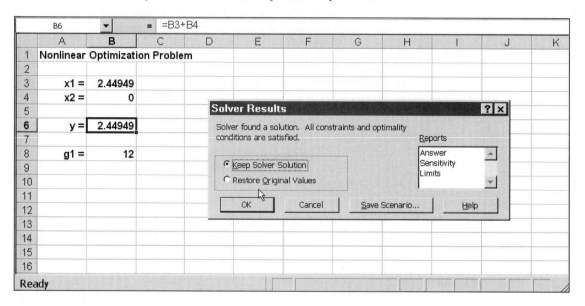

Figure 16.17 – The optimum solution

Figure 16.18 – The optimum solution after rounding

Once an optimum solution has been found, the Solver Results dialog box allows you to create three different reports, referred to as Answer, Sensitivity, and Limits, respectively. We have already discussed the Answer report and seen an example of an Answer report in Fig. 16.10. The Sensitivity report provides information concerning the sensitivity of the objective function to changes in the independent variables or changes in the constraints. For linear problems, the Sensitivity report also provides information concerning the relationships between changes in the independent variables and changes in the constraints.

The Limits report indicates the extent to which the independent variables can be changed before the constraints are violated. The accompanying changes in the objective function are also given.

Problems

16.5 Solve each of the linear optimization problems given in Prob. 16.1 with Excel's Solver feature, using the Assume Linear Model option. Compare each solution with the corresponding graphical solution obtained in Prob. 16.1.

16.6 Solve each of the linear optimization problems given in Prob. 16.2 with Excel's Solver feature, using the Assume Linear Model option. Compare each solution with the corresponding graphical solution obtained in Prob. 16.2.

16.7 Solve each of the linear optimization problems given in Prob. 16.2 with Excel's Solver feature, *without* the Assume Linear Model option. Compare your solutions with those obtained in Prob. 16.6.

16.8 Solve each of the nonlinear optimization problems given in Prob. 16.3 using Excel's Solver feature. Compare each solution with the corresponding graphical solution obtained in Prob. 16.3.

16.9 Solve each of the nonlinear optimization problems given in Prob. 16.4 using Excel's Solver feature. Compare each solution with the corresponding graphical solution obtained in Prob. 16.4.

16.10 Solve each of the following linear optimization problems using Excel's Solver feature.

(*a*) Maximize

$$y = 4x_1 + 2x_2 + 3x_3$$

subject to the following constraints:

$$3x_1 + 4x_2 + 6x_3 \leq 24$$

$$10x_1 + 5x_2 - 6x_3 \leq 30$$

$$x_2 \leq 4$$

$$x_1 \geq 0, x_2 \geq 0, x_3 \geq 0$$

(b) Minimize

$$y = x_1 + 2x_2 + 2x_3 - x_4$$

subject to the following constraints:

$$2x_1 - x_2 + 3x_3 + x_4 \leq 10$$

$$x_1 + x_2 - x_3 + 2x_4 \geq 1$$

$$-x_1 - x_3 + x_4 \leq 12$$

$$3x_1 + 2x_2 + x_4 = 5$$

$$x_1 \geq 0, x_2 \geq 0, x_3 \geq 0, x_4 \geq 0$$

(c) Maximize

$$y = 4x_1 + 5x_2 - x_3 - 2x_4$$

subject to the following constraints:

$$4x_1 - x_2 + 2x_3 - x_4 = 6$$

$$x_1 + 2x_2 - x_3 - x_4 = 3$$

$$-3x_1 + 3x_2 - 2x_3 + x_4 \leq 6$$

$$x_1 \geq 0, x_2 \geq 0, x_3 \geq 0, x_4 \geq 0$$

16.11 Solve each of the following nonlinear optimization problems using Excel's **Solver** feature.

(a) Maximize

$$y = x \sin x$$

within the interval

$$0 \leq x \leq \pi$$

(b) Minimize

$$y = (x_1 - 3)^2 + 9 (x_2 - 5)^2$$

within the region

$$x_1 \geq 0, x_2 \geq 0$$

(c) Minimize

$$y = 100(x_2 - x_1{}^2)^2 + (1 - x_1)^2$$

within the region

$$x_1 \geq 0, x_2 \geq 0$$

(d) Minimize

$$y = 3(x_1 - 1)^2 + (x_2 + 2)^2 + 5(x_3 - 4)^2$$

within the region

$$x_1 \geq 0, x_2 \geq 0, x_3 \geq 0$$

16.12 Steel is often blended with several different alloys, such as nickel and manganese, in order to give it special properties. In addition, the carbon content, which is present in all steel, will affect the properties of the steel.

Suppose a specialty steel company receives an order for 200 tons of steel containing at least 3 percent nickel and 2 percent manganese, with a carbon content not exceeding 3.5 percent. The steel will sell for $25 a ton. Assume the steel can be made from any combination of three different alloys, whose properties are given below.

	Nickel	*Manganese*	*Carbon*	*Cost per Ton*
Alloy 1	5%	8%	3%	$18
Alloy 2	4	2	3	15
Alloy 3	2	2	5	10

Develop a linear optimization model to determine the least expensive blend of alloys that will satisfy the requirements. (*Hint:* Let x_i be the number of tons of alloy i used to make the steel.) Solve the problem using Excel's Solver feature.

16.13 In the previous problem, suppose only 50 tons of alloy 2 were available. How would this affect the optimum blend?

16.14 A manufacturer of office furniture makes conference tables in two styles, Premier and Executive. Each style is available in either walnut or oak. Matching chairs are also manufactured in both styles, in either walnut or oak.

The material and labor requirements for each product are given below. Assume there is no distinction between walnut and oak in these requirements.

	Premier Table	Executive Table	Premier Chair	Executive Chair
Wood (sq ft)	100	140	16	20
Labor (hrs)	9	12	4	5

The selling prices for each product are given below.

	Premier Table	Executive Table	Premier Chair	Executive Chair
Walnut	$1200	$1500	$200	$250
Oak	1000	1200	180	220

Furniture-grade walnut costs $5 per square foot, and oak costs $4. Labor costs $18 per hour. Only 3000 hours of labor are available per week.

Develop a linear optimization model to determine the most profitable weekly production schedule. (*Hint*: There are eight different products: two walnut tables, two oak tables, two walnut chairs, and two oak chairs. Let x_i be the number of units of the ith product manufactured per week, where $i = 1, 2, \ldots, 8$.) Solve the problem using Excel's Solver feature.

16.15 Solve Prob. 16.14 with the additional restriction that at least four matching chairs must be manufactured for each table (i.e., at least four Premier walnut chairs for each Premier walnut table, etc.).

16.16 Solve Prob. 16.15 with the additional restriction that walnut cannot be obtained in quantities exceeding 10,000 sq ft per week.

16.17 A manufacturer of television sets has three factories, located in Philadelphia, St. Louis, and Phoenix. The company has four distribution centers, located in Atlanta, Chicago, Denver, and Seattle. All of the products are shipped from a factory to a distribution center, from which they are distributed to various customers across the country.

The plant capacities and distribution requirements (demands), in units per month, and shipping costs, in dollars per unit, are given below.

Philadelphia	St. Louis	Phoenix
5,000	8,000	12,000

Atlanta	Chicago	Denver	Seattle
6,000	8,000	5,000	6,000

From: →	Philadelphia	St. Louis	Phoenix
To: Atlanta	$100	$125	$300
Chicago	125	60	250
Denver	200	100	75
Seattle	325	225	175

Develop a linear optimization model to determine the least expensive way to supply the distribution centers with products from the factories. (*Hint*: Let $x_{i,j}$ represent the number of units shipped from factory i to distribution center j per month.) Solve the problem using Excel's Solver feature.

16.18 An oil refinery produces a stream of hydrocarbon base fuel from each of four different processing units. The processing units have capacities of C_1, C_2, C_3, and C_4 barrels per day, respectively. A portion of each base fuel is fed to a central blending station where the base fuels are mixed into three grades of gasoline. The remainder of each base fuel is sold "as is" at the refinery. Storage facilities are available at the blending station for temporary storage of the blended gasolines, if required.

Let N_1, N_2, N_3, and N_4 be the octane ratings of the respective base fuels, and let S_1, S_2, S_3, and S_4 be the profit derived from their direct sale. Let O_1, O_2, and O_3 be the octane numbers of the three grades of gasoline that must be blended on any given day to satisfy customer demands D_1, D_2, and D_3. Let R_1, R_2, and R_3 be the minimum octane requirements of the three grades of blended gasoline, and let P_1, P_2, and P_3 be the profit per gallon that is realized from the sale of each grade of blended gasoline. Finally, let x_{ij} be the quantity of the ith base fuel used to blend the jth gasoline.

The octane number of each gasoline can be expressed as the weighted average of the octane numbers of the constituent base fuels. The weighting factors are the fractions of the base fuels in each gasoline. Thus,

$$O_1 = \frac{x_{11}N_1 + x_{21}N_2 + x_{31}N_3 + x_{41}N_4}{x_{11} + x_{21} + x_{31} + x_{41}}$$

$$O_2 = \frac{x_{12}N_1 + x_{22}N_2 + x_{32}N_3 + x_{42}N_4}{x_{12} + x_{22} + x_{32} + x_{42}}$$

$$O_3 = \frac{x_{13}N_1 + x_{23}N_2 + x_{33}N_3 + x_{43}N_4}{x_{13} + x_{23} + x_{33} + x_{43}}$$

Develop a linear optimization model to determine how much of each base fuel should be blended into gasoline and how much should be sold "as is" in order to maximize profit. Solve the model using the following numerical data:

$C_1 = 13{,}000$ bbl/day	$N_1 = 82$ octane	$S_1 = \$0.90$/bbl
$C_2 = 7{,}000$	$N_2 = 95$	$S_2 = 1.05$
$C_3 = 25{,}000$	$N_3 = 102$	$S_3 = 1.25$
$C_4 = 15{,}000$	$N_4 = 107$	$S_4 = 1.60$
$D_1 = 13{,}000$ bbl/day	$R_1 = 87$ octane	$P_1 = 3.5¢$/gal
$D_2 = 25{,}000$	$R_2 = 89$	$P_2 = 4.5$
$D_3 = 18{,}000$	$R_3 = 93$	$P_3 = 6.0$

Note: 42 gal = 1 bbl

16.19 In Problem 16.18, how would the profit and the optimum policy be affected if the selling prices of the gasolines were changed to 3¢/bbl, 4.5¢/bbl (as before), and 5.5¢/bbl, respectively? What would be the consequences of decreasing D_2 by 7,000 bbl/day while simultaneously increasing D_3 by 2,000 bbl/day?

16.20 The cost associated with a set of electrical transmission lines is given by

$$C_1 = 10{,}000 \, n \, (1 + 2d^2)$$

where n is the number of lines in the set and d is the diameter of one line, in inches. In addition, the cost of transferring electrical power over the lines over their entire lifetime is given by

$$C_2 = 150{,}000/(nd^2)$$

To prevent major power disruptions in the event of a line breakage, it is required that $n \geq 10$.

Develop a nonlinear optimization model to determine the lowest overall cost $C = C_1 + C_2$. Assume that n is a continuous variable. Solve using Excel's **Solver** feature.

16.21 A pipeline is to transfer crude oil from a tanker docking area to a large oil refinery. The power required to pump the oil may be determined as

$$P = 0.4 \times 10^{-12} w^3/(\rho^2 d^5)$$

where w is the oil flow, in lb_m/hr, ρ is the density of the oil, in lb_m/ft^3, and d is the pipe diameter, in ft. The cost incurred, in dollars, is given by

$$C_p = 10{,}000d^2 + 170P$$

where the first term represents the cost of the pipeline and the second term represents the cost of the pump and the present value of the pumping power over the life of the refinery.

Suppose the oil flow is maintained at 10^7 lb_m/hr, and the oil density is 50 lb_m/ft^3. In order to keep sludge from settling in the pipeline, the oil velocity, v, must be at least 9 ft/sec. (*Note*: $w = 3600\pi d^2 \rho v/4$.) Develop a nonlinear optimization model to find the pipe diameter that will minimize the cost. Solve using Excel's **Solver** feature.

16.22 The oil carried by the pipeline described in Prob. 16.21 will be transported at an elevated temperature in an insulated pipeline. The heat loss is determined as

$$Q = \frac{500}{2.5 \times 10^{-4} + 0.0025t}$$

where Q is the heat loss in BTU/hr and t is the thickness of the insulation in ft.

The cost of the insulation is

$$C_i = 5 \times 10^5 td$$

where d is the diameter of the pipe in ft. The present value of the energy loss over the life of the refinery can be assumed to be

$$C_e = 0.5Q$$

Assuming the pipeline has not yet been built, develop a nonlinear optimization model to determine the pipe diameter and the thickness of insulation that will minimize the total cost, which is given by

$$C = C_p + C_i + C_e$$

Solve using Excel's **Solver** feature, with the restrictions that $d \geq 2$ ft and $v \geq 9$ ft/sec.

Appendix

EXCEL FUNCTIONS COMMONLY USED IN ENGINEERING APPLICATIONS

Function	*Purpose*
ABS(x)	Returns the absolute value of x.
ACOS(x)	Returns the angle (in radians) whose cosine is x.
ASIN(x)	Returns the angle (in radians) whose sine is x.
ATAN(x)	Returns the angle (in radians) whose tangent is x.
AVERAGE($x1, x2, \ldots$)	Returns the average of $x1, x2, \ldots$.
CONVERT($x, u1, u2$)	Converts x from unit $u1$ to unit $u2$.
COS(x)	Returns the cosine of x.
COSH(x)	Returns the hyperbolic cosine of x.
COUNT($x1, x2, \ldots$)	Determines how many numbers are in the list of arguments.
DEGREES(x)	Converts x radians to degrees.
EXP(x)	Returns e^x, where e is the base of the natural (Naperian) system of logarithms.
FORECAST($x, range1, range2$)	Returns the value of y corresponding to the given value of x. Based upon a linear trendline, with x values given in $range1$ and y-values in $range2$.
FV(i, n, x)	Returns the future value of n payments of x dollars each at interest rate i.
IF($e, v1, v2$)	Places value $v1$ in the active cell if logical expression e is TRUE. Otherwise (if e is FALSE), $v2$ is placed in the active cell. *Note*: $v1$ and $v2$ may be numbers, strings, formulas, or other IF statements.
INT(x)	Rounds x down to the closest integer.
IRR($x1, x2, \ldots$)	Returns the internal rate of return for a series of cash flows.
LN(x)	Returns the natural logarithm of x ($x > 0$).
LOG10(x)	Returns the base-10 logarithm of x ($x > 0$).
MAX($x1, x2, \ldots$)	Returns the largest of $x1, x2, \ldots$.
MEDIAN($x1, x2, \ldots$)	Returns the median of $x1, x2, \ldots$.

EXCEL FUNCTIONS (CONTINUED)

Function	*Purpose*
MIN($x1, x2, \ldots$)	Returns the smallest of $x1, x2, \ldots$.
MDETERM(*range*)	Returns the determinant of a nonsingular square matrix defined by *range*.
MINVERSE(*range*)	Returns the inverse of a nonsingular square matrix defined by *range*.
MMULT(*range1, range2*)	Returns the matrix product of two matrices defined by *range1* and *range2*, respectively. The matrices must conform to the rules of matrix multiplication.
MOD($x1, x2$)	Returns the remainder after $x1$ is divided by $x2$.
MODE($x1, x2, \ldots$)	Returns the mode of $x1, x2, \ldots$.
NPER(i, x, P)	Returns the number of payment periods for a loan of P dollars with constant payments of x dollars each at interest rate i.
NPV($i, x1, x2, \ldots$)	Returns the net present value of a series of cash flows at interest rate i.
PI( )	Returns the value of π. (The empty parentheses are required.)
PMT(i, n, x)	Returns the periodic (e.g., monthly) payment for an n-payment loan of x dollars at interest rate i.
PV(i, n, x)	Returns the present value of a series of n payments of x dollars each at interest rate i.
RADIANS(x)	Converts x degrees to radians.
RAND( )	Returns a random value between 0 and 1. (The empty parentheses are required.)
RANDBETWEEN($n1, n2$)	Returns a random integer between $n1$ and $n2$.
RATE(n, A, P)	Returns the interest rate for series of n equal payments of A dollars each with present value P.
ROUND(x, n)	Rounds x to n decimals.
SIGN(x)	Returns the sign of x. (Returns $+1$ if $x > 0$, -1 if $x < 0$.)
SIN(x)	Returns the sine of x.
SINH(x)	Returns the hyperbolic sine of x.
SQRT(x)	Returns the square root of x ($x > 0$).
STDEV($x1, x2, \ldots$)	Returns the standard deviation of $x1, x2, \ldots$.
SUM($x1, x2, \ldots$)	Returns the sum of $x1, x2, \ldots$.
TAN(x)	Returns the tangent of x.

EXCEL FUNCTIONS (CONTINUED)

Function	Purpose
TANH(x)	Returns the hyperbolic tangent of x.
TEXT(x, f)	Converts numerical value x to text with number format f. (*Note*: number formats are listed under Format/Cells…/Number.)
TRUNC(x, n)	Truncates x to n decimals.
VAR($x1, x2, \ldots$)	Returns the variance of $x1, x2, \ldots$.

INDEX

Abbreviations, unit, 206-207
Abscissa, 90
Absolute addressing, 405
Absolute and relative cell addresses
 mixing, 72
 toggling between, 72-73
Absolute cell address, 72
Acceleration, gravitational, 247-248
Accumulating compound interest, 426-427
Accuracy of a solution, 12
Accuracy, trapezoidal rule, 372-373
Active cell, current, 30
Addition, matrix, 331, 332, 339
Address
 absolute, 72
 cell, 5, 13, 19
 mixed, 72
 relative, 72
Addressing
 absolute, 405
 relative, 405
Adjusting column widths, 79-80
Aggregate, data, 242, 249
Algebraic equations
 graphical solution, 300-301
 nonlinear, 297-298
 numerical solution, 302-308
 solution of, 16, 302-308
Algorithm, simplex, 468
Alloy blending, 357, 489
Alphanumeric data, 138
Analysis toolpak, 203, 221
Analysis, data, 217
Analyzing a data set, 221-224
Annulus, heat exchanger, 49-50
Arguments, Excel functions, 51
Arithmetic coordinates, 97
Arithmetic operators, 38
Arrays, 338
Arrow keys, 30, 31
Assessing a curve fit, 255
Auditing toolbar, formula, 42
Autocalculate, 66
AutoCorrect, 79, 82
AutoSum, 220
Auxiliary conditions, 463
AVERAGE function, 218
Axis
 category, 90
 value, 90

Bar graphs, 126-130
 creating in Excel, 127-130
Beam deflection, 323
Best function, when curve fitting, 279-289
Bisection, method of, 303-306
Block of cells
 clearing, 67
 copying, 70-71
 moving, 71
 selecting, 65-67
Boolean expressions, 389-390
Branching operations, 389

Capacitor, voltage drop across, 98-102
Cartesian coordinates, 97
Cash flow diagram, 440
Cash flows
 irregular, 447-452
 uniform multipayment, 439-444
Category axis, 90
Catenary, 358
Cause-and-effect, 242
Cell address, 5, 13, 19
 absolute, 72
 mixed relative and absolute, 72
 relative, 72
Cell addressing, within a macro, 404-408
Cell formulas, displaying, 86-88
Cell names, 44
Cell reference, 19
Cell
 copying, 67-69
 currently active, 13, 14
Cells, 5, 13, 19, 23
 clearing a block of, 67
 copying, 70-71
 inserting and deleting, 76-77
 moving, 71
 naming, 44-45
 selecting a block of, 65-67
Changes, undoing, 71-72
Characteristics of a good graph, 90-93
 engineering optimization problems, 470-471
 nonlinear algebraic equations, 298-299
 optimization problems, 464-471
Chart toolbar, 96
Chart wizard, 93-96

Chart, altering its size or location, 96, 97
Charts, 90
 line, 122-126
 scatter, 97-107
 XY, 97-107
Circuits, RLC, 85, 325
Circular references, 40-42
 using Goal Seek, 310
Classical solution procedures, 17
Clearing a block of cells, 67
Column widths, adjusting, 79-80
Commonly used Excel functions, 494-496
Comparing two economic alternatives, 437-438
 in Excel, 451-452
 using IRR, in Excel, 458-460
Comparison operators, 38, 389
Complex conjugates, 299
Compound interest, 47, 248, 396-397, 424-434
 accumulating, 426-427
 frequency of compounding, 428-430
Conjugates, complex, 299
Conservation laws, 9
Constant
 numeric, 19, 32
 text, 19, 32
Constraints, 351, 463, 464
Constructing a cumulative distribution, 233-234
Constructing a histogram, 226-227
Convergence considerations using Goal Seek, 312-315
Convergence criterion, Newton-Raphson method, 306
Convergence using Goal Seek, 310
Convergence using Solver, 318
Conversions, unit, 396
CONVERT function, 202, 203
Converting temperature differences, 210-211
Converting temperatures, 209-211
Coordinate systems, when curve fitting, 280-282
Coordinates
 arithmetic, 97
 cartesian, 97
 logarithmic, 110-111
Copying a block of cells, 70-71
Copying a cell, 67-69

Copying data directly to Word, 182-185
Copying formulas, 72
Correcting errors, 36
Cost function, 463
Cost of a loan, determination of, 440-441
Creating a graph in Excel, 93-97
Cumulative distribution
 constructing, 233-234
 drawing inferences from, 237-238
 generation in Excel, 234-236
 plotting, 236-237
Cumulative distributions, 233-238
 in Excel, 234
Currently active cell, 13, 14, 30
Curve fit, assessing, 255
Curve fitting, 16
 in Excel, 256-262
 multiple functions, 283-284
 scaling the data, 287-288
 selecting the best function, 279-289
 straight line, 279-282
 substituting other variables, 284-286
 using various coordinate systems, 280-282

Data
 aggregate, 16
 alphanumeric, 138
 copying directly to Word, 182-185
 entering, 32-35
 exporting to a text file, 173-176
 filtering, 147
 graphical display, 7
 importing from a text file, 167-172
 input, 5
 linking with Word, 190-193
 nonadjacent, 107
 numerical, 138
 output, 5
 paired, 96, 126, 242
 retrieving in Excel, 147-152
 RTF, 186
 scatter, 16
 single-valued, 96, 122, 126
 sorting, 142
 source, 93
 transferring to PowerPoint, 194-198
 transferring to Word, 182-194
Data aggregate, 242, 249
Data analysis, 16, 217
Data items, formatting, 81
Data scaling, when curve fitting, 287-288
Data scatter, 91, 242
Data set
 adding data, 102
 analyzing, 221-224

fitting a logarithmic equation, 268-269
fitting a polynomial, 273-277
fitting a power equation, 270-273
fitting a straight line, 251-255
fitting an exponential equation, 264-268
Data spread, 219
Database operations, 138-139
Decision variables, 464
Deflection of a beam, 323
Degree of a polynomial equation, 298
Deleting individual cells, 76-77
Deleting rows and columns, 73-75
Delimiters, 170
Dependent variables, adding a new set, 102-105
Dependent variables, adding to an X-Y graph, 105-106
Descriptive statistics, 221
Design requirements, 463
Deviation, standard, 220
Dialog box, form, 141
Disabling smart tags, 79
Disbursements, 439
Displaying cell formulas, 86-88
Distributions, cumulative, 233-238
Divergence, Newton-Raphson method, 308
Drawing inferences from a cumulative distribution, 237-238

Economic alternatives, comparison of, 437-438
Economic analysis, engineering, 16
Editing a graph, 98-102
Editing a macro, 418-421
Editing a worksheet, 83
Editing shortcuts, 81-82
Editing, Excel worksheet, 65-83
Electrical circuit, 9-12, 47-48
 spreadsheet solution, 13-15
Embedding a worksheet within Word, 185-188
Embedding RTF data within Word, 186
End key, 31
Engineering economic analysis, 16
Engineering fundamentals, 9
Engineering optimization problems, characteristics of, 470-471
Entering data, 32-35
Entering Excel, 20
Equation of state, van der Waals, 297
Equations
 exponential, 112
 linear, 327-328
 nonlinear, 298-299
 polynomial, 298-299

power, 116
quadratic, 397-398
Equilibrium, 9
Equivalence factor, unit, 202
Error
 least squares, 249
 standard, 223
Errors, correcting, 36
Evaluating a polynomial, macro for, 405-408
Evaluating integrals, 16
Evaluating trigonometric functions, 53-54
Exam scores, student, 52-53, 56
Excel functions, 50-56
 commonly used in engineering, 494-496
 list of, 51, 494-496
Excel, 4, 20
 creating a graph in, 93-97
 entering and leaving, 20-25
 leaving, 28-29
Executing a macro, 401-404
Existing data set, adding data, 102
Exponential equation, 112
Exponential function, 264
 fitting to a set of data in Excel, 266-268
 least squares equations, 265
Exporting data
 to a text file, 173-176
 to an HTML file, 180-181
Expressions, logical, 389-390
Extensions, text file, 168

Feasible region, 466
Fibonacci numbers, 396
Fick's law, 18
Filtering a list, in Excel, 147-152
Filtering data, 147
Financial functions, 434
Fitting a function to a set of data
 logarithmic function, 268-269
 polynomial, 273-277
 power function, 270-273
 straight line, 251-255
  in Excel, 256-259, 279-282
 exponential function, 264-268
Flux, 9
Force, spring, 108-109
FORECAST function, 245
Form, dialog box, 141
Format, number, 33
Formatting a number, macro for, 402-404
Formatting data items, 81
Formatting toolbar, 21, 23
Formula auditing toolbar, 42

Formulas, 7, 13, 19, 37
  copying, 72
  moving, 73
  using, 37-45
  writing, 39-40
Four-bar linkage, 324
Fourier's law, 9
Frequency of compounding, 428-430
Frequency, relative, 225
Friction factor, 324
Functional relationships, 17
Functions
  Excel, 50-56, 494-496
  financial, 434
  list of, 51, 494-496
  selecting, 55-56
Fundamentals, engineering, 9
Furnace wall, 356-357
Future sum of money, present value of, 437
Future value of a series of uniform payments, 443-444

Gas, ideal, 47, 108
Gasoline blending problem, maximum profit, 491-492
General problem-solving techniques, 8-9
Goal Seek
  circular reference, 310
  convergence, 310, 312-315
  solving algebraic equations with, 309-316
Good graph, characteristics of, 90-93
GoTo key, 31
GPA, 46
Grade point average, student, 46
Graph
  bar, 126-130
  characteristics of, 90-93
  creating in Excel, 93-97
  editing, 98-102
  line, 122-126
  log-log, 116-120
  pie, 131-135
  semi-log, 110-115
  transferring to Word, 193-194
Graphical solution of algebraic equations, 300-301
Gravitational acceleration, 247-248

Heat exchanger, annulus, 49-50
Help, 22, 25-30
  answer wizard, 26
  contents, 25, 26
  index, 26
  online (web-based), 28
  What's This?, 28

Histogram, 225-231
  constructing, 226-227
  generation in Excel, 227
  plotting, 230-231
Histogram fundamentals, 225-226
History, time, 242
Home key, 31
HTML, 167
HTML data, transferring, 179-181
HTML file, exporting data to, 180-181
Hyperlinks, 82

Ideal gas, 47, 108
Identity matrix, 333
IF function, 390
  nested, 391-394
If-then-else, 389
Importing data, from a text file, 167-172
Inferences, from a cumulative distribution, 237-238
Input data, 5
Inserting and deleting individual cells, 76-77
Inserting and deleting rows and columns, 73-75
Integrals, evaluating, 16, 360-361
Integrand, 361
Interest, 425
  compound, 47, 248, 396-397, 424-434
Interest rate, 425
Internal rate of return, 455-461
  as the root of a polynomial, 460-461
  in Excel, 456-457
Interpolation, 16
  linear, 242, 243-246
Interval halving, method of, 303-306
Interval reduction techniques, 303-306
Interval width, effect of (trapezoidal rule), 371-373
Intervals, number of (trapezoidal rule), 372-373
Inverse matrix, 333
Investment opportunities, comparison of using IRR, in Excel, 458-460
Investment opportunities, comparison of, in Excel, 451-452
Investment, present value of proposed, in Excel, 441-443
IRR, 455
Irregular cash flow, present value of, in Excel, 447-449
Irregular cash flows, 447-452
Iterative techniques, 306-308

Keyboard shortcut, 399, 400
Kirchoff's laws, 10, 353, 354
Kurtosis, 223

Labels, 13, 19, 34-35
Least squares curve fitting, in Excel, 256-262
Least squares equations
  for a logarithmic function, 269
  for a power function, 270
  for a straight line, 252
  for an exponential function, 265
  for polynomials, 273
Least squares, method of, 243, 249-251
Leaving Excel, 22, 28-29
Legend, 91
Line graph, creating in Excel, 122-125
Line graphs, 122-126
Linear algebraic equations, simultaneous, 334, 335, 340-342, 344-349
Linear equations, 327-328
Linear interpolation, 242, 243-246
  in Excel, 245-246
Linear optimization problems using Solver, 477-481, 482
Linear programming, 467
Lingage, four-bar, 324
Linking, with Word, 189-193
List, 138
  creating in Excel, 139-141
  filtering in Excel, 147-152
  sorting in Excel, 142-143
Loan
  cost of, 440-441
  single-payment, 427-428
Loan repayment, 322
Local optima, 470
Logarithmic coordinates, 110-111
Logarithmic function, 268
  least squares equations, 269
Logarithmic scale, 111
Logical expressions, 389-390
Logical operators, 390
Log-log graphs, 116-120
  creating in Excel, 118-120

Macro, 399
  cell addressing, 404-408
  editing, 418-421
  executing, 401-404
  recording, 399-401
  saving, 409-411
  security level, 401
  viewing, 413-417
Macro to evaluate a polynomial, 405-408
Macro to format a number, 402-404
Manometer, 50

Mathematical solution procedures, 16-17

Matrix, 328
  identity, 333
  inverse, 333

Matrix addition, 331, 332
  in Excel, 339

Matrix form of a system of simultaneous equations, 329, 331

Matrix inversion, in Excel, 340-342, 344-352

Matrix multiplication, 329, 330

Matrix notation, 328-335

Matrix operations, in Excel, 338-339

Matrix subtraction, 332
  in Excel, 339

Max, 219

MAX function, 219

Maximum profit gasoline blending problem, 491-492

MDETERM function, 340

Mean, 218

Measured data, numerical integration of, 383-385

Median, 218

MEDIAN function, 218

Menu bar, 20, 21

Menus, long and short, 22

Method of bisection, 303-306

Method of least squares, 243, 249-252
  fitting a straight line, 251-255

Microsoft Office, 20

Microsoft PowerPoint, 167
  transferring data to, 194-198

Microsoft Word, 167
  copying directly to, 182-185
  embedding a worksheet within, 185-188
  transferring data to, 182-194

Min, 219

MIN function, 219

Minimization problem, nonlinear, in Excel, 482-486

Minimum cost shipping problem, 490-491

Minimum-weight structure, 468-469

MINVERSE function, 340

Mixing relative and absolute cell addresses, 72

MMULT function, 340

MOD function, 390

Mode, 218-219

MODE function, 218-219

Money
  present value of a future sum, 437
  time value of, 437-438

Monthly compounding
  single-payment loan, 430-431
  single-payment loan, in Excel, 431-434

Mouse, 30

Mouse pointer, 30

Moving a block of cells, 71

Moving around the worksheet, 30-32

Moving formulas, 73

Multipayment cash flows, uniform, 439-444

Multiple functions, when curve fitting, 283-284

Multiplication
  matrix, 329, 330
  scalar, 332, 339

Name box, 23

Naming cells, 44-45

Nested IF functions, 391-394

Newton-Raphson method, 306-308
  divergence, 308
  stopping condition, 306

Nonadjacent cells, selecting, 66

Nonadjacent data, 107

Nonadjacent rows, 98

Nonlinear algebraic equations, 297-298
  characteristics of, 298-299
  solution in Excel, 349-352

Nonlinear minimization problem, in Excel, 482-486

Notation, matrix, 328-335

NPV function, 449-452

Number of intervals, trapezoidal rule, 372-373

Numeric constant, 19, 32

Numerical data, 138

Numerical expressions, 37

Numerical integration, 361

Numerical integration of measured data, 383-385

Numerical methods, 17

Numerical values, 32-34
  appearance, 33

Objective function, 463

Objects, oscillating, 83-84, 325, 359

Office, Microsoft, 20

Ohm's law, 9, 10

Oil pipeline problem, minimum cost, 492-493

Operations
  branching, 389
  database, 138-139

Operator precedences, 43-44

Operators, 38
  comparison, 389
  logical, 390

Optima, local, 470

Optimization problems, 463
  characteristics of, 464-471

Optimization techniques, 17

Ordinate, 90

Oscillating objects, 85-86, 325, 359

Outliers, data, 242

Output data, 5

Page Up, Page Down keys, 31

Paired data, 96, 126, 242

Parameters, 5

Performance criterion, 463

Performance index, 464

Pie charts, 131-135
  creating in Excel, 132-135

Pivot chart, 163-164

Pivot table bar, 163-164

Pivot tables, 154-164
  creating within Excel, 156-163
  rearranging, 163-164

Plotting a cumulative distribution, 236-237

Plotting a histogram, 230-231

Pointer, mouse, 30

Policy variables, 464

Polynomial equations, 298-299
  degree of, 298
  graphical solution, 300-301
  solution using Goal Seek, 310-312
  solution using method of bisection, 305-306
  solution using the Newton-Raphson method, 307-308

Polynomial
  evaluating (macro for), 405-408
  fitting to a set of data in Excel, 273-277

Polynomials, 273
  least squares equations, 273

Potential, 9

Power equation, 116

Power function, 269
  fitting to a set of data in Excel, 270-272
  least squares equations, 270

PowerPoint, Microsoft, 167
  transferring data to, 194-198

Precedences, operator, 43-44

Prefixes, unit, 208

Present value of a future sum of money, 437

Present value of a proposed investment in Excel, 441-443

Present value of an irregular cash flow, in Excel, 447-449

Principal, 425

Printing a worksheet, 60-64

Problem, preparing to solve, 9-12

Problem-solving techniques, 8-9
Production scheduling, 465-468
   in Excel, 477-481
   maximum profit, 489-490
Profile, spatial, 242
Profit function, 464
Programming, linear, 467
Projectile, trajectory, 4
Proposed investment, present value of,
   in Excel, 441-443

Quadratic equations, 397-398

r^2, equation for, 254
RANDBETWEEN function, 395
Range, cell addresses, 52
Rate of return, internal, 455-461
Rate phenomena, 9
Recalculation, worksheet, 7, 15
Receipts, 439
Record, 138
Recording a macro, 399-401
Reference, cell, 19
References, circular, 40-42
Regression, in Excel, 260, 261-262
Relationships, functional, 17
Relative addressing, 405
Relative and absolute cell addresses,
   mixing, 72
Relative and absolute cell addresses,
   toggling between, 72-73
Relative cell address, 72
Relative frequency, 225
Resistive circuits, 353, 354
Retrieving a worksheet, 57-59
Retrieving data, in Excel, 147-152
Reynolds number, 324
RLC circuits, 85, 325
Root of a polynomial and internal rate
   of return, 460-461
Rows and columns, 5, 13
   inserting and deleting, 73-75
Rows, nonadjacent, 98
RTF data, embedding within Word, 186

Saving a macro, 409-411
Saving and retrieving a worksheet, 57-59
Scalar multiplication, 332
   in Excel, 339
Scale, logarithmic, 111
Scatter charts, 90, 97-107
Scatter, data, 16, 91, 242
Scheduling production to maximize
   profit, 465-468
   in Excel, 477-481
Scroll bars, 21, 24, 31
Scroll button, 24

Security level, macro, 401
Selecting a block of cells, 65-67
Selecting functions, 55-56
Selecting nonadjacent cells, 66
Selecting the entire worksheet, 66
Semi-log graphs, 110-115
   creating in Excel, 114-115
Series of uniform payments, future
   value of, 443-444
Shortcut key, 399, 400
Shortcuts, editing, 81-82
Simplex algorithm, 468
Simpson's rule, 374-380, 393-394
   comparison with trapezoidal rule,
   375-378
Simultaneous algebraic equations,
   solution of, 16
Simultaneous equations
   matrix form, 329, 331
   solution, in Excel, 340-342, 344-352
Simultaneous linear algebraic equations,
   solution of, 334, 335,   340-
   342, 344-349
Single-payment loan, 427-428
   with monthly compounding, 430-431
   with monthly compounding, in
   Excel, 431-434
Single-valued data, 96, 122, 126
Skewness, 223
Smart tags, 78-79
   disabling, 79
Solution of algebraic equations, 16
Solution of simultaneous algebraic
   equations, 16
Solution of simultaneous linear
   algebraic equations, 334, 335, 340-
   342, 344-349
Solution of simultaneous nonlinear
   algebraic equations in Excel, 349-352
Solution procedures
   classical, 17
   mathematical, 16-17
Solution, assessing the accuracy of, 12
Solver, 476-477
   convergence, 318
   for linear optimization problems,
   477-481, 482
   reports, 487
   solving algebraic equations with,
   316-321
   solving simultaneous algebraic
   equations with, 344-352
Solving algebraic equations
   numerically, 302-308
   using Goal Seek, 309-316
   using interval reduction techniques,
   303-306

   using iterative techniques, 306-308
   using Solver, 316-321
Solving equations graphically, 300-301
Solving optimization problems in Excel,
   476-487
Solving simultaneous equations, in
   Excel, 340-342, 344-352
Sorting a list in Excel, 142-143
Sorting data, in Excel, 142
Source data, 93
Spatial profile, 242
Spread, data, 219
Spreadsheet, 4, 19
Spreadsheet, overview, 4-7
Spring, force exerted by, 108-109
SSE, equation for, 253
SST, equation for, 254
Standard deviation, 220
Standard error, 223
Standard toolbar, 21, 23, 58, 70
Statistics, descriptive, 221
Status bar, 21, 24
STDEV function, 220
Steady-state, 9
Stop Recording bar, 400, 405
Stopping condition, Newton-Raphson
   method, 306
Straight line
   fitting to a set of data in Excel, 256-
   259, 279-282
   least-squares equations, 252
String, 19, 34-35
String operator, 38
Structure, minimum-weight, 468-469
Student exam scores, 52-53, 56
Student grade point average, 46
Substituting variables, when curve
   fitting, 284-286
Subtraction, matrix, 332, 339

Table, transferring from Word, 194
Tables, pivot, 154-164
Tabs, worksheet, 21, 23
Task pane, 21, 24
Temperature differences, converting, 210-211
Temperatures, converting, 209-211
Text constant, 19, 32
Text file
   exporting data to, 173-176
   extensions, 168
   importing data from, 167-172
Text import wizard, 168
Time history, 242
Time value of money, 437-438
Title bar, 20, 21
Toggling between relative and absolute
   cell addresses, 72-73

Toolbar
  chart, 96
  formatting, 21, 23
  formula auditing, 42
  standard, 21, 23, 58, 70
Toolbars, 21, 23
  adding or removing, 24-25
Trajectory, projectile, 4
Transferring a graph to Word, 193-194
Transferring a table from Word, 194
Transferring HTML data, 179-181
Transportation problem, 490-491
Trapezoidal rule, 361-373
  comparison with Simpson's rule,
    375-378
  equally spaced data, 367-373
  equally spaced data, in Excel, 369-
    373
  unequally spaced data, 363-367
  unequally spaced data, in Excel,
    366-367
Trendline, 91, 258-260
Trigonometric functions, evaluating,
  53-54
Truss, forces in, 355

Undo icon, 72
Undoing changes, 71-72
Uniform multipayment cash flows, 439-
  444
Uniform payments, future value of a
  series, 443-444

Unit abbreviations, in Excel, 206-207
Unit conversions, 396
  complex, 211-214
  in Excel, 203-209, 211-214
  simple, 202-203
Unit equivalence factor, 202
Unit prefixes, in Excel, 208
Using formulas, 37-45

Value axis, 90
Values, numerical, 32-34
van der Waals equation of state, 297
VAR function, 220
Variable substitution, when curve
  fitting, 284-286
Variables, decision, 464
Variance, 219
VBA, 413-421
Viewing a macro, 413-417
Visual Basic for Applications, 413-421
Voltage drop across a capacitor, 98-102
Volumetric flow, 49

What if, 7, 19
Width, column, 79-80
Wizard
  chart, 93-96
  text import, 168
Word, Microsoft, 167
  copying directly to, 182-185
  embedding a worksheet within, 185-188
  transferring data to, 182-194

Workbook, 23
Worksheet
  editing, 65-83
  embedding within Word, 185-188
  linking with Word, 190-193
  moving around, 30-32
  printing, 60-64
  recalculation, 7, 15
  saving and retrieving, 57-59
  selecting, 66
Worksheet cell, active, 23
Worksheet tabs, 21, 23
Worksheet window, adding or deleting
  items, 24-25
Writing formulas, 39-40

XY charts, 90, 97-107
X-Y graphs, 97-107
  adding data, 102-104
  adding dependent variables, 105-106